High-Performance Thin-Layer Chromatography in Food Analysis

HPTLC

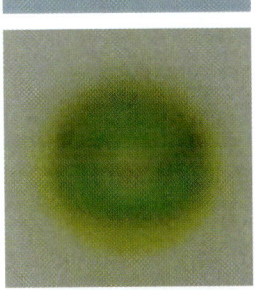

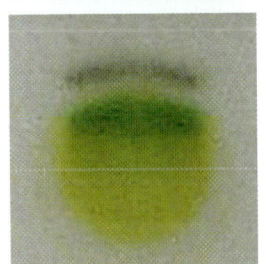

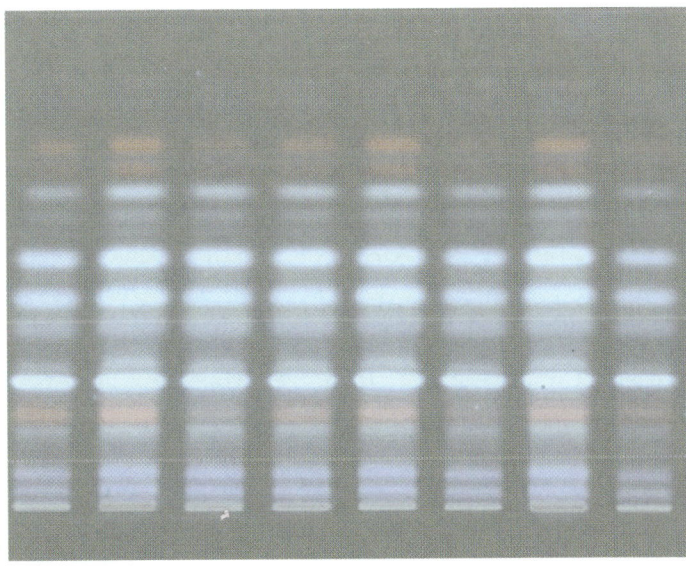

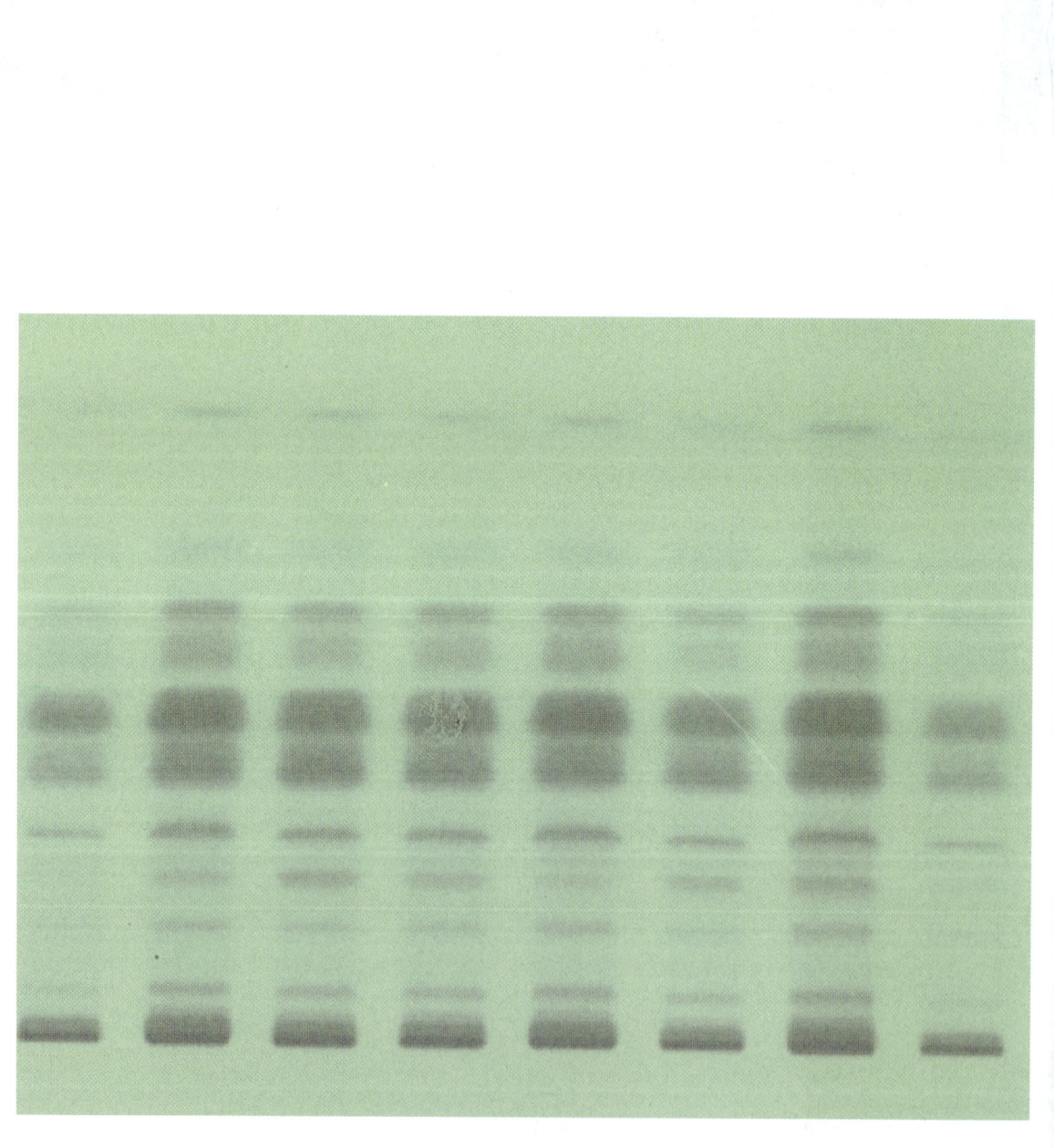

High-Performance Thin-Layer Chromatography in Food Analysis

HPTLC

Prem Kumar Jaiswal

MSc (Analytical Chemistry)
DPhil (Chemistry), DSc (Chemistry)

University of Allahabad
Allahabad (UP)

CBS

CBS Publishers & Distributors Pvt Ltd

New Delhi • Bangalore • Pune • Cochin • Chennai

High-Performance Thin-Layer Chromatography in Food Analysis

ISBN: 978-81-239-1837-2

First Edition : 2010

Published by Satish Kumar Jain and produced by Vinod K. Jain for

CBS Publishers & Distributors Pvt Ltd

Head off: CBS Plaza 4819/XI Prahlad Street, 24 Ansari Road, Daryaganj, New Delhi 110 002, India.

Website: **www.cbspd.com**

Ph: 23289259, 23266861/67   Fax: +91-11-23243014   e-mail: delhi@cbspd.com; cbspubs@vsnl.com; cbspubs@airtelmail.in.

Branches

• Bangalore: Seema House 2975, 17th Cross, K.R. Road, Banasankari 2nd Stage, Bangalore 560 070, Karnataka
  Ph: +91-80-26771678/79   Fax: +91-80-26771680   e-mail: bangalore@cbspd.com

• Pune: Shaan Brahmha Complex, 631/632 Basement, Appa Balwant Chowk, Budhwar Peth, Next To Ratan Talkies, Pune 411 002, Maharashtra
  Ph: +91-20-24464057/58   Fax: +91-20-24464059   e-mail: pune@cbspd.com

• Cochin: 36/14 Kalluvilakam, Lissie Hospital Road, Cochin 682 018, Kerala
  Ph: +91-484-4059061-65   Fax: +91-484-4059065   e-mail: cochin@cbspd.com

• Chennai: 20, West Park Road, Shenoy Nagar, Chennai 600 030, TN
  Ph: +91-44-26260666, 26202620   Fax: +91-44-45530020   email: chennai@cbspd.com

Printed at Paras Offset Pvt. Ltd., C-176, Naraina Industrial Area Phase-I, New Delhi

to

---

my parents
late Shri MP Jaiswal and
Smt AK Jaiswal

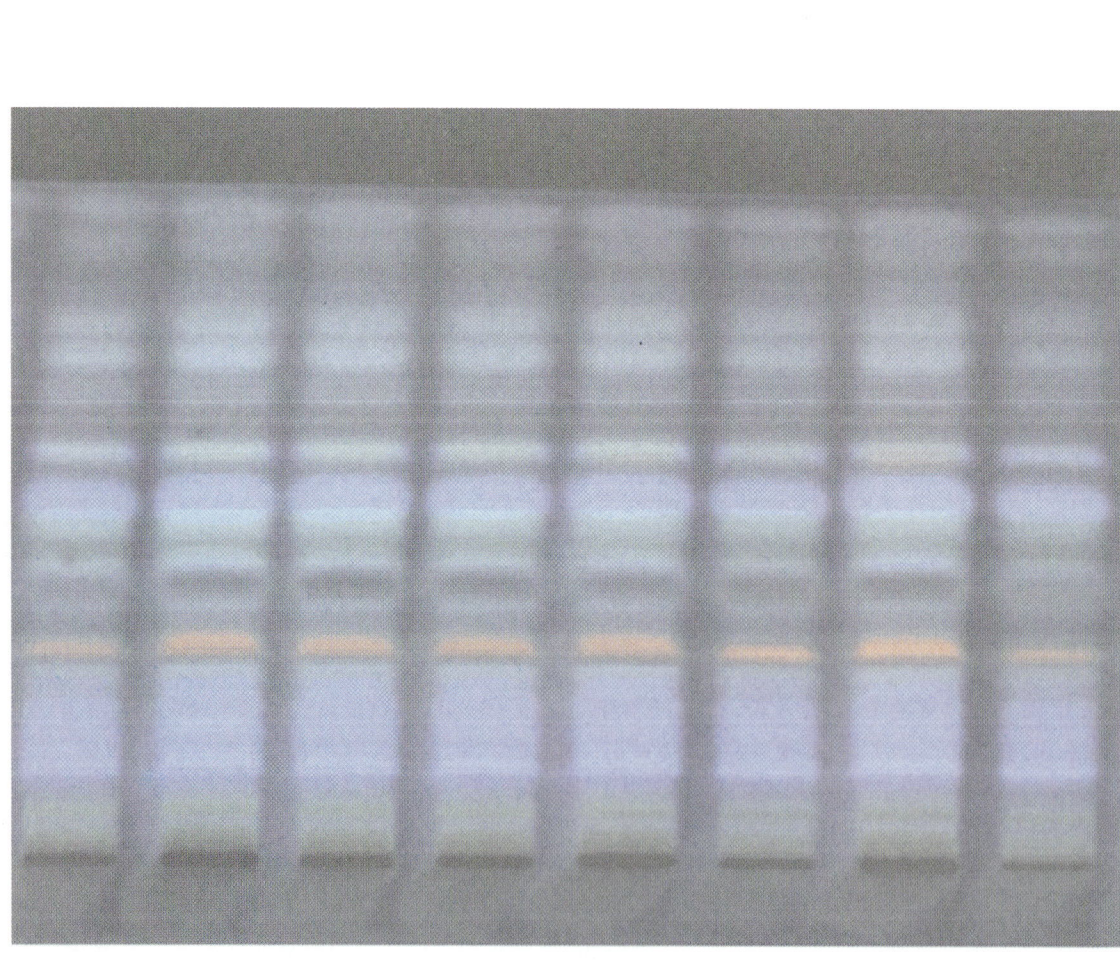

Preface

ecently quality assurance in food has been a challenge to analysts. Due to continual enforcement of stringent safety standards in food and expectations to provide safe foods to consumers, the job of analyst has become more responsible. Liberalization of world trade in food has opened up new vistas of growth. International standards and quality have played an important role in protection of health and safety of consumers. Adoption of various systems like ISO 22000, HACCP, laboratory accreditation, with analytical protocols to prevent hazards in food, enhance capability and reliability of analysis to meet the needs of the quality assurance.

Dynamic advancement in instrumentation and computerization has facilitated the laboratories to handle automated testing equipment with higher accuracy, reliability and ruggedness of analysis. But this has increased the burden for validating the methods and continual need for reliable analytical methods for use in compliance with national and international regulations in the areas of food quality and safety. In fact the old conventional and official methods used in food analysis need to be reviewed and replaced with modern instrumental techniques.

Food varies widely in chemical composition. Use of various additives in food, contaminations and adulteration in food in the context of specified safety and quality standards have always been a challenge to test the skills of food analysts.

In the past, thin layer chromatography (TLC) has been without any doubt one of the versatile and widely applicable separation techniques in chromatography used for establishing quality and safety of a food. But this has its own limitations in view of the present requirements of detection and quantitation at a much lower level. It is still a preferred method as it is simple, inexpensive and requires minimum sample pretreatment. Due to very low level of detection/determination to establish identity, purity, quality and presence of contamination and additives, high performance thin layer chromatography has emerged as a modern separation technique accepted worldwide and specially distinguished by its flexibility, reliability and cost efficiency. The method is quite comparable in its analytical capabilities with HPLC and GC. In many cases HPTLC offers more suitable solution and is often used as a confirmatory/alternative/screening method. It has its own advantages due to the off-line principle, which offers tremendous flexibility, simultaneous and parallel separation of many samples at high speed, visual evaluation of sample components with unsurpassed clarity, simple and convenient sample preparation and possibility of multiple evaluation of plate with different parameters, etc. In view of the versatile application of HPTLC as a preferred method, it has found place in pharmaceutical and clinical analysis; industrial chemistry; environmental toxicology; food, water, inorganic and pesticide analysis; plant and herbal analysis; purity of dyes and cosmetics; and forensic materials, etc.

An attempt has been made by me to elucidate the importance and value of HPTLC analysis for food and food products, the cost effective technology of the twenty-first century and versatile application of this instrument in food quality assurance. The book gives the methods of analysis for pesticide residues, flavours, organic acids, mycotoxins, antibiotics, additives, food colours, carbohydrates, proteins, pigments and dyes, and different adulterants in food, etc. The contents in this book also explain how HPTLC can be integrated into quality assurance programs with a high degree of compliance of analytical protocol. Requirements of laboratory accreditation like validation of method, different qualifications, good laboratory practices and documentation, etc. have been discussed at length. I have tried to design the description in this book in such a way that the users would be benefited not only in the field of food analysis but also in the other domains with a little change in application techniques. This book has taken special care to cover the syllabus related to food analysis, mycotoxins, relevant portion of instrumentation (HPTLC), chromatographic methods, safety and quality evaluation of additives and ingredients, natural and synthetic colours, food toxicants and contaminants, detection and estimation of adulterants, analysis of various foods, pesticide residues, etc. in the B Tech and M Tech courses prescribed by a number of Universities. Therefore, a significant portion of this book shall be useful to the students of these courses in understanding the different applications of HPTLC.

I have put limited theoretical description about HPTLC and the main emphasis has been that the book will facilitate the analysts to widely use this technique in different applications.

In the end I wish that the contents in this text would be a guide to generate many methods of analysis suiting their applications in different fields since the applications of HPTLC are unlimited.

Dr. Prem Kumar Jaiswal

Acknowledgements

This book could not have been written without strong technical support and motivation from my friends and encouragements from my family.

I am grateful to Mr Dilip Charegaonkar, Camag Application Laboratory, Mumbai, India and Switzerland, and Merck for providing technical inputs whenever required for the benefit of analytical scientists.

I would like to place on records his sincere gratitude to the different authors on HPTLC whose books have been constant source of inspiration to write on HPTLC related to food analysis.

My special thanks to all my family members who allowed our personal time meant to be spent with them towards writing this book.

I also express my gratitude to the publishers CBS Publishers & Distributors Pvt Ltd for cooperation and timely publication of this book. My special thanks to Mr YN Arjuna, Senior Director, CBS P&D, for his keen interest in improving the content and presentation.

Dr. Prem Kumar Jaiswal
e-mail:prem1948@yahoo.co.in

Contents

APPENDIX

Introduction

C hromatography is one of the oldest and widely used techniques as a method of separation of components in a mixture by a variety of phenomena such as partition, adsorption, ion exchange, etc.

Runge, Schoenbein and Goppelsroeder carried out considerable work during the period from approximately 1850 to 1900. However, the importance of the method was not fully appreciated.

The Russian botanist Tswett[1, 2] discovered adsorption chromatography, but this remained unnoticed until 1931. He observed that plant pigments like chlorophylls could be separated by filtering their solution in petroleum ether using a column containing calcium carbonate. It was noticed that yellow and green zones were formed on the column. He remarked "like rays of light in a spectrum, the various components of a mixture of pigments are separated in a definite order and can then be determined qualitatively and quantitatively". He separated leaf extracts into coloured bands using an inulin column. Separation of coloured oils when percolated through earth was observed by Day.[3] However, the former is usually credited, by and large, with the discovery and description of chromatography. The term "chromatography" which literally means "description of colours" was suggested because first separations involved, of necessity, coloured components.

However, in 1938, separation on thin layer chromatography was achieved by Izmailov and Shraiber[4], when looking for a simpler technique requiring less sample and sorbent, separated plant extracts using a glass plate coated with aluminium oxide. Slurry of sorbent of about 2 mm thickness was made on a microscope slide and the sample was applied as droplets to the layer. Methanol as a solvent was added dropwise over the spots and a series of circular rings were obtained of different colours, thus giving birth to circular TLC (thin-layer chromatography), but Izmailov and Shraiber named this as "drop chromatography".

In 1941, Martin and Synge[5] used an inert support (e.g. Kieselguhr) to hold a liquid phase and passed the sample in an immiscible solvent. This gave birth to liquid - liquid partition chromatography. It appears that the early developments in chromatography were monopolized and in recognition to their work, Noble Prize for Chemistry in 1952 was awarded to Martin and Synge for their contribution in this field.

Consden et al[6] in 1944 made a different approach applying principle of partitioning, which lead to invention of paper chromatography (later a well-known technique). Till such time there was no advancement in TLC.

Meinhard and Hall[7] in 1949 applied starch as binder to give some firmness to the layer for separating inorganic ions which was described as surface chromatography. Advancements were further made in 1951 by Kirchner et al[8] that used ascending method with silicic acid as sorbent for separation of terpene derivatives. They described the plates used as "chromatostrips". Reitsema[9] in 1954 could achieve separation of several mixtures in one run by using a broader plate.

Credit should go to Stahl for the development of thin layer chromatography in the 1950s'. In an attempt to develop micro-column chromatography, Izmailov and Shraiber[4] used "open" micro-column instead of closed one, which is nothing but a thin layer. Although a number of subsequent publications described about the use of this type of chromatography, these passed unnoticed until Stahl started search of highly sensitive micro-separation technique in connection with work on plant cell constituents. Stahl[10, 11] described this technique as "Dunnshicht Chromatography" which literally translated is "thin-layer chromatography".

Kurt Randerath[12] published a book on TLC in 1963 which was followed by Stahl[13] with a book entitled *Thin Layer Chromatography—A Laboratory Handbook* in 1965 and then expanded greatly in 1969. Kirchner[14] published a book *Thin Layer Chromatography*. These authors demonstrated the versatility of TLC and its application in the field of separation by which the technique gained wide acceptance throughout the world. Stahl[15] highlighted the importance of parameters like uniformity and thickness of layer, binder level and standardization of sorbents for specific surface area and particle size in achieving better reproducibility and separation.

Partition chromatography is an analytical and separation technique on the basis of different affinities of the components present in a mixture in a system of two immiscible phases, i.e. between stationary and mobile phase. The difference in migration of components can also be due to differential adsorption, ion-exchange, molecular sieving, etc. of the substance to be separated. The stationary phase is usually a solid or a liquid. But the mobile phase may be a gas, a liquid or a super-critical fluid.

The appropriate chromatography technique needs to be adopted depending upon several factors, i.e. physical and chemical nature of the sample/components/mixture.

1.1 CLASSIFICATION OF CHROMATOGRAPHY

Depending upon the separation principles, chromatography is classified as:

1.1.1 Partition Chromatography (Fig. 1.1)

This is based on a thin film formed on the surface of a solid support by a liquid stationary phase. Solute equilibrates between the immiscible mobile phase and the stationary liquid. This technique is used to separate components of mixtures based on partition between mobile and stationary phases.

1.1.2 Adsorption Chromatography (Fig. 1.2)

A mobile liquid or gaseous phase is adsorbed on to the surface of a stationary solid phase. Separation principle is based on particular interaction of functional groups of solid surfaces with the functionality of sample components dependent on H-bond, dipole-dipole, acid-base, dispersion forces, etc. Adsorption chromatography is widely used in TLC.

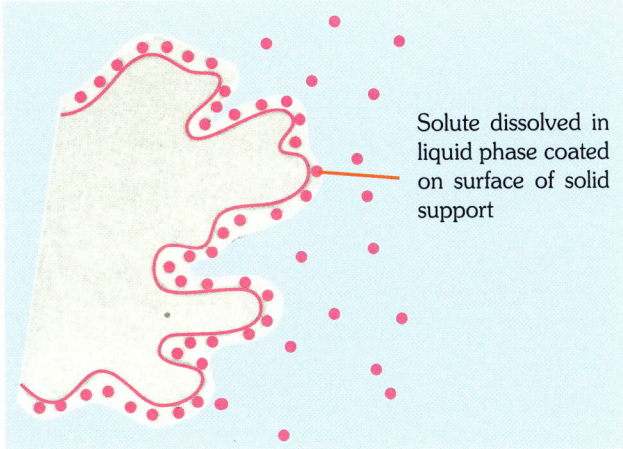

Fig. 1.1: Partition chromatography

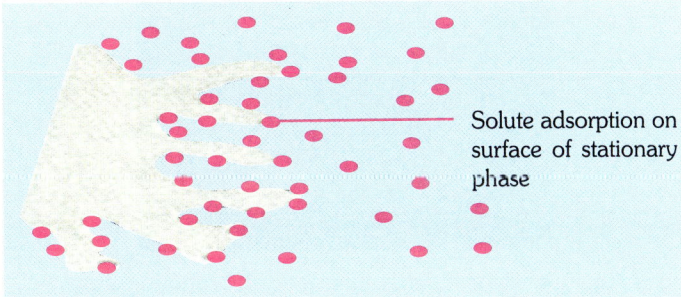

Fig.1.2: Adsorption chromatography

1.1.3 Ion-exchange Chromatography (Fig. 1.3)

In this technique, a resin (stationary solid phase) is used to covalently attach anions or cations on to it. Solute ions of the opposite charge in the mobile liquid phase are attracted to the resin through electrostatic forces.

This separation method is based on the electrostatic attraction between opposite charges on the stationary phase media and the molecules ion-exchange columns are categorized as either strong or weak. Strong ion-exchange columns are charged (ionized) at all pH levels while weak ion-exchange columns are only ionized in a certain pH range.

1.1.4 Hydrophobic Interaction Chromatography (HIC) (Fig. 1.4)

This involves retention on the basis of hydrophobic interactions.

HIC is water-eluted hydrophobic interaction chromatography. Loading is done in a high concentration of a structure-enhancing salt (cosmotrope). Elution is typically achieved with a gradient down to lower ionic strength buffer.

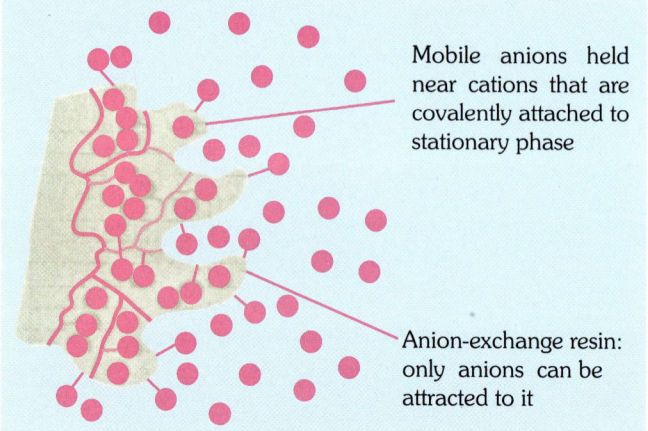

Fig. 1.3: Ion-exchange chromatography

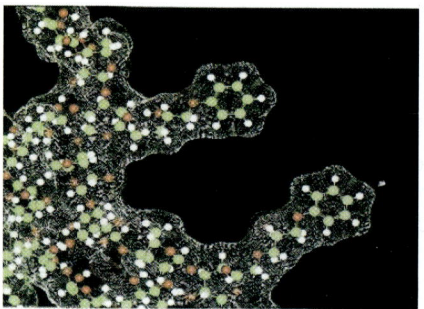

Fig. 1.4: Hydrophobic interaction chromatography

1.1.5 Affinity Chromatography (Fig. 1.5)

Affinity chromatography is based on the interaction between one kind of solute molecules and a second molecule immobilized on a stationary phase. Affinity chromatography uses a ligand attached to the media with binding affinity for either a specific molecule or a class of molecules, e.g. antibodies attached to bind one specific molecule or columns with antigenic proteins attached that bind a specific antibody.

Other types of affinity columns have lectins attached that will bind certain carbohydrates, but these are not used because most lectins are highly poisonous. The best type of ligand for affinity chromatography would be a synthetic molecule that is stable, safe, and inexpensive, has good specificity, and allows gentle elution.

1.1.6 Metal-chelate Affinity Chromatography

This form of purification is based on an immobilized metal ion that has affinity for a chain of histidines that are added to the target protein by recombinant molecular biological techniques to create a fusion protein. The protein is eluted with imidazole (the side chain of histidine) and then the chain of histidines is removed by an enzyme.

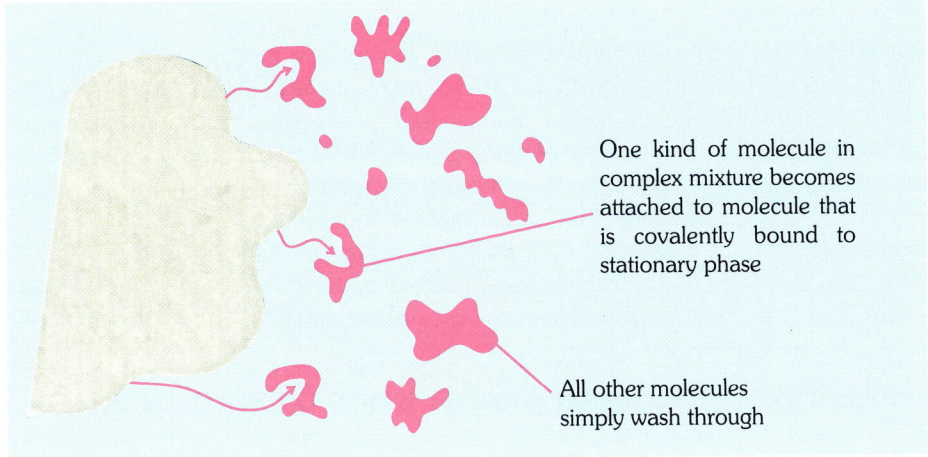

Fig. 1.5: Affinity chromatography

1.1.7 Size-exclusion Chromatography (SEC) (Fig. 1.6)

Separation is based on the size of the molecules (hydrodynamic volume is equal to volume created by the movement of the molecule in water) not the molecular weight. The difference between molecular weight and hydrodynamic volume is sharp. Proteins tend to be globular molecules while DNA or polysaccharides tend to be linear molecules.

While SEC is easy to understand and perform, this separation method gives the least resolution with the lowest capacity and largest dilution of the sample when compared to all other forms of chromatography. This is also known as gel permeation/gel filtration. It lacks an attractive interaction between the stationary phase and solute. The liquid or gaseous phase mobile passes through a porous gel which separates the molecules on the basis of its size. The pores are small and exclude the larger solute molecules, but allow smaller molecules to enter the gel, causing them to flow through a larger volume (therefore, longer distance); much cause the larger molecules to pass through at a faster rate than the smaller ones.

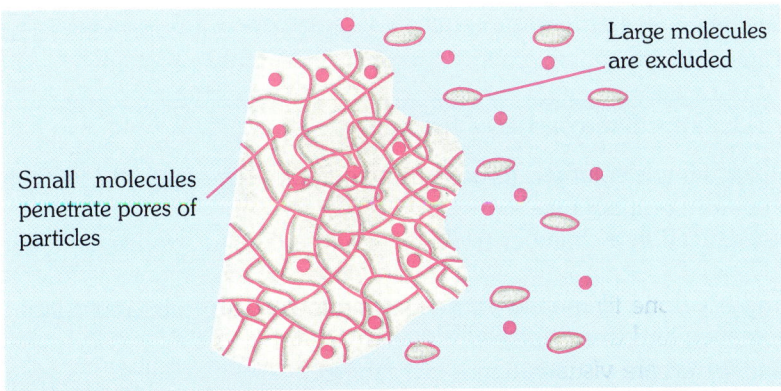

Fig. 1.6: Molecular exclusion chromatography

1.1.8 Supercritical Fluid Chromatography (SFC)

Supercritical fluids can be used as the mobile phase to separate analytes within chromatographic columns. As in supercritical fluid extraction (SFE), supercritical fluids can have solvating powers similar to organic solvents, but with higher diffusivities, lower viscosity, and lower surface tension. The lower viscosity allows higher flow rates compared to liquid chromatography, and the solvating power can be adjusted by changing the pressure. A major advantage of supercritical fluid chromatography (SFC) is that it offers the advantage of liquid-like solubility, with the capability to use a non-selective gas phase detector such as flame ionization detector. Analytes that cannot be vaporized for analysis by gas chromatography, yet have no functional groups for sensitive detection with the usual liquid chromatography detectors, can be separated and detected using SFC.

1.2 CLASSIFICATION OF CHROMATOGRAPHY BASED ON STATE OF PHASE

Based on stationary phase (solid or liquid) and mobile phase (gas, liquid or supercritical fluid), chromatography can be classified as given in Table 1.1.

S.no.	Chromatography	Mobile phase	Stationary phase
Table 1.1: Classification of chromatography according to state of phases			
1.	LLC/TLC	Liquid	Liquid
2.	GSC	Gas	Solid
3.	GLC/GC	Gas	Liquid
4.	LSC/LC	Liquid	Solid
5.	SFC	Supercritical fluid	Liquid (Solid)

1.3 PLANAR CHROMATOGRAPHY

Thin-layer chromatography is a mode of liquid chromatography, wherein the sample is applied as a spot or streak near the origin of a thin sorbent layer having support on thin glass, plastic or metal support. The mobile phase moves through stationary phases by capillary action and separation of components takes place in the open bed. The separation is done for components, having the same migration time, but different migration distances. The mobile phase may be single or mixtures of solvents.

The modern TLC procedure consists of following steps.

1. Standard and sample solutions to be chromatographed are applied at 8 mm away from the bottom edge or compact bands.
2. The plate is placed in a chamber (closed) with a measured quantity of mobile phase at the bottom.
3. Development is done till mobile phase rises up to 6–7 cm from application position.
4. Plate is removed and dried quickly.
5. Fractions migrated are visualized in a UV cabinet.
6. The plate is then quantitatively evaluated using a scanner, which records UV-VIS spectra as well.

7. Photodocumentation is carried out in UV (254 and 366 nm) and visible.
8. Post-chromatography derivatization may be done, if required.

The above procedure can meet GLP norms, if performed instrumentally.

Commercialization of TLC technique began in 1965 with availability of pre-coated TLC plates and sheets and the technique became very popular with about 450–500 research publications per year. TLC became a recognized technique for separation of mixtures, detection and semi-quantitative determination of adulterants, contaminants, etc. Silica gel also became the most popular sorbent with an average pore size of 60 Å. However, modifications to the silica gel started with silanization to produce reversed phase layers. A wider range of separation possibilities opened up based on modified silica gel layers, although even today more than 90% analysis is done using silica gel.

Interest in TLC has increased in recent years which is a result of the increasing instrumentation and automation in the technique. This has not only reduced the experimental error, but has also increased the efficiency of the method. New separation techniques have been developed (gradient and forced flow systems) and coupling with spectroscopic methods (TLC-UV/VIS, TLC-FTIR, TLC -Raman, TLC -MS). Manual operation has considerably been reduced, thus removing many sources of error. TLC is an inexpensive, simple method requiring little instrumentation, whereas modern TLC (normally termed as high-performance thin-layer chromatography or HPTLC) is a sophisticated instrumental technique for accurate quantitative determination. HPTLC is capable of giving fast results, high resolution separations, more accurate and precise quantitative results often rivaling those of other most widely used popular techniques such as gas chromatography and high performance liquid chromatography, with many advantages. TLC is an off-line technique in which various stages are carried out independently.

Schematic diagram of the steps in a high-performance thin-layer chromatographic analysis is given in Fig. 1.7.

With the introduction of high-performance thin-layer chromatography (HPTLC), the technique achieved quantification of separated components with high repeatability and reproducibility. Halpaap[16] was the first, in 1973 to recognize the advantage of smaller particle size of silica gel (about 5–6 μm) for TLC plates. Effect of particle size on development time, Rf values and plate height was demonstrated for improving the techniques. HPTLC during mid 1970's gave a new direction as the precision was improved ten times with reduction in analysis time, and use of less mobile phase with reduction in development distance. The technique is now fully instrumental and software based and is quite comparable to HPLC for much analysis.

The first commercial plates for HPTLC were called as "nano-TLC" by E. Merck the manufacturer. Halpaap et al[17] in 1980 reported reversed phase HPTLC and this then became commercially available as pre-coated plate. Hauck and Jost[18] in the year 1982 reported an amino modified HPTLC plate. During 1980's, improvements in computer controlled densitometry scanner were made, thus including the optimum of peak purity and measurement of spectra for all components after separation. Burger[19] made gradient HPTLC possible with automated multiple development (AMD). This made possible for increase in number and resolution of separated components. Subsequent developments saw the chiral separation field using chiral selectors and chiral stationary phases.

Today the entire HPTLC analysis process can be computer controlled and use of highly sensitive charged couple device (CCD) cameras has given an opportunity to electronically store images of "chromatograms" for future reference. Commercially available plates with extra pure 4-5 μm spherical silica gel have value addition to the technique reducing the background interference along with improved resolution, making it possible to use hyphenated techniques. However, the use of hyphenated techniques either on-line or off-line for quality assurance is a subject matter of challenge and validation of technique.

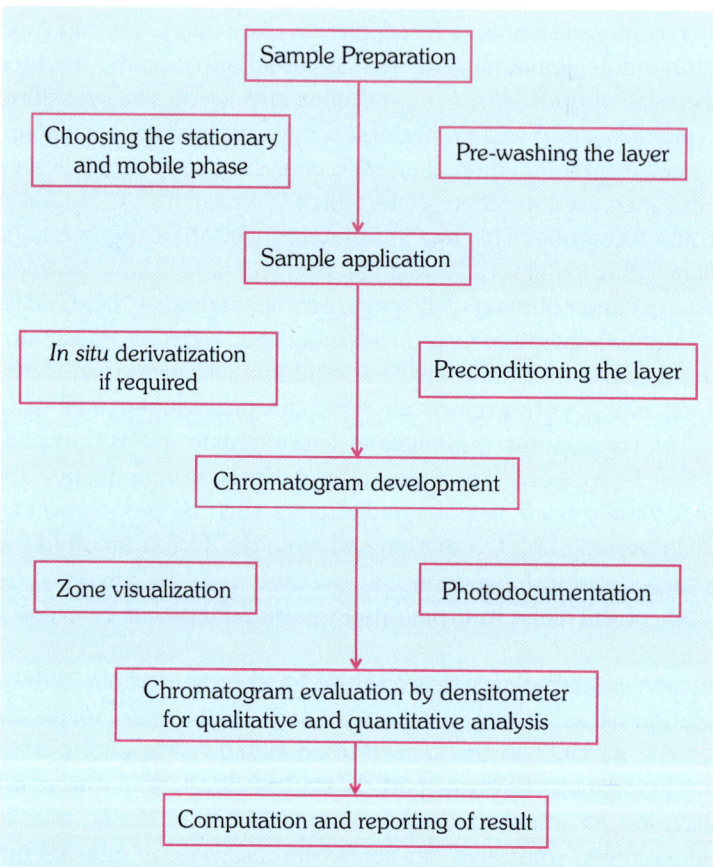

Fig. 1.7: TLC/HPTLC procedure
(Adopted from figure supplied by D. Haenchen, Camag Muttenz, Switzerland).

Sharma[20] has described advantages of this arrangement using an open disposable layer. HPTLC has the possibility of separating a large number of samples and standards even up to 60-70 at a time simultaneously on a single plate, thus making it high output, low cost analysis, with ability to construct calibration curves from standards chromatographed under identical experimental conditions. It analyses samples with minimum clean up, including those present as component that remain sorbed to the origin of the layer or suspended micro-particles. The sample can be analyzed using multiple separation steps and static post-chromatographic detection procedures with various universal or specific visualization reagents because sample components are "stored" on the layer without any chance of "loss". TLC is also highly selective and flexible due to availability of great variety of layers.

HPTLC started around 1975 with the introduction of commercially precoated plates with highly efficient uniform thinner layers (0.1-0.2mm), composed of small diameter particles (i.e. 5 µm average) and developed over a shorter distance of 3-7 cm. Other advantages of HPTLC are smaller sample volume (0.1-0.2 µl), starting spot diameter (1 or 1.5 mm), diameter of separated spots (2-6 mm), developing time (3-20 min.) and detection limits of up to low ng for absorption and low pg for fluorescence.

In planar chromatography, there is a different approach than column chromatography. Stationary phase in planar chromatography is a flat (thin) layer coated on an inert support (glass/aluminium/polyester). Whereas in column chromatography, column (glass, capillary glass, stainless steel) is packed with small particles. The term "thin-layer chromatography" has evolved accordingly. There are similarities between TLC and HPLC as both are liquid chromatography techniques. By and large same stationary and mobile phase can be used in both techniques. But difference in the method is way of defining retention of solutes.

In column chromatography, separation is based on individual time of different components to pass the column and thus variable retention times and fixed distance. But in planar chromatography mixtures migrate for a certain time and separation is based on variable migration distance and fixed time. Capacity factor "K" is used in column chromatography but R_f value to define location of separated fractions in planar chromatography. Mobile phase in column chromatography is maintained at optimum velocity through pressure. However, in case of planar, capillary forces drive the mobile phase.

1.4 BENEFITS OF HPTLC

HPTLC has several advantages over other chromatographic techniques which are discussed below keeping in mind the adulterants in food and analysis/quantification of contaminants, additives, preservatives, residual level analysis, etc. HPTLC is a rapid method most suitable for screening purpose due to its properties, i.e. easy to handle and maintain, cost-effective, and simultaneous analysis, etc.

1.4.1 Visible Results

Image is an additional feature of HPTLC, as real chromatogram (analog curves) on-line with other chromatographic techniques is readily available from densitometric evaluation. Different colours of the zone work as an additional dimension, whereas three-dimensional computer plots do not provide such information. On the basis of data obtained from Diode-array or multi-wavelength scan, all the samples could be compared on one wavelength or one sample at all wavelengths.

1.4.2 Simple Technique

No other chromatographic technique is comparable to HPTLC for a technical solution at appropriate level of sophistication. The instrument is good laboratory practice compliant, good manufacturing practice compliant and varies from manual to computer controlled devices offering high reliability and utmost versatility.

1.4.3 Simultaneous Analysis of Several Samples

Parallel analysis of samples up to 72 (approx.) on a 20×10 cm plate (36 on each side) is possible on off-line principle. Visual examination in tandem permits very rapid screening under the same chromatographic conditions. Such situations are difficult to maintain in a sequential process. There is a considerable savings of mobile phase in HPTLC analysis. Time required per sample can be as low as 1 minute and solvent consumption, about 0.15 ml in the best situation.

1.4.4 Quick Analysis

Besides parallel analysis of several samples leads to quick analysis, HPTLC does not require start-up time. Several different analyses can be done in parallel, each on a different plate, e.g. while A sample is being spotted; B can be developed; C can be scanned; and D can be photo documented, etc.

1.4.5 Flexibility in Method

The HPTLC based method of analysis can be very flexible. Many parameters can be used for a better separation. There is an unlimited choice of the mobile solvents in combination using different stationary phases. There are no restrictions with regards to interference with detections because chambers and TLC plates are inert. Mobile phase is completely removed from plate before detection, thus making it possible to use UV-absorbing solvents. Dedication of independent instruments for each step of chromatography provides greater flexibility. In HPTLC, various separations can be performed by simple way of changing the mobile phase. The instruments for sample application and detection have no bearing on mobile phase combination.

1.4.6 Easy to Handle

In column chromatography, column is cleaned thoroughly to avoid contamination, whereas in HPTLC, plates once used are disposable. Separation in HPTLC can be optimized for a particular portion of the sample without taking care of irreversible adsorption of matrix components at the application point while examining the plates. All components of samples can be seen at one time as part of chromatogram.

1.4.7 Multiple Detections and Examinations of Chromatogram

The "developed" plates with separated components work like a storage device, which can be subjected to repeated performances for detection and evaluations under same or different conditions. The plates can be subjected to examinations with and without derivatizations by use of various light sources. Sequential derivatizations are combined with multiple images/scans. Scanning densitometry can be performed before and after derivatization step. A wealth of information and data from a single plate can be collected about the samples.

1.4.8 Cost-effective

HPTLC is a state of art instrument providing a number of options and sophistication. Initial cost depends upon the requirements of analysis and depending upon the specifications. The operating cost and recurring expenditure in analysis is much less in HPTLC.[21] There is very low expenditure on maintenance of equipment and use of solvents.

1.4.9 Inevitable Progress

HPTLC will continue to grow as a highly selective, quantitative, rapid automated technique for analysis of samples.[22]

1.5 PRECAUTIONS

No doubt HPTLC offers a versatile analysis in different domains, but precautions need to be taken due to its limitations.

- HPTLC being an open system is exposed to environment and climatic conditions, hence due to precautions for achieving reproducible results should be taken especially in case of semi-volatile and photo/oxygen sensitive samples.
- Although computer controlled system for each step is now available for the HPTLC, but off-line principle asks for a manual intervention between two steps which may lead to problem in reproducibility.
- HPTLC may not be suitable for samples that must remain in solution, as the sample is dried on the plate. Separations requiring largely aqueous mobile phases may be difficult to use due to long development times.
- Parameters like humidity and composition of gas phase in the chamber can affect the separation. Modern automated developing chambers overcome this limitation.
- The separation power of HPTLC may be little lower than other chromatography techniques specially for complex samples. It may be cumbersome to achieve separation of all components.

1.6 EFFICIENCY IN SEPARATION

Better efficiency in separation can be achieved by taking proper precaution and optimizing the following conditions.

- By using HPTLC pre-coated ready made plates.
- Proper selection of the mobile phase.
- High quality of solvents used.
- Selection of a developing chamber of an appropriate size and shape.
- Adequate saturation of chamber.
- Using band system for sample application.
- Optimizing the relative humidity and temperature.
- Gradient development.
- Flow rate of solvent, separation distance and mode of development.
- Rapid and uniform drying of the plate after development.
- Keeping development distance as short as essential, usually 7 cm maximum.

1.7 COMPARATIVE EVALUATION OF HPTLC

Comparative evaluation of HPTLC with other chromatography has been enumerated in Tables 1.2, 1.3 and 1.4.

S. no.	HPTLC	HPLC
	Table 1.2: Comparison of technique used by HPTLC and HPLC	
1.	Flexible	Limited flexibility
2.	Several samples can be analyzed at a time.	Only one sample at a time
3.	Easy to learn and operate	Trained personnel are required.
4.	Low cost of analysis and consumes less time	With total automation it is costly. However, takes more time.
5.	Low maintenance	High maintenance
6.	Fully automated	Fully automated
7.	Sample preparation is easy.	Sample preparation is laborious, time consuming and expensive.
8.	Choice of solvent for dissolving sample is not critical.	Solvent selected for dissolving should be compatible with column and mobile phase.
9.	Sample clean-up is easy.	Sample requires proper purification and clean up so that column is not damaged.
10.	*In situ* derivatization is very simple.	*In situ* derivatization is not practical.
11	Solvents need no prior treatment like filtration and degassing.	Solvents have to be pre-treated.
12.	Solvent of analytical grade may also be used for routine purpose.	Solvent of chromatographic grade is recommended for use.
13.	Consumption of mobile phase is very low.	Consumption of mobile phase is high.
14.	No possibility of contamination of the previous analysis as every time a new plate is used, hence no carry over.	Since the columns are used several times, it is necessary to remove the contaminants and proper cleaning of columns. Possibility of the carry over of the contaminants.
15.	Mobile phases having pH 8 and more can be used.	Mobile phases having pH more than 8 not recommended.
16.	Visual detection is possible.	Visual detection is not possible.
17.	Multiple scanning of several samples can be done at a time.	Not possible
18.	Resolution limited.	Resolution is superior.
19.	Separation by adsorption common	Separation by partition common
20.	Selectivity is high.	Selectivity is poor.

S. no.	HPTLC	HPLC	GC
	Table 1.3: Comparison of technique for HPTLC, HPLC and GC		
1.	Small amount of mobile phase required.	Large amount of mobile phase required.	Large amount of mobile phase required.
2.	Temperature and pressure control not required.	Temperature and pressure control required.	Temperature and pressure control required.
3.	Off-line system	On-line system	On-line system
4.	One can visualize the sample.	One cannot visualize the sample.	One cannot visualize the sample.
5.	It's a multiple sample process.	At a time, only one sample can be processed.	At a time, only one sample can be processed.
6.	Economical	Expensive	Expensive
7.	Sample preparation is minimal.	Sample preparation is tedious.	Sample preparation is tedious.
8.	It's the fastest technique.	Comparatively slow.	Comparatively slow.
9.	Low maintenance	High maintenance	High maintenance

S. no.	Feature	HPTLC	TLC
	Table 1.4: Comparison between HPTLC and TLC		
1.	Technique	Instrumental	Manual
2.	Layer	Pre-coated	Lab made/pre-coated
3.	Shape of applied sample	Rectangular, e.g. (6 mm L × 1 mm W)	Circular (2-4 mm dia.)
4.	Sample transfer by	Spray-on	Contact
5.	Vol. precision	Within 1-2%	Depends (2-10%)
6.	Precise positions of sample application	±0.1 mm	±1 mm
7.	Sample application position, from bottom	8 mm	10 to 15 mm
8.	Micro-preparative chromatography	Ideal	Crudely possible
9.	GLP	Compliant	Not compliant
10.	Vol. range	0.1 TO 500 μl	1 to 10 μl
11.	PC connectivity	Yes	No
12.	Method storage	Yes	No
13.	Validation/qualification	Yes	No
14.	Sample holder	Syringe	Capillary/pipette
15.	Qualitative analysis	Yes	Yes
16.	Quantitative analysis	Yes	No
17.	Wavelength range	190-800 (Monochromatic)	254 or 366 nm, visible
18.	Spectrum measurement	Yes	No
19.	Evaluation by	Scanner	Analyst
20.	Chromatogram	On plate-scan/image	On plate image only
21.	Layer thickness	100/200 μm	250 μm
22.	Efficiency	High due to smaller particle size	Lower
23.	Separations	3-7 cm	10-15 cm
24.	Analysis time	2-10 min.	15-120 min.
25.	Solid support	Wide choice of stationary phases	Wide choice
26.	Development chamber	New type that require less mobile phase(twin trough)	More amount (Flat bottom)
27.	Sample application	Automatic	Manual
28.	Scanning	Use of UV/visible/fluorescence scanner scans the entire chromatogram qualitatively and quantitatively. Scanner is an advanced type of densitometer	Visual assessment in UV cabinet or after derivatization

References

1. Tswett M. *Proc Warsaw Soc Nat Sci Bio Sec* **14**, No. 6; 1903.
2. Tswett M. *Ber Dent Botan Ges* **24**; 316, 384: 1906.
3. Day D T. *Science* **17**; 100 7: 1903.
4. Izmailov N A, Shraiber M S. *Farmatisiya* **3**: 1; 1938.
5. Martin A J P, Synge R L M. *J Biochem* **35**: 1358; 1947.

6. Consden R, Gordon A H, Martin A J P. Partion chromatography with paper. *J Biochem* **38:** 224; 1944.

7. Meinhard J E, Hall N F. *Anal Chem* **21**: 185; 1949.

8. Kirchner J G, Miller J M, Keller G E. *Anal Chem* **23**: 420; 1951.

9. Reitsema R H. *Anal Chem* **26**: 960; 1954.

10. Stahl E. *Pharmazie* **11**: 633; 1956.

11. Stahl E. *Chemiker-Ztg* **82**: 323; 1958.

12. Randerath K. *Thin-Layer Chromatography*, Academic Press, London, U.K.1963.

13. Stahl E. *Thin-Layer Chromatography—A Laboratory Handbook*, Springer-Verlag, Berlin, Germany, 1965.

14. Kirchner J G. *Thin-Layer Chromatography*, 2nd edn., Techniques in Chemistry, vol. XIV, Wiley-Interscience, Chichester, UK, 1978.

15. Stahl E. *Thin-Layer Chromatography*, 2nd edn, E. Stahl (ed), Springer-Verlag, Berlin, Germany, **5**, 1969.

16. Halpaap H. *J Chromatogr*, **78**: 77-78; 1973.

17. Halpaap H, Krebs K.F, Hauck H E, J.HRC and CC, 3, 215-240, 1980.

18. Hauck HE, Jost W. *Instrumental High Performance, Thin-Layer Chromatography, Proceedings of 2nd International Symposium*, R. E. Kaiser (ed), Interlaken, Switzerland, 25-37, 1982.

19. Burger K Z. *Anal Chem* 318, 228, 1984.

20. Sharma J. *J Assoc of Anal Chem* **74**: 435-437; 1991.

21. Renger B. Benchmarking HPLC and HPTLC in pharmaceutical analysis. *J Planar Chromatography* **12**: 58-62; 1999.

22. Sharma J. *Anal Chem* **74**: 2653; 2002.

2

Theoretical Concept of
Thin-Layer Chromatography

Chromatography theory discusses about the physico-chemical relationship governing the separations. TLC separation is basically the interaction of the compounds for separation (solutes) with the stationary and mobile phase. A number of books are available on theoretical concept of the planar chromatography[1-4] which can be referred to for details about the theory of chromatography. In this chapter, a brief discussion has been made about the theoretical concept.

2.1 RETENTION MECHANISMS

TLC is based on the affinity of sample with the stationary and mobile phase. *Partition coefficient* (K) is defined as the ratio of substance concentration in the stationary phase (C_s) and substance concentration in the mobile phase (C_m)

$$K = C_s/C_m \qquad \text{(Eq. 2.1)}$$

A molecule moves with the flow in the mobile phase, whereas it remains static on stationary phase. Therefore, the molecule is retained in respect to mobile phase flow. Retention mechanism is based on different properties. Normally, three main physico-chemical processes, i.e. adsorption, partition, complex formation and ion exchange are mainly responsible for determination to retention capacity and selectivity.

2.2 ADSORPTION CHROMATOGRAPHY

Stationary phase is normally a solid and a variety of interactions takes place at the surface. Adsorption takes place between polar sub-structures of solutes and different adsorptive centres of the sorbent. Solute having high adsorption capacity tends to bind more strongly to this sorbent. This results in enhanced retention, whereas solute with lower adsorption strength will result in more easy flow. The character of adsorptive forces engaged in solute-sorbent interaction is complicated issue. Interaction is influenced by solute sub-structure with permanent dipole-movements, with inducible dipole-movements and with capacity to form hydrogen bond. The degree of retention is based on the number and type of functional groups of the molecule of the sample and on its steric structure.

In normal/straight phase chromatography, the mobile phase is organic and less polar than adsorbent. Since mobile phase molecules are adsorbed in the stationary phase and so are in competition with the sample molecule. Various approaches are made for mathematical description of adsorption.[5–7] One approach is that adsorbed solvent molecules are displaced from adsorption sites on stationary phase by sample molecules, whereas the other speaks of retention mechanism as adsorption process highlighting sample-mobile phase interaction. Both processes occur alternatively based on composition of mobile phase. For adsorption chromatography, the activity is stated as a certain property of adsorbent affecting retention. Activity of the stationary phase is also of fundamental importance, as it may be influenced by the relative humidity to which stationary phase was exposed before development of chromatogram.

2.3 PARTITION CHROMATOGRAPHY

It is governed by the molecular hydrophobicity which takes place between the solute molecules and mobile phase. It is expected that solute having more preference for the mobile phase (in context of their high solubility) will elute prior to the compounds with lower solubility in the eluent. Therefore partition chromatography is based on different solubility of sample compounds in immiscible liquid phases. The stationary liquid phase is immobilized on its solid porous support either by adsorption or by chemical bonding.

When component of mobile phase is much stronger adsorbed, or wets the surface better than other components, then partition system can be generated dynamically during the process of chromatography. This is called as solvent generated stationary phase, e.g. water forms multilayer on silica gel/cellulose when water saturated mixtures of organic solvents are used as mobile phase. This is suitable for separation of water soluble polar solutes. Acid/base/buffer may also be added to mobile phase to control protolysis and to separate non-dissociated acidic or basic compounds. It is easier to achieve reproducible results in partition system as compared to true adsorption system. Whereas it is difficult to interpret relative retention of a given set of compounds (e.g. monosaccharides or amino acids in mixtures). Separation optimization may be trial and error method. The stationary phases for chromatography are chemically bonded on to silica gel. Reversed phase (RP) and polar bonded phased (PBP) are looked as a monomolecular liquid film on a solvent support. RP system comprises of non-polar stationary phase and polar aqueous mobile phase. The reason for retention in such cases is because of their at least partial hydrophobic nature; organic molecules are "expelled" from aqueous mobile phase and then "incorporated" into lipophilic stationary phase. On a reverse phase, retention depends on aspects of lipophily such as steric hindrance, absence of branches or length of carbon chain. Description of PBP such as cyano-, amino-, etc. is more complicated, because they tend to interact with the functional group of the samples. Therefore in such cases, adsorption on the surface of such phases needs to be considered.

2.4 COMPLEX FORMATION

It is based on the reversible formation of coordination compounds between Lewis acids and bases, e.g. in case of separation of unsaturated or polyaromatic compounds, the stationary phase is impregnated with Lewis acids such as metal ions, e.g. Ag^+, Zn^{2+}, or strong electrophilic organic compounds, e.g. caffeine, picric acid, etc. Separation of compounds depends upon the degree of unsaturation and according to steric effects. Chiral compounds can also be separated through complex formation.

2.5 ION-EXCHANGE CHROMATOGRAPHY

In this type of chromatography, stationary phase is able to stoichiometrically replace its counter ions and with ions of the analyte. Cation exchange is having functional groups like sulfonate, phosphate or carboxylate on resins, dextranes or cellulose support. Anion exchangers feature with ammonium, quaternary alkyl ammonium salts. *Retention* is described as a function of charge density and polarizability of the concerned ions.

2.6 ION-PAIR CHROMATOGRAPHY

It is a simple way to separate electrolytes. The stationary phase is C_8 or C_{18} bonded phase. Ion pairs are formed by adding to the mobile phase ion-pairing reagents such as sodium alkyl ammonium bromides for acidic samples. These are separated like and together with neutral compounds in the same sample. The principal retention mechanism is based on the partitioning of the ion pair between stationary and mobile phase. Another aspect assumes that ion-pairing reagent may be part of the stationary phase whether by the process of impregnation before chromatography or through absorption out of mobile phase during chromatography. Thus retention may be visualized as dynamic ion exchange.

2.7 SIZE-EXCLUSION CHROMATOGRAPHY

This is based on separation of molecules according to their sizes. The specific pores of the stationary phase permit penetration by molecules up to specific size (normally expressed as molecular weight). If the molecule is smaller, it will diffuse deeply into the pore and is retained longer. Hence molecules with larger size will elute first. Silica gel, synthetic resins, cross-linked dextranes are by and large, used as stationary phase.

2.8 INCLUSION CHROMATOGRAPHY

It is based on formation of inclusions as in specific cases like cyclodextrins make cages of particular dimension and chemical viability in which the sample molecules may enter. Separations of optical and structural isomers are possible with this technique.

2.9 EFFICIENCY, DIFFUSION, SELECTIVITY AND RESOLUTION

Regarding distribution of sample between stationary and mobile phase, it can be assumed that during the time taken to establish the corresponding equilibrium of concentrations, mobile phase advanced a certain distance which is called as theoretical plate. Normally, the height equivalent to a theoretical plate (HETP, or H) is in the range of 10 to 100 µm. In a different approach to separation, H is called efficiency which has the ability to generate sharp peaks in chromatographic system.

The chromatographic system should be selective for separation of compounds because compounds are retained to a different extent. This can be achieved if the partition coefficients of the compounds are different and selectivity factor (α) is more than 1. It is prescribed as

$$\alpha = K_1/K_2 \ (K_1 > K_2) \tag{Eq. 2.2}$$

where α is selectivity factor (greater than 1) and K is partition coefficient.

Ideally, several theoretical plates must be passed for quantitative separation. But whenever the molecule of the sample is in mobile phase, it is subjected to diffusion which counteracts the efficiency.

These can be discussed into more details regarding their calculation and empirical determination.

H (HETP) and plate number (N)

This can be defined as efficiency of chromatographic system.

Plate number (N) is expressed as

$$N = L/H \tag{Eq. 2.3}$$

where 'N' is plate number, 'L' is length of system, and 'H' is the height equivalent to a theoretical plate.

The plate number can also be calculated directly from the chromatogram assuming a Gaussian peak (Fig. 2.1). Peak base width (W_b) or width at half height (W_h) may be used. Width is defined in time units so that N has no dimension.

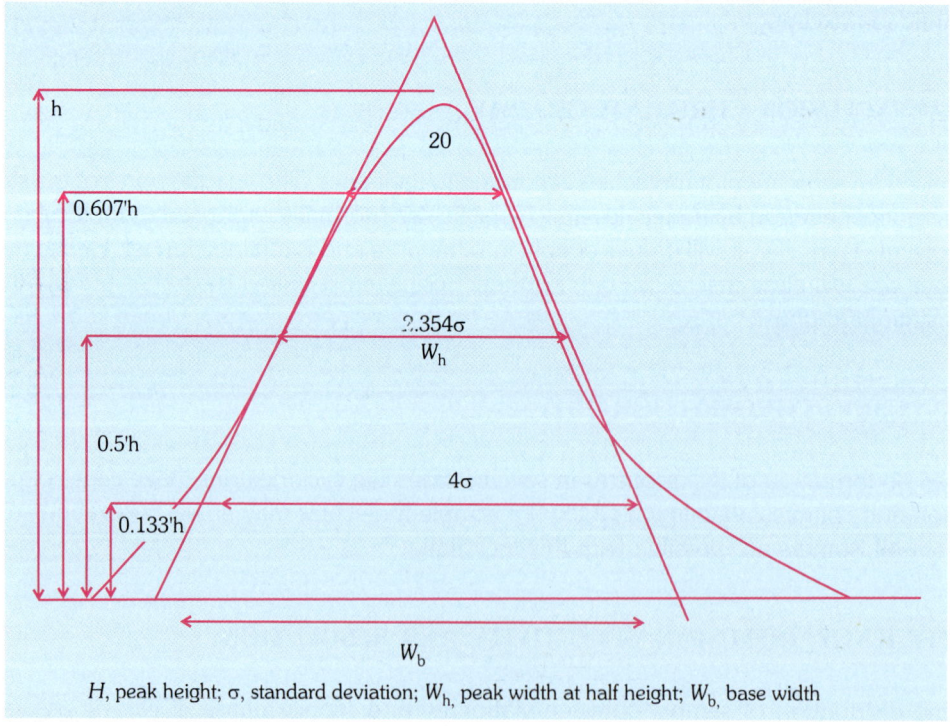

H, peak height; σ, standard deviation; W_h, peak width at half height; W_b, base width

Fig. 2.1: Characteristics of a Gaussian peak

$$N = 4(t_R/W_h)^2 \qquad \text{(Eq. 2.4)}$$
$$N = 16(t_R/W_h)^2 \qquad \text{(Eq. 2.5)}$$

where N is plate number, t_R is retention time, W_h is peak width at half height, and W_b is base width.

It is always desirable to obtain large theoretical plate numbers to ensure sharp peaks and for separation of several compounds with baseline resolution. Large plate numbers can be achieved theoretically with low values for H or developing distances. But practically there are physical limitations.

2.9.1 Broadening of Band

Broadening of band occurs due to diffusion significantly, when the sample is injected as a band into a stationary phase through which the solvent is flowing. This is known as longitudinal diffusion. This occurs due to heat movement of molecules in solution and is important at low flow rates.

Molecules adopt several different paths when mobile phase flows through bed of heavily packed small particles. It is based on different length of individual paths and different flow rates in paths of different width. This is called "eddy diffusion" and is controlled by the quality of the packing of chromatographic bed. Various effects such as mobile phase mass transfer, stationary phase mass transfer also occur.

Combination of all these effects causes the peak to get broader the longer the separation path is on TLC. Therefore, retention is a major cause of band broadening whereas it is the requirement of separation. The key factor is flow rate of mobile phase. If the rate is very low, longitudinal diffusion has major role and in extreme cases there is no separation if mobile phase is stagnant. When flow rate is high, the stagnant mobile phase and stationary phase mass transfer become dominant. If one looks at efficiency (H), it is seen as the ability of chromatographic system to provide narrow bands or to minimize band broadening.

The Van Deemter equation links different causes for band broadening with the efficiency H.

$$H = A + B/u + Cu \qquad \text{(Eq. 2.6)}$$

where H is height equivalent to a theoretical plate, A is constant (eddy diffusion, mobile phase mass transfer), B is constant (longitudinal diffusion), C is constant (stagnant mobile phase, stationary phase mass transfer), and u is flow velocity (cm/s).

2.9.2 Resolution

The separation of two compounds is expressed as resolution (R_s) between the corresponding peaks in the chromatogram.

$$R_s = (t_2 - t_1)/1/2(W_{t1} + W_{t2}) \qquad \text{(Eq. 2.7)}$$

where R_s is resolution, $t_{1 \text{ resp } 2}$ is retention time of peaks, and $W_{t1 \text{ resp } 2}$ is base width of peaks.

A resolution of $R_s = 1.0$ is required for separation of two compounds to at least 98%, whereas baseline resolution, i.e. 99.5% can be achieved at $R_s = 1.25$.

A number of factors will affect the resolution and therefore should be used to control it. In these contexts, the following equation may be given.

$$R_s = \underbrace{\frac{1/4(\alpha - 1)}{a}}\, \underbrace{\frac{N^{1/2}}{b}}\, \underbrace{\frac{(k'/1 + k')}{c}} \qquad \text{(Eq. 2.8)}$$

where R_s is resolution, α is selectivity factor, N is plate number, and k' is capacity factor.

'a' of the equation is "selectivity" and easy to control by proper combination of stationary and mobile phase which are basically responsible for good resolution.

The efficiency term 'b' is normally affected by alteration in the length, flow rate of the mobile phase or packing material diameter and porosity.

The term 'c' basically depends on the solvent strength of the mobile phase. Weaker solvents can give larger k'. However, when k' is initially less than 1, increasing it will improve R_s. However, when k' is greater than 5, the increase may be marginal due to excessive band broadening.

2.10 QUANTITATIVE EVALUATION

As a matter of analytical interest, it has to be watched that all components in an analyte should be detected. Identification in a qualitative analysis can be done by use of retention time/migration distance of the corresponding substance by the position of peak maximum in a chromatogram. The height and/or area of the individual peaks are quantitatively evaluated as given in Figure 2.2.

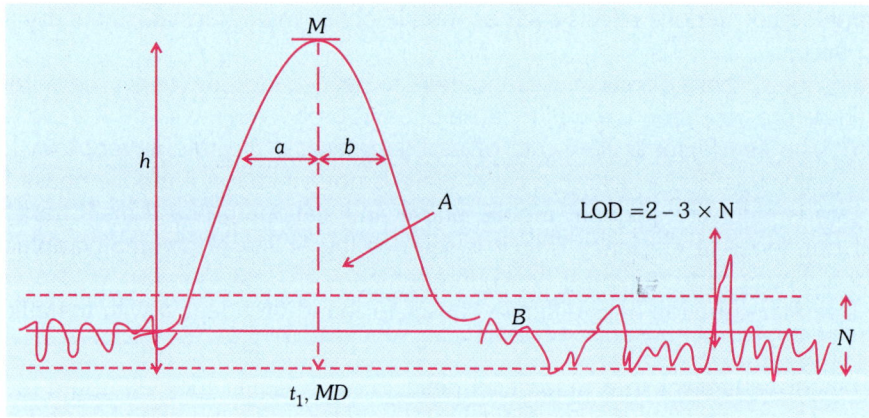

Fig. 2.2: Evaluation of peaks

The signal (detector output) is due to either sensitivity of the concentration which corresponds to a change of mass per unit volume, or mass sensitive due to change of mass per unit time. A reference standard of known mass/concentration in the same proximity is needed for evaluation.

The sensitivity of a mass sensitive detector can be defined as the quotient of peak area and related sample amount. Baseline (B) is a signal output by a detector when sample is absent and when plotted against time/position. Normally, baseline is flat; however, it depicts small fluctuations which are called noise (N). Every individual compound has a fixed limit of detection (LOD) in a given system which can basically be defined as the amount of signal generated significantly (approximately three times larger than baseline noise). The dynamic range of the detector is obtained by plotting detector signal against increasing sample quantity. It is between the range that creates a discernible change in the signal at the lower end (noise) and upper end (saturation). The peak asymmetry parameter (a/b) describes tailing (more than 1) or leading (less than 1) peaks.

2.11 MISCELLANEOUS ASPECTS

The principle used in HPTLC is "offline". It allows a critical examination of chromatogram as a sequence of individual or by and large independent steps. It is possible to characterize by many parameters, e.g. which are not available in online separations.

2.12 MOBILE PHASE FLOW

The capillary action is responsible for mobile phase flow. TLC is dependent on capillary forces. It is not possible to control them in general. The surface of the stationary phase when covered with mobile phase molecules lowers the energy of the small particles which have very high surface energy on account of micropores, and cavities between particles of layer packing. This constitutes a reason for flow. Another reason is from solvent flow which tries to bring down its energy by entering a capillary cavity. The resulting energy change ΔE is defined by the following equation.

$$\Delta E = \gamma V_m / D \tag{Eq. 2.9}$$

where ΔE = energy change, γ = surface tension, V_m = molar volume, and D = pore diameter.

Detection of the true solvent front is a difficult task due to establishment of frontal volume ingredients because smaller capillaries of particles are filled first and larger pores close to the migration solvent front tend to remain empty for a longer time.

The aggregate velocity of a mobile solvent decreases with increasing distance of development, because resistance against the flow grows simultaneously assuming that capillary forces in a particular sorbent is constant with respect to a specified solvent system. Prior to solvent entering into an empty capillary at solvent front, it has to migrate through a bed of increasing length of wet particles.

The movement of mobile phase front position in context of immersion line is described by following equations.

$$Z_f = \sqrt{kt} \tag{Eq. 2.10}$$

$$K = 2K_0 d_p (\gamma/\eta)\cos\theta \tag{Eq. 2.11}$$

where Z_f = mobile phase front position, k = flow constant, K_0 = permeability d_p = average particle size, γ = surface tension, η = viscosity, θ = contact angle.

Actual velocity of a mobile phase during development is difficult to predict as it is affected by several factors such as extent of saturation of development chamber, evaporation of components of mobile phase, resistance developed due to increase in migration distance.

In reverse phase plate, the front velocity slows down as compared to normal plates due to large contact angle (θ) of water on the non-polar surface of the capillary. The flow constant (K) becomes low when increasing water content (40%), θ reaches to about 90°. At this stage, no movement is seen.

It can thus be concluded that mobile phase tends to slow down with increasing development distance, eventually coming to a halt. Irrespective of R_f value, resolution invariably exhibits a maximum between 5-7 cm development distance, with best at 6 cm.

$$R_s = \frac{1}{4}(\alpha - 1)(R_F N)^{1/2}(1 - R_F) \tag{Eq. 2.12}$$

where R_s = resolution, α = selectivity factor, N = plate number, R_F = retardation factor.

2.13 EFFECTS OF GASES

The gases present in chamber interact with stationary as well as mobile phase, besides sample components. On one hand, it can be interpreted as advantageous because these can be used to improve the separation. On the other hand, it is impossible to control and interpret the results of separation in context of mathematical expressions.

2.14 EFFECT OF HUMIDITY

Ambient air contains certain amounts of water vapor which is expressed as humidity. *Absolute humidity* is defined as amount of water in grams per cubic meter of air. Whereas *relative humidity* is expressed as the ratio of absolute humidity and saturation humidity (max. amount of water vapor that is contained in air at a given temperature).

$$\%RH = (H_{abs}/H_{sat}) \cdot 100 \qquad\qquad\qquad \text{(Eq. 2.13)}$$

where H_{abs} = absolute humidity (g H_2O/m^3 air), H_{sat} = saturation humidity (g H_2O/m^3 air).

Silica gel forms adsorption equilibrium with relative humidity. The silica gel becomes less active with higher humidity. Activity decrease results in increase of R_f values. However, this effect is less with increasing polarity of the mobile solvent. It is worthwhile to indicate that amount of absorbed water on sorbent layer also affects the selectivity of the separation, position of secondary fronts, and shape of the chromatographic zone. Thus, a good separation may be achieved at any activity, but the results may be susceptible to changes.

2.15 REACTIONS IN DEVELOPMENT CHAMBERS

The equilibrium established between components of the mobile phase with their vapor is called chamber saturation. The gas phase composition may differ from that of developing solvent depending on the vapor pressure of the individual components.

The time taken to reach saturation will be on the ratio of solvent volume, surface of chamber volume and geometry.

Secondly, preconditioning occurs when dry stationary phase will be exposed to the solvent present in gas phase. When plate is not in contact with the developing solvent (e.g. in a sandwich chamber), equilibrium will eventually reach which is called sorbtive saturation. The surface of stationary phase is covered by thin layer of solvent molecules. Thus, composition of such layer depends on the concentration of adsorptive interaction.

During the course of development, mobile phase rises through the layer of the plate. Part of solvent tries to evaporate in the chamber unless the chamber is fully saturated. This evaporation may be controlled by vapor pressure as in saturation and by adsorption equilibriums of the mobile phase components. The evaporation becomes important the higher the mobile phase rises in case of unsaturated chamber. This process inhibits the migration of the mobile phase and tries to change its composition. During migration process, the components of the mobile phase may separate due to stationary phase on account of different degree of interaction with the individual solvents that make up the mobile phase. These effects can be minimized by saturating the chambers with a thick filter paper for at least 20 min and pre-conditioning the stationary phase.

TLC occurs in a non-equilibrium condition between stationary, mobile and gas phases. Efforts should be made to keep constant as far as possible all external parameters for reproducible chromatographic results.

2.16 DEVELOPMENT MODES

Normally, ascending development of a plate in a "normal all glass tank" is the technique widely used in planar chromatography in past and now. But there are other techniques used quite successfully due to their different aspects.

2.16.1 Single Development

This mode is widely used when the plate is vertical. While in case of horizontal chamber, the plate is located horizontally with the layer facing down. Theoretically there may not be any difference between these two modes.

2.16.2 Circular Development (Fig. 2.3)

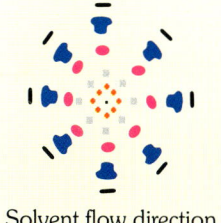

Solvent flow direction

Fig. 2.3: Circular development

In this mode, the solvent is introduced at the centre of a circle on the circumference of which sample is spotted. Area of the circle increases with time, the radial solvent throughput decreases. Zones of samples get stretched to the side after development. R_f values in this case are larger as compared to linear chromatography in low R_f range. This technique is useful for compound separation having low R_f values.

2.16.3 Anticircular Development (Fig. 2.4)

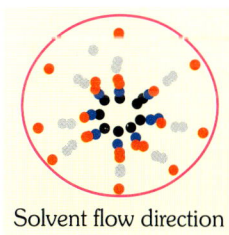

Solvent flow direction

Fig.2.4: Anticircular development

This mode requires a special chamber which is able to introduce mobile solvent on the periphery of a circular layer. In the process of moving towards centre, the velocity of mobile phase goes on increasing thus allowing the sample zones stretched towards the direction of chromatography.

R_f values in this case are generally lower than linear chromatography, thus permits better separation of sample components with high R_f values.

2.16.4 Continuous Development

The principle behind such technique is to produce a continuous flow rate of the mobile phase to separate difficult pairs of compounds at low R_f values on a short bed of adsorbent. This process is achieved by evaporation of mobile solvent of low solvent concentration at the top or some point of the plate. This technique consumes more time, i.e. about 1-2 hours.

2.16.5 Multiple Developments

Basically there are three types of multiple developments. Development with the same solvent repeatedly is done for improving the separation between two sample zones at low R_f values. The objective is to enhance the separation path beyond the extent of normal path available on the plate. But this should not make any difference for a sample where mobile phase front, in fact is as long as the flow is achieved. This approach is like continuous development. In this process, solvent front for every subsequent development attains the lower spot first, thus theoretical gain in resolution is lost to a certain extent. It is time consuming and may not enhance the results significantly.

Multiple developments with different solvents permit gradient development and thus provide the scope of separating mixtures for a wide polarity range. There can be a three-step manual method or automatic 25-step gradient development technique.

The important aspect of multiple developments is drying of plate for removal of solvent runs. Heating the plate may disintegrate the structure of sample component. Drying with cold air must take care to avoid contamination of the plate. These may affect the results of subsequent runs. Drying under vacuum may provide better results.

Two-dimensional separations are carried out to check method suitability (no additional peaks in second run) or stability of sample in chromatographic conditions. The sample in question is applied in one corner of a plate and developed. A second dimension development is done at right angles (after turning the plates through 90°). All separated fractions must come in one line.

2.17 RESULTS

Sample separation is evaluated on the basis of different retention of the each component of sample by using migration distance. The position of the sample zone on a plate can be described in a simple way by applying x and y coordinator on the lower left corner of the plate as origin. This approach can provide necessary information for routine qualitative and quantitative analysis. However, due care should be taken that position of the sample application and solvent should remain the same and all other conditions are kept constant.

2.17.1 R_F Value and other Parameters (Retardation Factor, Ratio to Front)

R_F value is a relative measure of sample position in chromatogram with reference to position of the solvent front and is explained by following equation.

$$R_F = Z_i/(Z_f - Z_0) \tag{Eq. 2.14}$$

$$R_F = Z_i/Z_f' \tag{Eq. 2.15}$$

where Z_i = migration distance (MD) of substance (mm), Z_f = migration distance of front measured from the immersion line (mm), Z_0 = distance between immersion line and sample application position (mm), $Z_f' = Z_f - Z_0$ ("front migration").

True R_F value (R_F') can be correlated with partition coefficient (K) as given below.

$$R_F' = 1/[1 + K(A_s/A_m)] \tag{Eq. 2.16}$$

$$R_F' = 1/(1 + k') \tag{Eq. 2.17}$$

where R_F' = "true" retardation factor, K = partition coefficient, A_s/A_m = phase cross-section (= phase ratio), k' = capacity factor.

The true R_F takes into consideration effects of gas phase and preconditioning. It differs from observed R_F ($R_{F(obs)}$) by a factor of 1.8 to 2.2. It has been observed that measured R_F value is generally low and can therefore be corrected by following formula.

$$R_F' = \xi R_{F(obs)} \quad \xi = 1.8 \dots 2.2 \tag{Eq. 2.18}$$

For obtaining the reproducible results, it is essential that sample application position and the level of developing solvent in a chamber should always be kept constant.

R_F value is explained in two significant digits, i.e. $R_F < 1 R_F > 0$. However, it is a normal practice to use hR_F value which is described as below.

$$hR_F = 100 \cdot R_F \qquad \text{(Eq. 2.19)}$$

In case, position of a substance zone is stated relative to that reference standard, the relative retardation factor is defined by following equation with the sample zone in determining distance of reference compound.

$$R_{rel} = Z_i/Z_{st} \qquad \text{(Eq. 2.20)}$$

where R_{rel} = relative retardation factor, Z_i = migration distance of substance (mm), Z_{st} = migration distance of reference compound (mm)

For theoretical work, e.g. the elaboration of structure retention relationship, the following equation is useful.

$$R_M = \log(1 - R_F)/R_F \qquad \text{(Eq. 2.21)}$$

where k' (capacity factor) $R_M = k'$

2.17.2 Resolution Optimization

Resolution is calculated by using peak data from analog curve of the chromatogram or by using the following equation.

$$R_s = \underset{a}{\underbrace{1/4(\alpha - 1)}} \; \underset{b}{\underbrace{(R_F N)^{1/2}}} \; \underset{c}{\underbrace{(1 - R_F)}} \qquad \text{(Eq. 2.22)}$$

where R_s = resolution, α = selectivity factor, N = plate number, R_F = retardation

This equation is the theoretical basis for any improvements. If two peaks are not properly separated, the peaks (i) can be shifted from each other, selectivity (a) (ii) can be made sharper, efficiency (b), (iii) can be shifted to a optimum position (R_F) (c) [Fig. 2.5].

Selectivity (a) can be altered by changing composition of the mobile phase without changing polarity for improving resolution.

Efficiency (b) is enhanced by use of a stationary phase of homogenous and small packed particles. R_F values of the components in sample should be significantly high to possible extent for obtaining increased efficiency. But the last one (c) has an opposite effect. It decreases with higher R_F. Two extremes are observed when resolutions of two compounds are impossible. First is $R_F = 1$, when efficiency is large but (c) is zero, as sample moves in the solvent front and not separated.

Second is $R_F = 0$, then $c = 1$, and efficiency (b) becomes zero. This results that samples do not migrate at all and no separation is possible to achieve.

Therefore, it is necessary to optimize the position of sample components in the chromatogram between the two extremes.

2.17.3 Separation Number

Resolution is considered an important factor in separation of components. But this cannot be treated as the only one in case of multicomponent mixtures. The issue may be that how many substances can be separated with particular resolution in a chromatographic system. So the separation number (SN) also known as spot capacity, defines such problems, based on $R_s = 1$

$$SN = (Z_f/W_0 + W_1) - 1 \qquad \text{(Eq. 2.23)}$$

SN = separation number, Z_f = migration distance of front, W_0 = width of peak at half height for $R_f = 0$, W_1 = width of peak at half height for $R_f = 1$.

Typical separation numbers in HPTLC are approx. 15 on 6 cm development distance. Hence maximum 15 substances may be separated in practice by isocratic HPTLC.

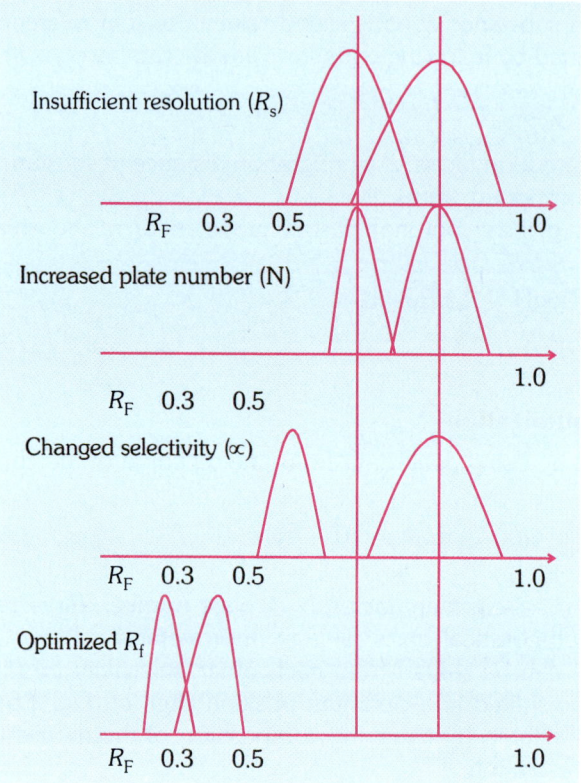

Fig. 2.5: Figures illustrate the statements

2.18 DENSITOMETRY

Densitometry is a technique for generating analog curves of the chromatogram for qualitative and quantitative analysis. It is a measurement of optical density (intensity) of spots or bands on a plate. Densitometer is also known today as "scanner". However, unlike the traditional densitometer, the modern scanners have UV and fluorescence measurements as well as spectrum recording facilities. In this technique, a beam of monochromatic light scans the individual tracks of the plate and the diffusely reflected light from the plate to detector is evaluated and noted as a function of position. Light falling on a surface of a plate is absorbed, reflected or scattered.

A light beam of defined shape falls vertically on the plate. Since the detector is placed at 45° angle, only a small but constant portion of light reaching the detector can be measured of all light approaching the plate. This mode of measurement is known as "Reflection". But when plate is underneath the light beam, the detector records a baseline as a function of the position. Therefore baseline corresponds to 100% reflection. For plotting, the signal is inverted to ensure that "baseline" is at "zero".

Substance zone absorbed at the wavelength of incident light generates a lower signal in the detector. A peak is observed in the plot after inversion. The peak maximum corresponds to highest absorption. This mode is known as "**absorption measurement**".

Most measurements in HPTLC are made in reflection/absorption.

When it is possible to excite a substance by ultraviolet light to fluoresce, measurement in such mode is called "**fluorescence measurement**". The monochromatic light (mercury or deuterium lamp) fall on the plate but the amount reflected in the direction of the detector is blocked quantitatively by a universal UV light blocking filter thus resulting baseline signal as "zero". If the beam passes a fluorescing zone, (visible) light so emitted with a longer wavelength will pass the filter to reach the detection. A peak is observed in the chromatogram.

The plate can be measured in transmission (when light is not absorbed by the plate support), when photo detector is placed underneath the plate. This is rarely done in practice.

2.19 QUANTITATIVE DETERMINATION

For quantitative measurement, it is necessary to have a functional relationship between measured signal and amount of substance in a separated zone. In TLC, sample is adsorbed in the layer but not in a homogenous manner throughout the depth. Absorption phenomena are very complex in HPTLC, as light is reflected not only from surface of the layer but also from deeper areas. There is no linear relationship between reflection and concentration as is evident from following equation.

$$(1 - R)^2/2 \cdot R = 2.303\varepsilon \cdot C/S \hspace{3cm} \text{(Eq. 2.24)}$$

where R = reflected light, ε = extinction coefficient, C = substance concentration, S = scatter coefficient.

However, there is an increase in signal with increase in concentration which can be predicted. The chromatographic data is evaluated with mathematical expression producing peak area and height of substance zone. With the help of calibration standards, calibration function is calculated. In general, data suits from absorption measurements by polynomial functions. Linearization by Kubelka-Munk formula may not improve the accuracy of the result.

In fluorescence measurement, fluorescence signal is proportional to the intensity of light used for excitation. Mercury lamp is most suitable because of emission of spectral light of very high intensity due to its line spectrum. This should be seen that emission wavelength selected should match the excitation wavelength of substance. The following linear equation will apply.

$$F = k \cdot I_0 \cdot \varepsilon \cdot C \hspace{3cm} \text{(Eq. 2.25)}$$

where F = fluorescence, I_0 = irradiation intensity, ε = extinction coefficient, C = substance concentration, k = constant.

In case of absorption and fluorescence measurements, it is necessary to select the wavelength, where the sample components have highest or maximum absorption for ensuring sensitive detection.

References

1. Geiss F. *Fundamentals of Thin Layer Chromatography*, Heidelberg: Huthig; 1987
2. Poole CF, Poole SK. *Chromatography Today*, Amsterdam: Elsevier Science; 1991

3. Snyder LR, Kirkland JJ. *Introduction to Modern Liquid Chromatography*, 2nd ed. New York: John Wiley & Sons; 1979.
4. Frey HP, Zieloff K. *Qualitative and Quantitative Dunnschichtchromatographie*: Grundlagen and Praxis. Weinheim, Germany: VCH Publishers; 1992.
5. Synder LR. *Principles of Adsorption Chromatography*. New York: Marcel Dekker; 1968.
6. Soczewinski E. Solvent composition effects in thin-layer chromatography systems of the type silicagel-electron donor solvent. *Anal Chem* **41**: 179-182; 1969.
7. Scott RPW, Kucera P. Solute-solvent interactions on the surface of silica *gel. J Chromatogr* A **149**: 93-110; 1978.

HPTLC Stepwise Analysis

3.1 SAMPLE PREPARATION

Before preparation of the sample solutions for application on plate, various steps such as obtaining a representative sample, making it homogeneous and then chemical treatment for enrichment/isolation of minor components, removal of other interfering substances if required, may be necessary. The sample preparation procedure should be so accurate that recovery of the required ingredients for analysis is maximum and consistent for accurate quantitative determination. The solution for analysis should be adequately reduced and concentrated so that analyte, even if present in traces, could be detected and it can be separated as a discrete compact spot or zone. Pure samples or their concentrated extracts can also be directly spotted on HPTLC plate for analysis. However, if the analyte is present at low level in a complex food sample, solvent extraction clean up (purification), derivatization, isolation of analyte if interfering agents are present, concentration procedures should be adopted. Once ready, the samples must be chromatographed immediately. Prepared samples should normally be kept in refrigerated conditions. Sometimes it may be possible to apply crude sample for separation. Impurities, which migrate with the analyte, adversely affect its detection or distort the zone. Therefore streaking or trailing causing components must be eliminated prior to HPTLC. Common clean up procedures include liquid-liquid extraction, solid-liquid extraction, defatting, desalting, deproteinization, and decolouring, etc. A pre-chromatography derivative of the analyte can be made in solution or *in situ* at the origin by over spotting of a reagent to improve resolution on plate, e.g. amino acids or bind a volatile component, e.g. formaldehyde with dimedone.

Due to introduction of commercially available HPTLC plates, pre-treatment of samples before application on plates has become simpler. Fractions of interest can be distinguished from impurities and sample contaminants and so do not cause a problem by and large. Sample pre-treatment is simplified in HPTLC as compared to HPLC, where contaminants can adversely affect the column performance. On application of samples to the sorbent layer, both the components of interest and the interfering impurities are deposited together. However, as the development starts, the contaminants are separated while all the components migrate in the direction of flow of the solvent. In case the sample solvent is mainly aqueous or viscous, then dilution with an organic polar solvent like methanol, ethanol or acetonitrile, will aid application to the layer. Thus the sample solution will wet the surface and penetrate into the sorbent effectively. Use of mobile phase as the sample solvent is recommended where possible. This eliminates artifacts if any, caused by other solvents. Filtering of the sample solution may be necessary. Sometimes, the volume of the sample is too small. Filters of 0.45 or 0.2 μm can be used. A variety of such filters are commercially available based on cellulose, cellulose acetate and nitrate, alumina, PES, polypropylene or PTEE. If contaminants other than components of interest may interfere with the chromatogram, then sample clean up procedure is necessary, e.g. when the desired components are present in traces or recovery has to be increased for quantitative determination, then it is desirable to remove the contaminants as far as possible.

Basically foods can be classified, for the purpose of removal of the contamination, into a fatty, starch or sugar, sugar-based, protein-based, or mixed composition. It is considered necessary that extractions of the desired components are validated for each specific type of food to get the optimum uniform and consistent recovery.

Solid phase extraction (SPE) cartridges with appropriate vacuum manifold can be used for extraction. SPE sorbents can be specific, exploiting the different physico-chemical properties of the sorbents. These include diatomaceous earth, silica gel, C_2, C_8, C_{18}, CN, Diol, NH_2, phenyl-bonded silica gel, and cation- and anion-exchangers based on silica and also various polymers. The basic principle is that a sorbent is chosen that can either absorb the analyte but not the impurities or vice-versa. When analyte is absorbed, it is necessary to wash the extraction column with a solvent that removes the impurities without any effect on the analyte. Next, the analyte is eluted by careful selection of solvent at appropriate concentration. It is possible to use SPE columns for sample concentration. Proteins can be removed by precipitation. Sample pre-treatment may be eliminated by pre-derivatization of analytes in a sample in a few cases or by use of concentration zone TLC plates.

3.2 SOLID SAMPLE EXTRACTION

Sometimes extraction of analyte of interest by use of suitable solvent may be necessary by a procedure more than simply shaking appropriate solvent and keeping it overnight for better extraction, as the sample may contain a number of components differing in their polarity. Therefore, extraction procedure should be such that necessitates isolating the components of different groups. Boiling or soxhlet extractions are more effective using appropriate solvent on certain occasions but can also extract a lot of other unwanted substances.

A general procedure for a sample extraction of major compounds from aqueous based samples is given in Figure 3.1.

Spices based products containing active polar ingredients or their dilutions with water should be boiled in methanol/ethanol to completely dissolve analyte of interest. Subsequently, their fractionation based on polarity of different groups can be carried out. pH adjustment is an selective process in ether extracts. At high pH by use of sodium hydroxide, or sodium bicarbonate, organic acids and phenolics can be made water soluble as these are converted into salts. Amines and amides by use of mineral acid at low pH will also form salts and can be extracted similarly. The aqueous extracts are neutralized and organic matter is extracted with an immiscible solvent like ether and then concentrated for application on plates or large volumes can be applied instrumentally.

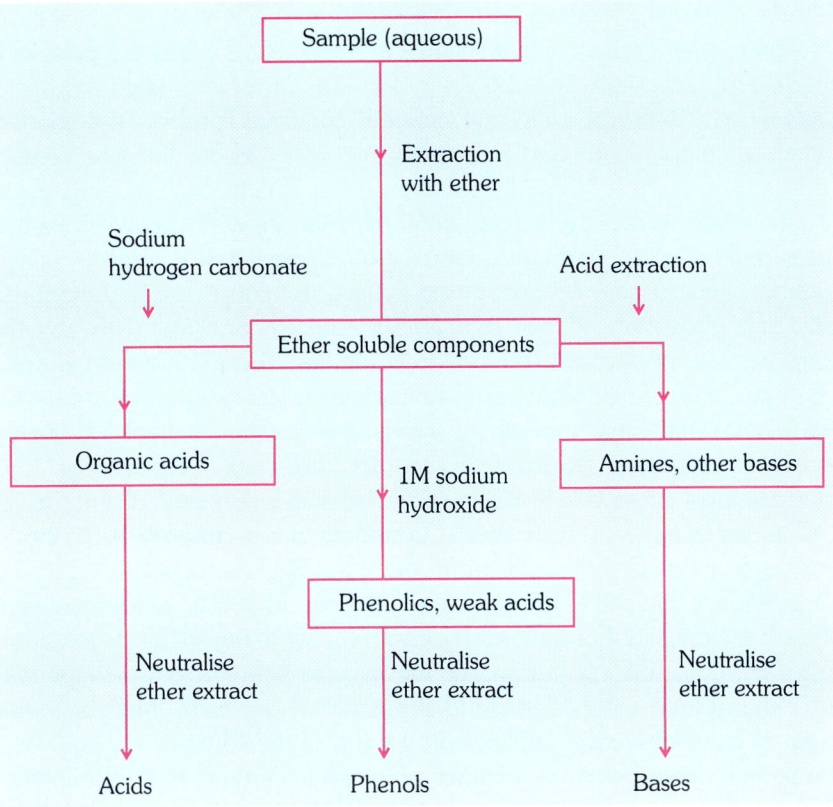

Fig. 3.1: Sample extraction

3.3 SOLID PHASE EXTRACTION (SPE) SYSTEMS

Diatomaceous earth which is a natural siliceous deposit of diatom frustules; kieselguhr, celite, bentonite of different varieties is used as filtering aids or clearing agents. High purified grades of kieselguhr are used as absorption materials for sample pre-treatment in chromatography. Normally, short columns of dry sorbent are prepared using a glass or plastic tube with a taper at one end. Sample is poured directly into the column and allowed to stand for 10-15 minutes. A water immiscible solvent is used to elute the analytes, evaporated to low volume and reconstituted with a suitable solvent for application in TLC.

3.4 BONDED SILICA GELS AND POLYMERS

Highly specific bonded phases designed to cope with a different variety of samples from a different ranges[7] find use in sample pre-treatment. Use of a particular cartridge is made which is based on selection of retention for the analytes in sample, sample matrix and composition of samples. For such analysis, specifically designed commercial pre-packed bonded phase columns(s) are available, e.g. reversed-phase cartridges are used for non-polar compounds in aqueous solutions. Methanol, acetonitrile, or either solvent mixed with a suitable buffer solution which are polar in character are used as eluent. When the sample analytes are polar, sample solvents used are non-polar, e.g. n-hexane, iso-octane or dichloromethane, normal SPE cartridge is used. Diol-bonded silica gel is considered better as compared to cyano or amino or silica gel 60. In some cases, low polarity solvents are selected as eluents. Normally, medium polar eluents such as ethyl acetate, acetone, etc. are more often used. When the sample contains analytes with positively charged functional groups and the sample solution is aqueous, then cation-exchange SPE cartridge which is based on sulphonic or carboxylic acid as functional group is used. Elution is made using buffers based on citrate, acetate, and phosphate. When the sample contains analytes with negatively charged functional groups such as organic acids in aqueous solution, then SPE cartridge used is based on amino or quaternary amine functional group. Phosphate or acetate buffers are used as eluents. It is worth while to note that ionic strength of the sample solution is kept low for optimum ion-exchange capacity in cartridge, because strong solution will result in ion leakage.

The support material of the sorbent is first conditioned with an organic solvent such as acetonitrile or methanol. This will solvate the CH chains bounded to the support. After this, a volume of solvent used to dissolve the sample is passed through the SPE cartridge.

The sample solution is injected at low vacuum in the cartridge. The analyte concentrates on the sorbent, whereas it is expected that undesirable components of the solute matrix are not absorbed and elute from cartridge, but the volume of the sample solution varies depending upon the concentration of the desired components and sensitivity of detection. More volumes are used if there is necessity to concentrate the analytes on the top of the sorbent.

Water or buffers or water/methanol mixtures are used to remove the undesirable components and left behinds for reverse phase and ion-exchange/SP cartridges. However, it is necessary to maintain the pH at the same value as that of original sample solution, because any change in pH may make premature elution of analytes.

Solvent is selected to elute the analytes from the sorbent. Selection of solvent is made on the basis that it will weaken the strength of the interaction between the analytes and the sorbent so that analytes are eluted, but should leave behind more strongly bound matrix components. The analytes are concentrated prior to application on the TLC plates. Sometimes the complete analyte is made to dry and then reconstituted in a solvent which gives better wetting of TLC plate.

3.5 IN SITU CLEAN-UP

The disposable nature of the HPTLC plate can be put to an advantage, i.e. clean-up of all samples on the plate itself, before chromatography, e.g. removing fatty matrix of extracts of nuts in the analysis of aflatoxins. A 20×20 cm aluminium foil supported plate is used. Samples are applied at 110 mm from lower edge. The plate is then placed in a developing chamber which contains diethyl ether up to 100 mm. This development moves the fatty matter to the top of the plate. After development, the plate is dried and cut at 120 mm from bottom edge, i.e. 10 mm above the original point of application. The cut plate is then turned around 180° and chromatographed in the normal way.

References

1. Lautie JP, Stankovic V*J Planar Chromatogr* **9**; 113-115, 1996
2. Jacob K, Egeler E, Hennel B, Neumeier D Fresenius-J. *Anal Chem* **330**(**4-5**); 386-387, 1988
3. Ikai Y, Oka H, Kawamura N, Yamuda, M, Harada Suzuki M. *J Chromatogr*, **411**; 313-323, 1987
4. Kessel S, Hauck HE. *Chromatographia*, **43**(**7-8**); 401-404, 1996
5. Imrag T, Junker-Buchheit, A. *J Planar Chromatogr*, **9(2)**; 146-148, 1996
6. Fischer W, Bund O, Hauck HE. Fresenius-*J Anal Chem*, **354(7-8)**; 889-891, 1996
7. Wall PE. *Thin-Layer Chromatography—A Modern Practical Approach*, 2005

4

Techniques for Sample Application

4.1 SAMPLE APPLICATION

As in all chromatography methods, HPTLC too requires that the physical dimension of the sample at the time the separation begins should be as compact as possible. Sometimes poor separations are blamed on the quality of the layer or the application techniques, or the processing of the sample. The optimum amount of the sample to be applied on the sorbent layer is important too. In case of manual application of the sample, care should be taken that the application process does not damage the plate or the upper surface of the layer. This may result into uneven spots or irregular chromatogram. Devices like pipette, capillary, etc. can damage the sorbent layer when the spots or bands are applied on it. Manual band application can be uneven and differ in concentration along the band. It is desirable that the constant temperature and humidity are maintained and atmosphere is free of chemical fumes and solvent vapors. Therefore HPTLC analysis should be carried out in laboratories free of vapors and chemical fumes, etc. These fumes and vapors may affect the sample, or may get absorbed on the layer and affect stationary-mobile phase equilibrium during development. There should not be any dust or other particles on the sorbent layer. Before sample application, the sorbent layer can be cleaned with a dry air blower. Breathing on the layer should be avoided as this may add to the moisture.[1]

Sample application on the plate can be made manually with simple equipments or can be semi-automated or fully automated. Semi-automated or fully automated instruments are commercially available that has tremendously enhanced the quality of sample application. The objective in chromatography is to achieve precise and reproducible spots/zones on the plate without damaging them. Whereas manual application is suitable for qualitative separation, band applicators will give better resolution for improved quantification.

Fully automated sample application in HPTLC must be precise in positioning. Precise volumes are essential too. In manual application, the rate of delivery cannot be controlled and hence the spot diameter. In band application, the delivery rate is selectable and together with simultaneous gas drying, ensures a predictable band size. Semi-automated/full automated gives reproducible quantification.

Automated sample application devices are desirable for better resolution and especially in case of quantitative analysis using HPTLC. Generally, 0.5 to 5 µl of the sample is applied in the form of spot or 2 to 10 µl as narrow band. Varying volumes can also be applied depending upon the components present in the extract, linearity range of response, accuracy and precision. The amount of sample that can be applied to a TLC plate depends upon the components for separation, layer thickness and the development method. For screening purposes, disposable quantitative capillary pipettes or precision syringes are used. The syringes allow the intermittent application of more volume by using different combinations with the repeating dispenser incorporated in the spotting device. This helps in keeping starting zones, as small as possible, for efficient separation. The use of micro-syringe, wherein automated sample application is not possible, will aid in improving the resolution as compared to sample application by capillary for volume above 2 µl. High density solvents such as chloroform may leave the capillary before sample application. High viscosity solvents may not enter properly in the capillary. This is also possible that volatile solvents like acetone, diethyl ether, etc. may evaporate partially even before application. The spotting of the sample and standard should be made according to a spotting pattern (template). This works as a guide for spotting. The template acts as a support for the hand and shields the surface of the plate from damage besides being an aid to accurate positioning.[2] Automated devices overcome the above difficulties.

At the time of sample application, proper volume of concentration should be taken care of, depending upon the components for separation and their sensitivity to the method. In case larger quantity of the sample is applied, it may not absorb completely leading to over loading of the sample. The tailing of the zone starts and thus, leading to undesirable resolution. In HPTLC, samples and standards are always applied as "bands" (1 mm wide rectangles) due to advantages given below.

Automatic or semi-automatic sample applicators can "spray-on" the sample from a syringe, on to the layer, in a slow, controlled manner. An inert gas carries the sample pushed out from the syringe and deposits it on to the layer. Drying occurs at the same time.

1. Applied sample zone is as thin as possible for better resolution.
2. Sample zone size remains constant (and selectable) independent of the volume.
3. R_f of the samples and the standards are more consistent.
4. Interfering factors of the samples and other impurities present in analytes of interests cause very little disturbance in the chromatogram because distribution of the sample is over a larger area.
5. Densitometer gives a better response due to even distribution of concentration in the band.
6. Better concentration-response curve is obtained. There is no necessity to prepare solution with different concentration by and large.
7. Large quantity of the samples can be handled. Therefore, the application volume can be adopted as per convenience to suit the concentration of analyte and the sample is uniformly distributed over entire range of applied band. For quantitative analysis, this significantly improves the signal to noise ratio of a measurement.
8. Resolution of chromatogram is improved significantly in case of band rather than as a spot, because of counter diffusion in the middle of the band. They remain compact.
9. A larger volume and quantity can be applied, since the application area is bigger.
10. Silica gel is most widely used for HPTLC separations but it is not wetted easily. So aqueous samples are physically applied as a rectangle, e.g. 4 mm wide × 8 mm long allowing speedier application.
11. Bands travel in a straight line, compared to spots.

It may be noted that bands 2 or 3 mm long spread like circular spots. Bands 9 mm or longer allow few samples to be applied on a plate. 6 mm long bands for non-aqueous and 8 mm long for aqueous samples are usually suitable.

In the region of low R_f, the major difference has been noticed and band development gives better results. The separation resolution in low migration area is sharper for bands as compared to spots. Spot capacity or separated number can be calculated from the following equation.[3]

$$SN = \frac{Zf}{b_0 + b_1} - 1$$

(Eq. 4.1)

In case of low loading of sample, separation numbers are almost same but when concentration is increased, the separation number falls drastically for separation of spots as compared with bands.

Spot application requires less automation, it is inexpensive and normally consumes less time. Application of spot even with low polar solvent may still lead to concentration of solute in the centre of the spot with less concentration at the perimeter, thus leading to uneven distribution of sample. After development, further diffusion into the layer may take place. In such cases, for spectrodensitometric determination, a slit length suitable for the largest diameter spot has to be selected. But for bands, 4/5 of the band length is quite sufficient. If spots do not migrate accurately in vertical direction for ascending development, chromatogram adjustment is required on the scanning equipment. For the band scanning, it is rarely necessary. Therefore, for screening and identification purpose, spot techniques with minimum equipments may be suitable. But when precision, accuracy, and reproducibility are required, automated band application is much better especially, in cases where separated zones are very close to each other. Concentrated zone plate is also not suitable for sample band application except for qualitative analysis, because solute largely concentrates in the centre of the band. Thus, sample distribution is uneven along with the length of the band. Only instrumental techniques comply with SLP.

4.2 CONCENTRATION ZONE HPTLC

Commercially, plates are available that have two zones, e.g. a 20 × 20 cm concentration zone plate may have a 5 × 20 cm band of an inert material known as concentration zone. The remaining 15 × 20 cm zone is then used for separation. The advantage of such plate is that unevenly applied spots/bands can become uniform bands by the time the sample reaches the "separation zone". Depending on the adsorbent used as "concentration zone", sample clean-up can be aided.

Concentration zone HPTLC plates[4, 5] are suitable for application of larger amount of sample. The sample is applied to the concentration zone as dilute large spots in a desired location. Sample loading may vary depending upon the required resolution between separated components. If the sample is very dilute and concentration techniques are not feasible. Sample application can be repeated on the concentration zone keeping in view that drying is done between loadings so that solvent front migrates quickly through the concentration zone taking the sample with it. The sample spot may start to concentrate into the band. They become focused by the time they reach the end of concentration zone. It is possible that impurities are retained by absorption in lower zone; hence separation of sample may start in the upper zone on the plate. However, resolution may be better than normal spot development. It also depends upon the choice of mobile phase and low polarity less viscous solvent gives better result. But polar viscous solvent as mobile phase may result into the little improvement during separation. Reverse phase HPTLC plate is more suitable for increasing the potential of the concentration zones especially for wide range of application.

After application, it is desirable to dry the spots/bands prior to chromatogram development. A problem by "wet" application zone can be incompatible with the mobile phase, especially when sample contains non-volatile solvent like glycerin, etc. which will not evaporate during application or drying and is also not miscible with mobile phase.

4.3 APPLICATION METHODS

HPTLC, being a flexible method, allows sample application in a variety of ways too, unlike GC or LC. There are four well known sample application patterns, which depend on the aim of the experiment.

Sample Application

Mode 1: Quantitative Analysis

U1 S1 U2 S2 U3 S3 U1 S1 U2 S2

For quantitative analysis of the sample application, guidelines are as follows:
- Distance for sides–15 mm
- Distance from bottom edge 8 mm
- Sample and standards interspersed

Mode 2: *In situ* Clean-up

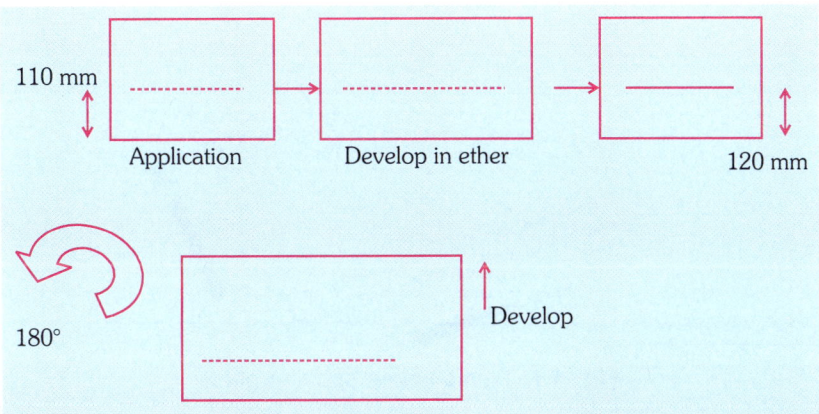

Samples containing matrix especially fatty extracts, e.g. extracts of nuts for aflatoxins can be cleaned up on the plate itself. This causes a big saving in money and time while improving the quality of results.

Sample bands are applied at 110 mm from bottom in the quantitative application pattern and then developed in a chamber where petroleum ether is filled up to 100 mm. During development, the fatty matter is moved to the top while the polar components remain behind. After drying, the plate is cut at 120 mm from lower edge, turned 180° and then developed in a suitable mobile phase.

Mode 3: Superimposition
Once the samples and standards are applied sometimes it may be essential or useful to superimpose upon it. The superimposition could be for the purpose of:

(a) Internal standard—as is commonly practiced in GC or LC.
(b) To improve LOQ by superimposing the fraction(s) of interest with the same substance at LOQ level.
(c) For pre-chromatography derivatizations by adding a reagent that will react with component(s) of interest, with a view to improve separation and/or detection

Mode 4: Micro-preparative Separation
For isolating fractions on a mg scale, micro-preparative chromatography is practiced. This requires a thick layer up to 2 mm and 500 μl volumes to be applied along the whole length of the plate, i.e. about 190 mm. Fractions at the same R_f can then be collected separately by scrapping the layer.

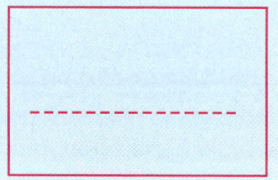

4.3.1 Manual Application
The simplest application apparatus is a capillary and spotting template or guide (Fig. 4.1). The capillary can be replaced by a pipette or syringe.

Fig. 4.1: Multipurpose spotting guide

The other easy method of sample application to the plate is glass capillary of fixed volumes (Fig. 4.2). Disposable capillary or capillary dispenser systems are better options. They give a reasonable degree of reproducibility.

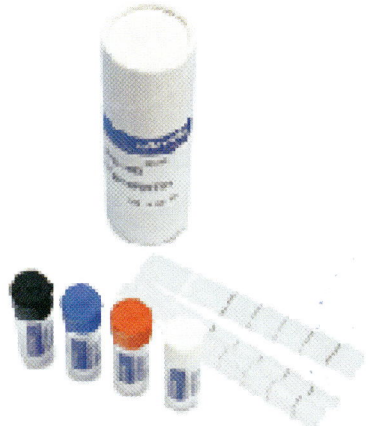

Fig. 4.2: Fixed volume and calibrated disposable glass capillaries

Fig. 4.3: Capillary dispenser unit

Micro-syringes can be used but they can often damage the layer. Extreme care is required. Band application with manual techniques is by and large impossible without some damage to the layer. Further, it is difficult to achieve uniform concentration for the rectangular band and straight line application.

Manual application instruments help in precise positioning of samples. Disposable fixed volume capillaries can ensure precise volumes but diameter of applied spot cannot be controlled.

4.3.2 Instrumental Techniques for Sample Application

For spot application, the sample may be introduced on the layer surface precisely at a desired location with negligible damage to the layer. A figure of instrument (Fig. 4.4) for fixed volume sample application is given below.

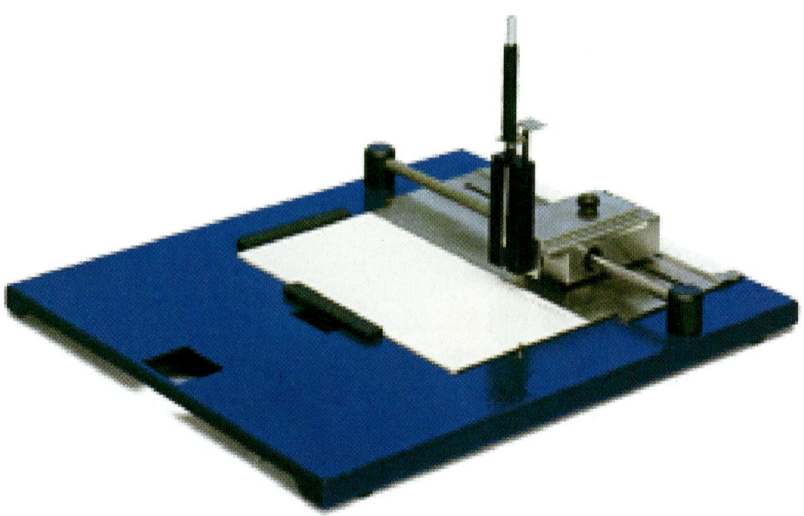

Fig. 4.4: A manual but instrumental sample applicator using capillaries

Most convenient and scientific way is semi-automated (Fig. 4.5) or fully automated (Fig. 4.6) sample applicator to apply sample solutions.

Fig. 4.5: A semi-automated sample applicator

4.3.3 Semi-automatic Application of Samples

Most effective for quantification is automatic technique to apply the samples as bands. A pre-set program on a computer with fully controlled sample application process for accurate quantitative determination of analyte at low detection level with automated instruments is recommended. Automated band application renders narrow application of sample, because syringe moves over a pre-determined distance of the plate at a constant speed with pre-set quantity of sample by spray technique. Constant flow of gas ensures a thin application zone. Further, the TLC plate never comes in contact with syringe, whereas application of 1-20 µl gives extremely thin line with negligible diffusion of sample. In semi-automatic applicators, the syringe is cleaned and refilled with each sample solution manually. Protocol of sample application for any number of methods can be stored in a computer.

Fig. 4.6: Fully automatic sample applicator

It is seen from Figure 4.7 that the result for band application is strongly dependent of the solvent of the sample, whereas for bandwise spray-on application this is not the case. Furthermore, the separation is significantly improved for the latter.

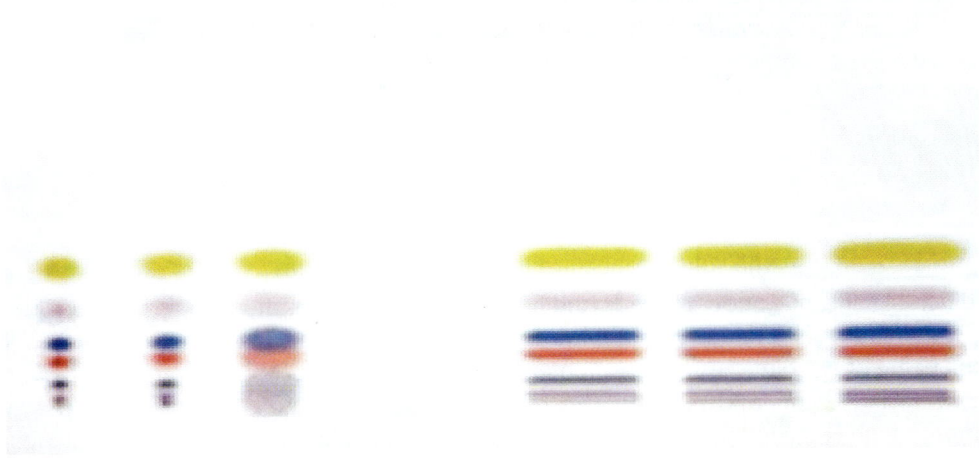

Fig. 4.7: Advantage of spray technique

In case of modern semi-automatic applicators, it is important to note that the transfer of sample from syringe to layer is fully automatic. The sample change over is manual. Removable sample syringe helps in proper cleaning and prevents sample carry over. Air bubbles are easily removed. Quantity of sample required is low. The needle tip of the syringe is conical. This allows the spray on inert gas to focus the sample and obtain a thin line. Semi-automatic applicators use 100 μl syringes which enable duplicate application of samples and multiple applications of standard without refilling the syringe. Sample application speed is controllable and must be optimized for the solvent being used. Too slow rate might dry the sample on the needle tip itself while too fast speed would cause band spreading as there is not enough time to dry the sample completely during application.

4.3.4 Automatic Application of Samples

Bands are applied with spray on technique as in case of semi-automated unit. However, spots in such cases are applied using this technique or by direct transfer (contact) process. The samples are kept in vials with septum seals. As per pre-set computerized program or analysis method, the robotic arm moves from vial to vial picking up sample solution in a syringe and delivering on a layer at pre-calculated position. Between sample solutions, the arm will move the syringe to a bottle containing fresh solvent for rinsing. Thus in the fully automatic sample applicator, the sample and standards pick-up, application, drying, etc. do not need manual intervention at all. However, they are more expensive.

4.4 OPTIMUM SAMPLE QUANTITY

It is necessary to standardize the quantity of the sample to be applied on the sorbent layer. This depends upon number of factors such as sample matrix, sorbent layer thickness, nature of the solvent, etc. As a thumb rule, the applications for HPTLC plate is 50-200 ng per spot, 1-4 μg per band as the sample application. For TLC plates, it is 0.1-2 μg per spot, 2-10 μg per 10 mm band application.

It is necessary to know why the sample overloading poses problem. Therefore, it is necessary to understand about the equilibrium ratio between solute in the stationary and mobile phase. As soon as development of sample occurs, equilibrium ratio also establishes between the solute in the stationary phase and mobile phase. Distribution coefficient (K) plays an important role and the same is defined as:

$$K = \frac{C_s}{C_m}$$

(Eq. 4.2)

C_s = the solute concentration in the stationary phase

C_m = the solute concentration in the mobile phase

If K differs significantly under a given condition, then sample separation would be good. However, greater the value of K, greater will be time of the solute retention in stationary phase. Whereas lower the value of K, solute resides for greater time in mobile phase. Hence separation of analytes with low K value occurs and retained into mobile phase which are carried through the layer as compared to high K values. Besides in the stationary phase, solute molecules do not move. Under ideal circumstances ratio between C_s to C_m should follow the linear correlation system (Fig. 4.8a) resulting scanned peak A (Fig. 4.8b). However, if the sample is overloaded, then relation between C_s to C_m will become non-linear (Fig. 4.8c) and result of scan is erratic peak (Fig. 4.8d). Sample overloading will result in little increase in R_f value. This leads to increase in loading concentration (Fig. 4.8e). Repeatability and reproducibility of the sample is also not consistent. A significant error thus occurs in quantification. It is necessary to locate the problem related to overloading of the sample and overcome such situation. With the availability of precision and high quality HPTLC plate, diffusion can be kept minimum and the sorbent layer could not be blamed. Positioning of the sample is also important and the sample application should be at a proper ideal distant from the one edge of the plate. It would be ideal to apply samples at least about 1 cm from the edge for proper development.

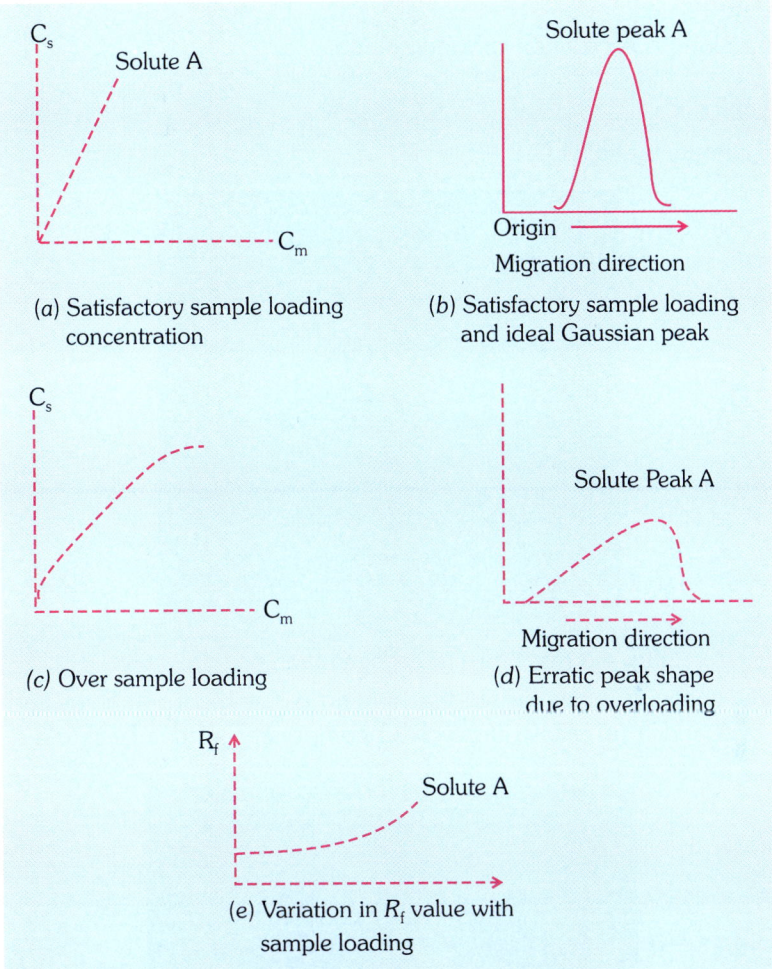

(a) Satisfactory sample loading concentration

(b) Satisfactory sample loading and ideal Gaussian peak

(c) Over sample loading

(d) Erratic peak shape due to overloading

(e) Variation in R_f value with sample loading

Fig. 4.8: Variation in R_f value with sample loading

Temperature and humidity do have effects on migration of the analytes and knowledge of the effects in a particular case is very important while optimizing the resolution. It is well known that moisture present in the layer and abnormal temperature would adversely affect the partition of the components between the stationary and mobile phases. Problems can occur due to the incomplete evaporation of solvent from sample. Due to evaporation of the solvent, localized cooling effect takes place. Thus moisture condenses on the layer and is absorbed. Therefore, it is advisable that area of application is dried for proper migration of the components of the sample. Automated development chambers ensure complete drying by blowing a large amount of air at room temperature.

4.5 EFFECT OF OVERLOADING

Overloading of the sample should be avoided and the volume should be optimized. Effect of overloading has been demonstrated in the following experiment (Fig. 4.9.a, b).

4.5.1 Concentration

Effect of concentration on separation of food colours.

4.5.2 Aim of Analysis

Profiling of food colours by HPTLC.

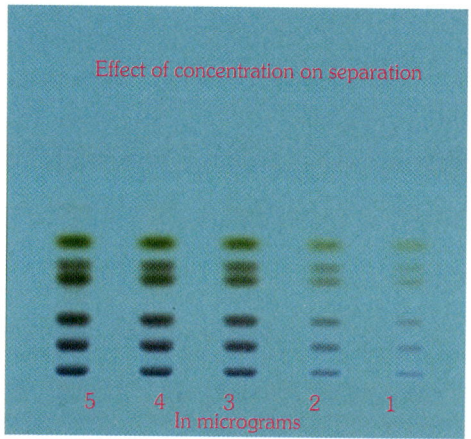

Fig. 4.9 (a): Effect of concentration on separation

From the above image, it is established that one μg is suitable for analysis and overloading of sample with 2, 3, 4 and 5 μg shows unresolved components and reflects the adverse effect of overloading.

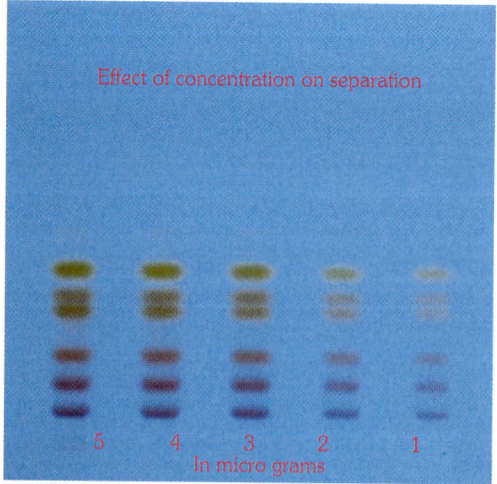

Fig. 4.9 (b): image@visible

4.6 CONCLUSION

Sample application in modern TLC requires:

- (*i*) Accurate volumes,
- (*ii*) Precise positioning,
- (*iii*) Compact starting zones,
- (*iv*) Instrumentation is available for manual, semi-automatic and automatic application of samples,
- (*v*) Depending on the aim and method of analysis, different modes of application are required, e.g. routine quantitative analysis, superimposition, clean-up on plate, micro-preparative chromatography, etc.

References

1. Sherma J. Basic techniques, materials and apparatus. In Sherma J, Fried B (eds). *Handbook of Thin Layer Chromatography*, Marcel Dekker, New York, **3**; 1991
2. Stock R, Rice CBF (eds). *Chromatographic Methods*, Chapman and Hall, London, UK, **279**; 1974
3. Kaiser RE. *HPTLC High-Performance Thin-Layer Chromatography*, Zlatkis A, Kaiser RE (eds), Elsevier, Oxford, UK, 21-32, 1977
4. Abbot DC, Thompson J. *Chem Ind* 310, 1965
5. Musgrave A. *J Chromatogr* **41**; 470, 1969

Stationary Phases, TLC Plates and Sorbents

HPTLC offers a wide selection of stationary phases and sorbents in terms of layer materials, their support and sizes. Silica gel is most widely used sorbent out of more than 25 sorbents used in thin layer chromatography. It is necessary to use an appropriate sorbent for separation purpose depending upon the components to be separated. As early as in 1973, Scot[1] evaluated many papers to establish the use of various sorbents by TLC. He noticed that silica gel was very popular followed by cellulose. Subsequently appreciable changes have been noticed with the introduction of chemically bonded phases, thus opening a new field for possibilities. Camag Bibliographic Service offers a computerized literature survey for a specific analyte. Analysts can get an idea for selection of sorbents for HPTLC for different classes of compounds from the Table 5.1.

Table 5.1: Selection of sorbents	
Sorbent	**Classes of compounds**
Silica gel	Generally for all classes
Aluminium oxide	Basic substances such as alkaloids, amines, terpenes, aliphatic hydrocarbons, steroid
Cellulose	Food dyes, carbohydrates, amino acids, antibiotics
Kieselguhr	Aflatoxins, carbohydrates, herbicides, antibiotics
Polyamides	Antioxidants, flavonoids, pesticides, sugars
Modified silica gel	
(a) Amino bonded	Carbohydrates, carboxylic acids, vitamins(B)
(b) Cyano bonded	Preservatives, pesticides
(c) Diol bonded	Steroids, hormones
(d) Reversed phase	Antioxidants, capsaicin, lipids, fatty acids
(e) Chiral modified	Amino acids, lactose, dipeptides
(f) Impregnated with silver nitrate	Lipids
(g) Impregnated with caffeine	Polyaromatic hydrocarbons
(h) Boric acid or phosphate	Carbohydrates

Contd…

Contd…

Sorbent	Classes of compounds
(i) Ethylene diamine tetra-acetic acid (EDTA)	Phospholipids, phenols
(j) Boric acid	Ascorbic acid derivatives, sugars, glycerides, lipids
(k) Transition metal salts	Amino acids
(l) Iron (III) salts	Phenolic acids

Precoated plates with silica gel 60 can be modified by immersing the plate in a solution containing modifier in a hydro-alcoholic solution or in alcoholic solutions, e.g. the plate is ready for use after-it is dried in an oven. Concentration of the modifier is fixed according to compounds to be separated and concentration of analytes. Adjustment of pH value is also done for separation, e.g. (i) inorganic acids (0.1-0.5N) for separation of phenols, acids, amino, (ii) sodium hydroxide or potassium hydroxide (0.1-0.5N) for separation of alkaloids, amines and basic compounds and (iii) buffer salts for separation of curcumine, sugars, heavy metals, etc. Such modifications are not available commercially.

5.1 PREPARATION OF HPTLC PLATES/SHEETS

Before the availability of commercially available plates/sheets, the glass plates were being prepared in advance of separation procedure. The procedure was to mix 30 gm of sorbent with 60 ml of ultrapure water and a slurry is made which is transferred to a plate coater apparatus and applied evenly over the glass plates in one continuous operation, allowed to dry for about half an hour and then activated for approx. 30 minutes in an oven at about 105° C. Gypsum binder (approx 12%) is added to improve the mechanical strength of the layer. Organic binders such as starch (1-2%) carboxyl methyl cellulose, polyvinyl alcohol (1-5%) have also been recommended for use as binder. Although these are strong binders, they suffer from draw back of their solubility in aqueous based mobile phases and charring after treatment with acids and heating. Home made plates/sheets are not suitable for HPTLC analysis, as they have bigger particles, not so homogenous and uniform leading to an inefficient separation and a noisy baseline during scanning. Factory made HPTLC layers are essential.

It is desirable to maintain the following conditions during preparation of plates (Table 5.2).

Design-ation (Plates)	Amt. of Sorbent (g)	Amt. of water (ml)	Adjusted layer thick-ness (μm)	Layer thick-ness after drying (μm)	Drying time in the air (min)	Activation time (min)	Temp (°C)	Notes
Table 5.2: Working conditions for the preparation of analytical TLC layers								
Aluminium oxide 60 G	30	40	300	250	60	30	120	
Aluminium oxide 60 GF$_{254}$	30	40	300	250	60	30	120	

Contd…

Contd…

Designation (Plates)	Amt. of Sorbent (g)	Amt. of water (ml)	Adjusted layer thickness (µm)	Layer thickness after drying (µm)	Drying time in the air (min)	Activation time (min)	Temp (°C)	Notes
Cellulose	25	100	250	100	60	30	120	60-90 sec. electric mixer at slow stiring rates
Florisil	25	60	300	250	60	30	120	
Silica gel 60G	30	60	250	200	30	30	105	Setting time of gypsum:2-4 min.
Silica gel 60 GF$_{254}$	30	60	250	200	30	30	105	Setting time of gypsum:2-4 min.
TLC Silica gel 60 G	30	65	250	200	120	30	105	Setting time of gypsum:2-4 min.
TLC Silica gel 60 GF$_{254}$	30	65	250	200	120	30	105	Setting time of gypsum:2-4 min.
Silica gel 60 H	30	65-70	250	200	120	60	120	
Silica Gel 60 HF$_{254}$	30	65-70	250	200	120	60	120	
Silica Gel 60 HF$_{254+366}$	30	65-70	250	200	120	60	120	
TLC Silica Gel 60 H	30	65	250	200	120	30	105	
TLC Silica gel 60 HF$_{254}$	30	65	250	200	120	30	105	
Kieselguhr G	40	100-110	300	250	60	30	120	
Polyamide 11	16	60	300	150	30	–	–	Suspending liquid: methanol. Shake 10minutes.

5.2 PRE-COATED TLC/HPTLC SHEETS AND PLATES

Pre-coated TLC/HPTLC plates/sheets with suitable polymeric binders are now available (Fig. 5.1). These are resistant to chemical elution and abrasion.

Fig. 5.1: Pre-coated plates

The binders are resistant to most mobile phases. Strong polar solvents and detection reagents can be used as needed. Gypsum is not used in pre-coated layers.

Details of the readymade plates have been described below.

Classical silica TLC plates made from silica gel 60 with polymeric binder result in an adherent and hard surface that will not crack or blister. The narrow particle size distribution and the smooth and dense layer surface ensure narrow bands for high separation efficiency with low background noise. Specifications of classical pre-coated TLC plates are given in Table 5.3.

Table 5.3: TLC pre-coated layers	
Specification	**Values**
Mean particle size	10 - 12 μm
Particle size distribution	5-20 μm
Layer thickness	250 μm, 200 μm
Typical plate height	30 μm
Typical migration distance	10 - 15 cm
Typical separation time	20 - 200 min.
Number of samples per plate	10

(ChromBook 06/07, E.Merck)

Silica gel has a share of nearly 80% of thin-layer chromatography applications for both adsorption and partition thin layer chromatography. A large range of substances such as aflatoxins, alkaloids, anabolics, benzodiazepins, carbohydrates, fatty acids, glycosides, lipids, mycotoxins, nucleotides, peptides, pesticides, steroids, sulfonamides, surfactants, tetracycline and many others can be separated and it is a unique plate suitable for analysis in foods.

Aluminium foil or plastic sheet backed flexible plates are suitable for cutting to desired size as shown in Figure 5.2.

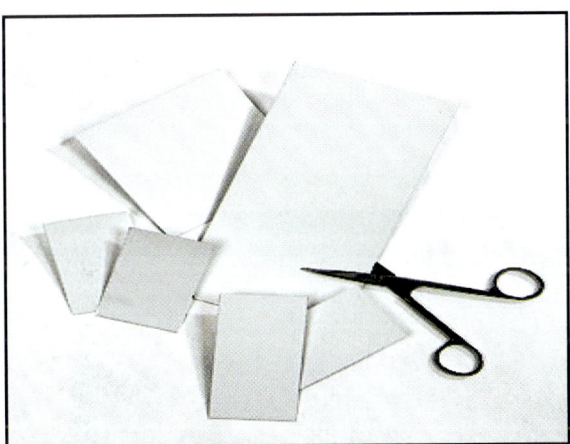

Fig. 5.2: Flexible plates

Cutting of sheets/plates should be done in a way that minimum amount of damage occurs to surrounding layers. Careless cutting may affect in a capillary crack between layer and food, thus causing mobile phase to migrate rapidly at the edges. The sharp cutting scissor must be held at an angle to the plate and not perpendicular. The plate to be cut must be held in an aluminium foil.

5.2.1 Aluminium Oxide TLC Plates for Basic and Neutral Compounds Using Different pH Conditions

Aluminium oxide plates give distinct separation features with regard to the pH range used. Under aqueous conditions, basic compounds can be separated on basic aluminium oxide plates, while neutral compounds are separated on neutral pH plates.

5.2.2 Kieselguhr and Mixed Layer Plates for Specific Applications

Kieselguhr, a natural diatomaceous earth can be used for the separation of polar or moderately polar substance. Mixed layer plates utilize a combination of classical silica gel 60 and kieselguhr providing good separation properties for special applications such as separation of inorganic ions, herbicides and some steroids.

The plates are available with layer thickness 0.2 mm in format of 20 × 20 cm.

5.2.3 High Performance (HPTLC) Silica Plates for Quick and Quantitative Analysis of Complex Samples

HPTLC unmodified silica plates provide higher speed and sensitivity than classical TLC plates and therefore better suited for higher resolution.

HPTLC plates use an optimized silica 60 sorbent with an average particle size of only 5-6 μm and 4-8 μm in general as compared to 10-12 μm used for TLC pre-coated layers. The smaller particles give a smoother surface and a superior separation power. Band diffusion is reduced leading to compact sample bands or zones after development. These features ultimately result in increased sensitivity and faster analysis. Spherical HPTLC silica gel 60 of optimum particle size is available which causes less resistance to flow of mobile phase.

Glass backed HPTLC silica plates are available in a variety of formats. In the classical range, two kind of fluorescent indicators are used: (i) the green fluorescing F_{254} and (ii) the blue fluorescing acid-stable F_{254s}. Both indicators fluorescence in UV light at an excitation wavelength of 254 nm.

Merck's HPTLC plates AMD with extra thin layer of only 100 μm have also been developed for automated multiple development (AMD) described later.

Superiority of HPTLC plate with that of TLC plate is evident from the following comparison given in Table 5.4.

Table 5.4: Comparative specifications of HPTLC and classical TLC plates		
Specification	**HPTLC**	**Classical TLC**
Mean particle size	5-6 μm	10-12 μm
Particle size distribution	4 - 8 μm	5 - 20 μm
Layer thickness	200 μm(100 μm)	250 μm(200 μm)
Typical plate height	12 μm	30 μm
Typical migration distance	3 - 7 cm	8 - 15 cm
Typical separation time	3 - 20 min	20 - 200 min
Number of samples per plate	< 36 (72)	< 10
Detection limits (absorption)	0.1-0.5 ng	1 - 5 ng
Detection limits (fluorescence)	5 - 10 pg	50-100 pg

Comparison of the particle size distribution of TLC, HPTLC and preparative layer chromatography is demonstrated in Figure 5.3.

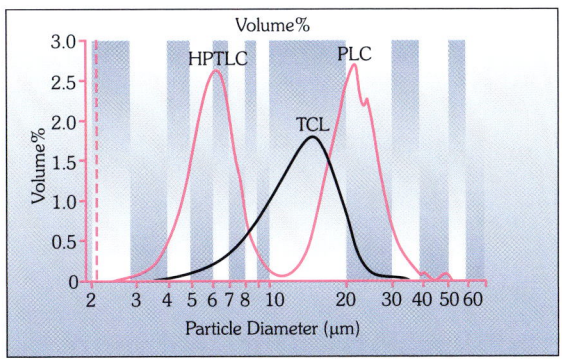

(ChromBook 06/07, E.Merck)

Fig. 5.3: Effect of particle diameter

5.3 APPLICATION OF HIGH PERFORMANCE SILICA PLATES (HPTLC)

HPTLC plates are ideal for quantitative analysis of:

- Drugs
- Medicinal plant extracts and formulations
- Trace analysis in food
- Automated applications for quantitative separations

Comparisons of the separation of dansyl amino acids as shown in Figure 5.4 (a) on a classical TLC silica gel 60 plates and Figure 5.4 (b) on an HPTLC silica gel 60 plate under identical conditions are cited.

The comparisons clearly demonstrate that the HPTLC plate delivers sharper zones with shorter migration distances and running times. Further the HPTLC plate is suitable to separate the double number of samples simultaneously.

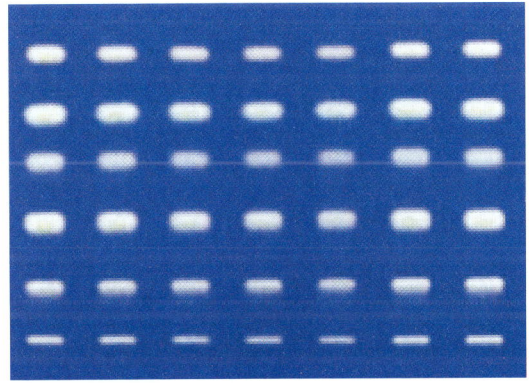

Fig. 5.4 (a): Separation on TLC layer   **Fig. 5.4 (b):** Separation on HPTLC layer

Table. 5.5 indicates the detail of experimental conditions for separation of dansyl amino acids.

Table 5.5: Experimental conditions		
1.	Compounds	1. N-alpha-dansyl-L-asparagine
		2. Alpha-dansyl-L-arginine
		3. Dansyl-L-cysteic acid
		4. N-dansyl-L-cysteic acid
		5. Dansyl-glycine
		6. N-N-didansyl-L-tyrosine
2.	Sample volume	HPTLC 0.3 µl; TLC 4 µl
3.	Mobile phase	Ethylacetate/methanol/propionic acid (22/10/3)
4.	Migration distance	HPTLC: 5 cm;TLC: 10 cm
5.	Analysis time	HPTLC: 14 min;TLC: 42 min

(ChromBook 06/07, E.Merck)

5.4 HPTLC PLATES WITH SPHERICAL PARTICLES

Unique HPTLC LiChrospher® plates are the first thin layer chromatography products based on spherical silica particles. They offer performance and speed to enable high throughput analysis of complex samples with 20% reduced running times, highly compact zones, lower detection limits and lower back ground scanning noise. This can be hyphenated to Raman spectroscopy for further identification and quantitation of unknown components.

Branded LiChrospher® plates are based on Merck's spherical-shaped silica 60 with a small particle size of 6-8 µm and narrow particle size distribution. Though it has similar selectivity, but plate height, separation numbers besides velocity constants are better. The plates are available in format of 20 × 10 cm and 20 × 20 cm size with backing of glass and aluminium. Electron micrograph of comparison of HPTLC spherical silica particles with silica plates are given in Figure 5.5.

Such HPTLC plates can be used for a wide range of separation, but especially for trace analysis of pesticides mixtures.

A comparison of separation of Hexazinone, Metoxuron, Monuron, Aldicarb, Azinphos-methyl, Prometryn, Pyridate and Trifluralin on LiChrospher with HPTLC silica plate is given below (Fig. 5.6). Sample volume taken is 50 µl with mobile phase as petroleum ether 40-60° C: acetone (7 : 8). Detection is done at 254 nm in UV range.

(a) A LiChrospher® plate   **(b) An HPTLC silica plate**

Fig. 5.5: Electron micrograph of plates

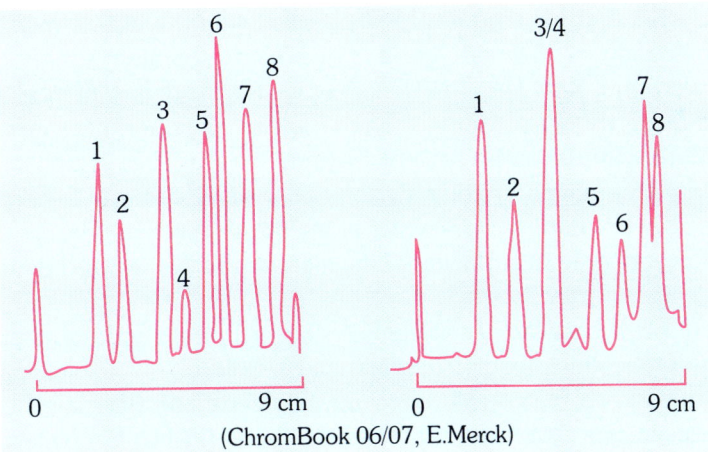

(ChromBook 06/07, E.Merck)

(a) Scan of pesticides separation on HPTLC LiChrospher® Si 60

(b) Scan of pesticides separation on HPTLC Si 60 plate

Fig. 5.6: Scans of pesticides separation

Analysis times on a LiChrospher® Si 60 F_{254} compared with an HPTLC Si 60 F254 is given in Table 5.6 for comparison.

Table 5.6: Comparison of plates			
Eluent	**Migration distance**	**LiChrospher silica gel 60 F_{254}**	**HPTLC silica gel 60 F_{254}**
Toluene	4 cm	4 min	5 min; 45 sec
Ethyl acetate / toluene (95:5)	5 cm	6 min	7 min; 50 sec
Methyl ethyl ketone / 1-propanol/water/acetic acid (40:40:20:5)	5 cm	20 min	26 min; 30 sec
n-hexane/ toluene/acetone 19 min. (70:20:10)	7 cm		13 min

(ChromBook 06/07, E.Merck)

5.5 RP-MODIFIED SILICA PLATES (TLC AND HPTLC) FOR SPECIAL SEPARATIONS

RP-modified silica layers are suitable for several separation problems that cannot be adequately resolved by unmodified polar silica. Such plates are suitable for separation of extremely non-polar and highly polar substance using aqueous solvent systems. Analysis of certain polar substances amenable for ion-pair chromatography is possible. Atmospheric humidity does not affect the layer activity significantly.

RP-plates (RP-2, RP-8 and RP-18) are based on silica gel 60 which is modified with aliphatic hydrocarbons of increasing chain length, i.e. 2, 8 and 18 carbon atom chains. Longer the carbon chain, greater the hydrophobicity. The chain length as well as the degree of modification determines the tolerance ability with the water of the solvent system. Migration time increases in the order RP-2, RP-8 and RP-18 with the same solvent. Separation sequence is unaltered too. The RP-2 shows higher polarity and affinity of aqueous solutions tolerating up to 80% water, whereas RP-8 and RP-18 can be run with up to 60% water in the solvent system. The special HPTLC RP-18 W with a defined lower degree of surface modification can be wetted and developed with 100% water.

RP phases do not show catalytic activity unlike silica gel plates and are, therefore, suitable for some unstable substances which tend to degrade by oxidation.

The RP-18 modified silica plates with concentrating zone are especially suited for the high-resolution separation of polycyclic aromatic hydrocarbons (PAH). RP plates are also used for separation of amides, antibiotics, and fatty acids. An example of separation of indeno-(1, 2, 3-c, d)pyrene (0. 05%); 3, 4-benzfluoranthene-(0.05%) and; fluoranthene(0. 05%) is given. Mobile phase used is acetonitrile: water (9 : 1) with the normal chamber without saturation and the migration distance is 5 cm. Detection is done with UV (366 nm). Influence of the hydrocarbon chain length on retention is demonstrated in Figure 5.7.

Separation of gallic acid and its esters viz. dodecyl gallate, butyl gallate, ethyl gallate, methyl gallate using RP modified layer is shown in Fig. 5.8. The excellent resolution shows the efficiency of the plate. Mobile phase as 1 N acetic acid:methanol (70:30). Migration distance is 5 cm and scanning is done at 265 nm.

5.6 CN-, DIOL-, NH₂- MODIFIED SILICA PLATES FOR TLC AND HPTLC

NH_2-, CN- and Diol-modified silica sorbents are less polar than classical silica phases and are thus suitable to separate hydrophilic or charged substances.

Silica gel 60 is modified with a cyano propyl group for CN-modified plate. Diol-modified plate is a silica surface modified by vicinal diol alkyl ether. These moderately polar plates with its intermediate properties bridge a gap in ranges of silica plates to be used for normal phase and reverse phase. CN plate permits unique two-dimensional separations by using normal phase mechanism in the first direction followed by reverse phase in the second direction.

Amino modified silica (NH_2 plate) has weak basic ion exchange characteristics. This enables separation of charged compounds such as nucleotides, purins, pyrimidines, phenols and sulfonic acids using simple eluent mixtures. It allows reagent free detection of certain chemical substances by thermo-chemical fluorescence activation, e.g. sugars.

The modified plates provide wide range of applications for:

(a) CN-silica: Benzodiazepine derivatives, pesticides, plasticizers, tetracycline antibiotics, gallic acid ester

(b) Diol-silica: Glycosides, anabolic steroids, aromatic amines and particularly dihydroxybenzoic acids

(c) NH_2-silica: Charged compounds

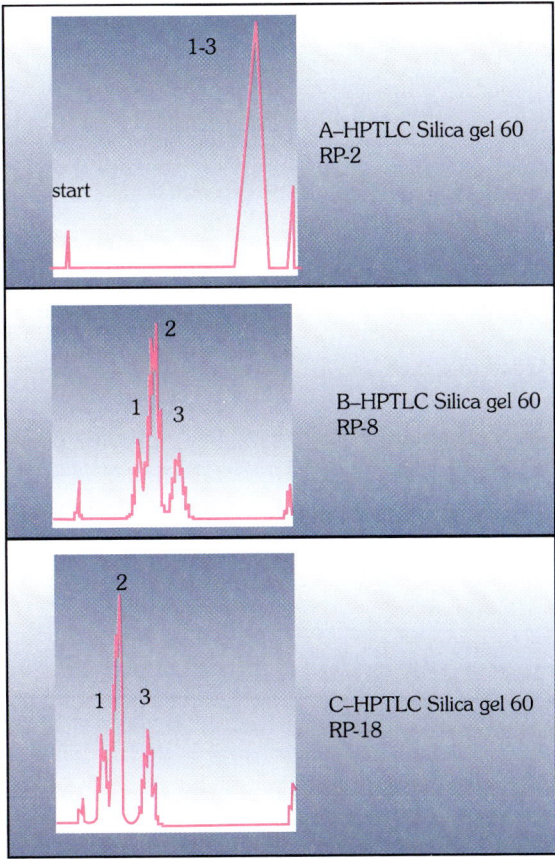

Fig. 5.7: Influence of hydrocarbon chain length on retention

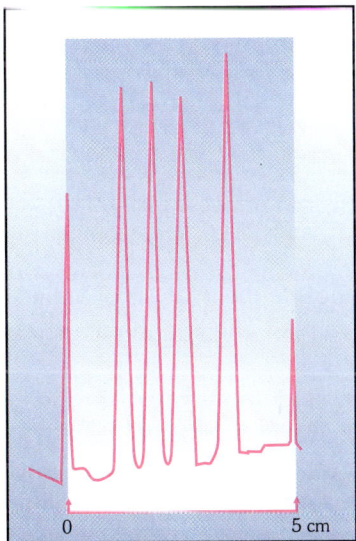

Fig. 5.8: Separation of gallic acid and its esters on HPTLC silica RP-18 WF$_{254}$.

Separation of oligo-nucleotides on an HPTLC NH_2-modified silica gel 60 plate is given in Figure 5.9 with the following experimental conditions.

Compounds	1. ApUpG	0.1%
	2. ApApU	0.1%
	3. ApApC	0.1%
	4. ApApA	0.1%
Sample	300 nl	
Mobile phase	Ethanol-water(60/40 v/v)plus 0.2 M lithium chloride	
Detection	UV 254 nm (TLC/HPTLC Scanner 2)	
Migration distance	7 cm	

5.7 CELLULOSE HPTLC PLATES

Pure (organic sorbent) is suitable for the separation of hydrophilic substances by partition chromatography. The HPTLC cellulose layers utilize high purity rod-shaped microcrystalline cellulose. These layers have low diffusion and so better separations.

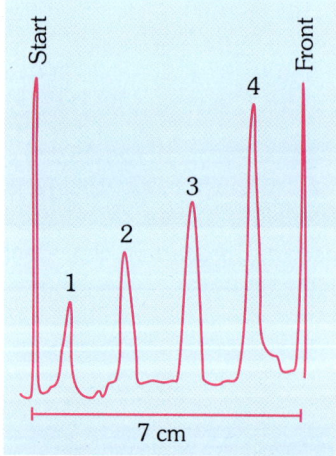

Fig. 5.9: Separation of oligo-nucleotides on an HPTLC NH_2-modified silica gel 60 plate

Applications of cellulose plates include the analysis of amino acids, carbohydrates, phosphates, nucleic acids and nucleic acid derivatives. The important applications are amino acids finger prints and metabolic studies.

Suitability of HPTLC cellulose is demonstrated through separation of polar compounds (phosphates) through Figure 5.10. The experimental conditions for separation are given below. Cellulose plates can be used to replace paper chromatography.

PEI cellulose plates are polyethyleneimine modified cellulose (strong basic anion exchanger) useful for analysis of nucleotides, nucleoside and nucleobases, vanadyl mandelic acid and sugar phosphates. The plates should be stored at 0–4° C to reduce deterioration.

Compounds	1. $(NaPO_3)_3$
	2. $Na_5P_3O_{10}$
	3. $Na_4P_2O_7$
	4. Na_2HPO_4
Sample volume	250 nl
Mobile phase	Dioxin sol.160 g TCA, 8 ml 25% ammonia in 1 lit. water; 70/30
Migration distance	7 cm
Detection	586 nm(TLC/HPTLC scanner)

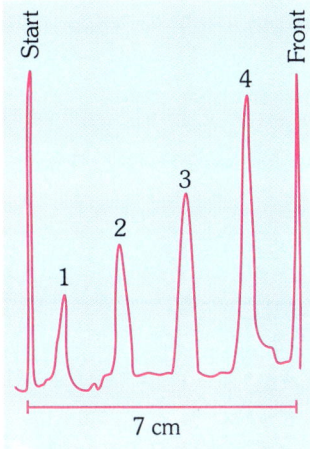

Fig. 5.10: Separation of polar compounds as demonstrated by the separation of phosphates

5.8 CONCENTRATING ZONE HPTLC PLATES

These plates are suitable for quick and easy sample application of large volumes of diluted samples. It provides high sample loading, better resolution due to uniform sharp bands and aids in sample clean-up. Concentration zone plates provide concentrating zones of 2 to 5 cm.

Concentrating zone plates facilitate sample application, hence used when manual sample application is desired. Separation of lipophilic dyes is given in Figure 5.11 showing the utility of concentrating zone plates.

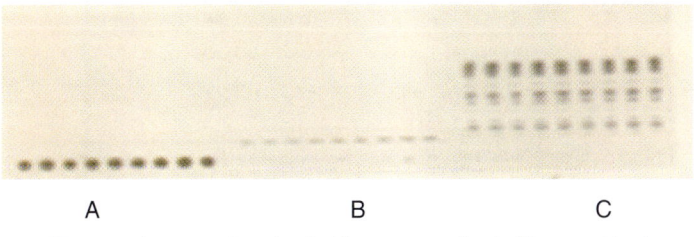

A   B   C

(Separation application) (Concentration) (Separation)

Fig. 5.11: Separation of lipophilic dyes

Note: Stages of the development of a preparative layer concentration zone plate silica gel 60 for separation of lipophilic dyes with toluene as mobile phase.

High resolution separation of poly aromatic hydrocarbons can be achieved too, using this technique.

5.9 ULTRA-THIN MONOLITHIC SILICA PLATE (UTLC)

This plate (Fig. 5.12) is suitable for ultra-fast and high sensitive analysis of trace analysis. It has all features of TLC technique such as easy and fast separations with the sensitivity and speed of ultra-thin monolithic layers. It provides highly reduced migration distances and short development times, very low sample quantities for costly samples, increased sensitivity due to reduced layer thickness, free from binder and stable in pure water.

The monolithic unmodified nano-SiO_2-layer of 10 µm has mesopores of 3-4 nm and macropores of 1-2 µm providing a 25-fold increased sensitivity compared to classical HPTLC. It enables to detect the analytes in the pg range.

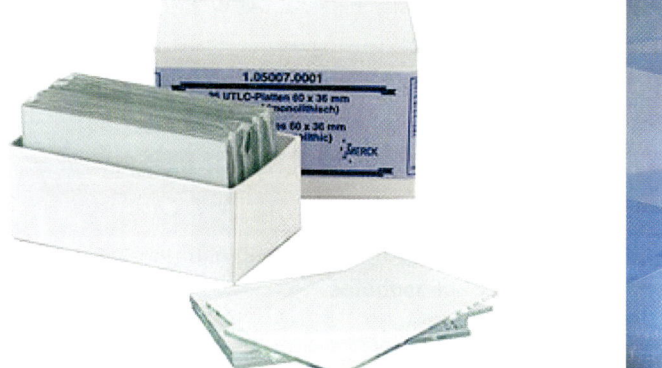

Fig. 5.12: Ultra-thin monolithic plates (ChromBook 06/07, E. Merck)

Specification of ULTC is provided in Table 5.7.

Table 5.7: Specification of ULTC plate	
Description	**Specification**
Plate format	60 × 36 mm
Layer thickness	10 µm
Stationary phase	Silica SiO_2
Additions	No binder
Sample volume	Spot wise: 5-20 nl, Band wise: up to 100 nl
Detection limit	10 pg
Migration distance	1-3 cm
Analysis time	1-6 min

(ChromBook 06/07, E.Merck)

5.9.1 Application of Ultra-thin Monolithic Unmodified Silica Plate (UTLC)

It is suited for small, simpler samples with low analyte concentration. It is suitable for separation of steroids, azepams, amino acids, phthalates and phenols.

An example of separation of Sico Fast Blue 50401N, Nitro Fast Blue 2B and Ceres violet I is given. The application volume is 10 nl (0:1% in toluene) and the spotting device used is syringe. Mobile phase used is toluene and ideal migration distance is 1.5 cm with saturated chamber having migration time as little as 165 seconds. Figure 5.13 shows the separation of lipophilic dyes on UTLC plate.

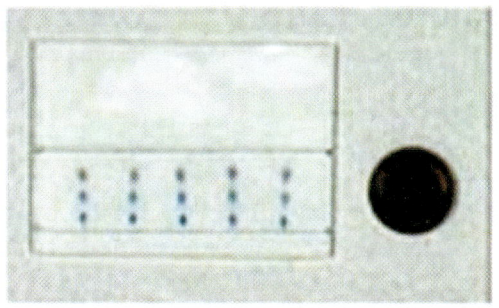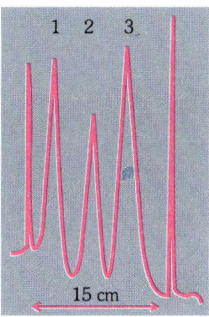

Fig. 5.13: Separation of lipophilic dyes on a UTLC plate

The experimental details of other compounds viz. 4-aminophenol, 2-aminophenol, biphenyldiol (2/2), 2-benzene-4-chlorophenol, separated on this plate is given below. Mobile phase is toluene: chloroform: methanol (80: 10: 10). Total migration distance is 2 cm (with chamber saturation) and migration time is 240s (approx.). Detection of separated components is done with UV 200 nm and the chromatogram is shown in Figure 5.14.

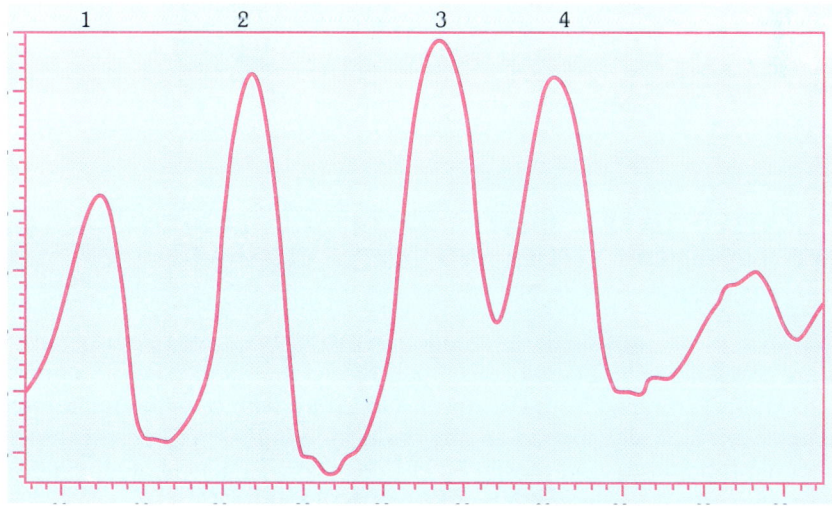

Fig. 5.14: Excellent separation of phenols

5.10 HPTLC PREMIUM PURITY

The new HPTLC premium purity plate is suitable for high performance, and contamination free separations. Layers are highly pure, exhibiting minimal background noise even with medium-polar solvent systems such as toluene: ethyl acetate (95:5).

The HPTLC premium purity plate is based on HPTLC Si 60 F_{254}. The special packing prevents any deposition of plasticizers such as phthalates from the wrapping material that could appear as unknown extra zone when using medium-polar mobile phase.

The HPTLC premium purity plate is suited for demanding HPTLC analysis for quantitative separations, impurity profiling of samples and other trace analysis.

5.11 GLP HPTLC PLATES FOR GLP APPLICATIONS

Laser coded GLP plates are available for working according to GLP requirements (Fig. 5.15).

The plates provide item, batch and individual plate number on the top of every plate for convenient back tracing of article, batch, and individual plate number. Every plate can be documented.

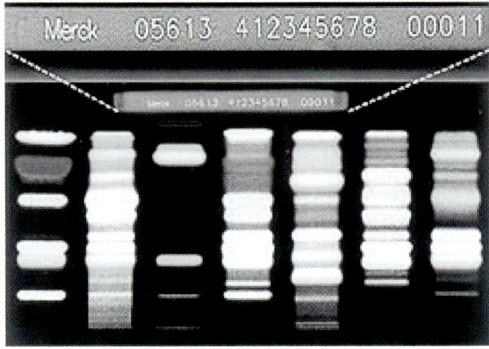

Fig. 5.15: GLP plates (ChromBook 06/07, E.Merck)

5.12 PREPARATIVE LAYER PLATES (PLC) FOR ENRICHMENT OF TARGET ANALYTES IN mg QUANTITIES AND SAMPLE CLEAN-UP

Preparative thin layer plates provide the separation of mg up to gm sample using up to 2 mm thick layers. Plates are available with layers of silica gel, RP 18-modified silica or aluminium oxide in several layer thicknesses, ranging from 0.5 mm up to 2 mm with or without fluorescent indicator.

In PLC, samples are typically applied as one band (about 190 mm) across the entire width of plate and components are visualized mostly by UV detection. The separated components may be isolated by extraction after scrapping the band from the layer.

Preparative layers find application in the final purification of synthesized organic chemicals. Phytochemicals can also be isolated and or purified. New drug discovery labs use PLC in a big way.

Thicker layers have lesser resolution than thin layers under identified conditions. PLC does not require special equipments, which is a big advantage.

Thick layers can also be prepared in the lab.

Reference

1. Scott RM. *J Chromatogr* **11;** 129, 1973

Preparation of Sorbents in HPTLC

6.1 SILICA GEL/KIESELGEL/SILICIC ACID

It is a white amorphous porous powder manufactured by precipitation from silica solution by reaction with acid. The primary particles appear and grow. Water is removed for gel formation. The process is not simple. Temperature and pH have to be well controlled. Colloidal particles are developed which are condensed and shrunk to form hydrogel. After washing and heating at approximately 120° C, an amorphous hard but porous gel popularly known as silica gel is formed. Quality specifications of silica gel is listed in Table 6.1

S. no.	Specifications	Silica gel 60 (40-63μm)	Silica gel 60 (63-200μm)
1.	Particle size range (μm)	40-63	63-200
2.	Mean particle size (μm)	50 (approx.)	105 (approx.)
3.	Specific surface area [m^2/g]	500 (approx.)	500 (approx.)
4.	Specific pore volume [Ml/g]	0.71-0.78	0.71-0.78
5.	Mean pore size [Å]	60	60
6.	pH (10% aqueous suspension)	7 ± 0.5	7±0.5
7.	Loss on drying [%] (3 hours/120° C)	<7	<7
8.	Bulk density [g/ml]	~0.45	~ 0.45
9.	Back pressure{bar} (column 250-4, n-heptane, 20 ml/min)	~2	~0.5
10.	Iron [%] (acid extraction)	<0.02	<0.02
11.	Chloride [%] (acid extraction)	<0.02	<0.02
12.	Theoretical plates [N/m]	~ 3000 (anisole)	~ 3000 (anisole)
13.	Relative retention 2-nitroanisole /4-nitroanisole	~ 1.73	~ 1.73

Table 6.1: Specification of silica gel

(ChromBook 06-07)

6.2 BONDED SILICA GEL PHASE

While the availability of bonded and other layers are very useful in certain analysis, plain round phase silica gel remains the work horse of adsorbents.

(*i*) **Reversed phase:** Paraffin oils or silicone is normally used for preparation of reversed phase plate. These are relatively easy to prepare in the lab.[2-5] Lab made plates have their own advantages and disadvantages. They may not be reproducible. Development of bonded silica gel plates for reversed phase chromatography is a boon to the modern chromatographer. Silica gel is bonded with organosilanes of various chain lengths. Dimethyl, ethyl, octyl, undecyl, octadecyl and phenyl have been used for commercial preparation to produce C_2, C_8, C_{12}, C_{18} and aromatic rings bonded to silica matrix.

(*ii*) **Amino bonded phase:** Silica gel 60 is chemically bonded with aminopropyl groups through siloxane linkages. Basic nature of amino group leads to typical application for separation of nucleotides, mono and polysulfonic acids, purines, and phenols, etc.[6-8]

(*iii*) **Cyano bonded phase:** It is made by chemical binding of cyanopropyl group via siloxane to silica gel. Klaus et al[9] reported first time such plates for separation of polynuclear aromatics. These plates have wide application[10-11] i.e., benzodiazepine derivatives, pesticides, plasticizers, tetracycline antibiotics, phenols, some estrogens, gallic acid esters, alkaloids and sorbic acid.

(*iv*) **Diol bonded phase:** Plates are modified to possible extent with a vicinal diol alkyl ether group bonded via siloxane to silica gel surface by silanisation procedure. Such plates are suitable for a number of separations especially digitalis glycosides, anabolic steroids, aromatic amines and particularly dihydroxybenzoic acids.[12-14]

(*v*) **Chiral bonded phase:** Wainer et al[15] reported in 1983 an ionically bonded chiral plate by use of N-(3, 5-dinitrobenzoyl)-R-(–)phenylglycine. Several efforts followed to make chiral bonded plate with other combinations. Commercial readymade plates are available. Such plates have facilitated resolution of amino acids, different groups of amino acids, dipeptides and catecholamines.[16-19]

6.3 OTHER SORBENTS

(*i*) **Cellulose:** Cellulose is a natural product, with a polymeric structure (glucopyranose units joined by oxygen bridge). It is an ideal phase for separation of hydrophilic substances like carbohydrates, amino acids, and inorganic ions, etc.[20-27]

In planar chromatography, normally two types of cellulose are used. First one with native fibres and typical polymerization between 400-500 glucopyranose units is appropriate for paper chromatography and TLC layer. The other one, i.e. microcrystalline form made by hydrolysis technique, is suitable for HPTLC. It has a degree of polymerization between 40-200 glucopyranose units. For preparation of plates, binders are not necessary. Resolutions are generally not sharp as compared to silica gel. This phase is used mostly for amino acids separations.

(ii) Cellulose bonded polyethyleneimine (PEI) modified phases: This works as a strong basic anion- exchanger. This phase has a specific use, e.g. in analysis of nucleotides, nucleosides and nucleo-bases, vanadylmandelic acid and sugar phosphate.[28] The plate requires storage at 0–4° C to prevent deterioration and discolouration which is a serious drawback.

(iii) Acetylated cellulose: It is prepared by reaction of hydroxyl groups on cellulose to produce a layer having reversed phase properties. Its major use was for separation of polyaromatic hydrocarbons. Lepri et al[29] used it as a chiral layer for separating optical isomers. Due to long development time (more than 2 hrs) such plates could not gain popularity.

(iv) Aluminium oxide: It occurs in different crystal forms. Aluminium oxide with a mean pore diameter of 90 Å is used for TLC. It is prepared by calcinations of aluminium hydroxide. It is a synthetic sorbent like silica gel. It is made in acidic, basic or neutral pH range for various types of applications. Under aqueous conditions, acidic compounds such as phenols, sulphonic, carboxylic and separation on acidic alumina. Basic compounds like amines, imines and basic dyes are separated on basic alumina. Neutral compounds, e.g. aldehydes, ketones and lactones are suitable for separation on neutral alumina. Since alumina is chemically more reactive than silica gel and also more hydroscopic so it could not gain much popularity in HPTLC. Specification and quality of aluminium oxide have been given in Table 6.2.

Table 6.2: Specifications of aluminium oxide						
Sorbent	Characteristics	Spec. surface area S_{BET} (m^2/g)	Pore volume V_P (ml/g)	Particle size d_P (μm)	pH	Activity
Aluminium oxide 60	Irregular particles of alumina medium pore size: 6 nm (60Å)	~160	0.3	63-200	4,7,9	I
Aluminium oxide 90	Irregular particles of alumina medium pore size: 9 nm (90Å)	~90-120	0.27	63-200	4,7,9	I, II-III
Aluminium oxide 150	Irregular particles of alumina medium pore size: 15 nm (150Å)	~60-90	0.27	63-200	4,7,9	I-II

(v) Kieselguhr: It is a natural diatomaceous earth comprising of skeletal remains of tiny marine organisms. It is basically silicon dioxide, but also contains varying amount of other oxides of aluminium, iron, magnesium, titanium, potassium and calcium as oxides, hydroxides and carbonates (about 10% in total).[30] It is popular as filter aid due to high porosity. In conjunction with 15% calcium sulphate as binder, kieselguhr was being used very commonly for separation of polar compounds.

(vi) Polyamide: It is produced from polycaprolactam(nylon 6), polyhexamethyldiaminoadipate (nylon 66) or polyaminoundecanoic acid (nylon 11). The separation on polyamide depends upon the hydrogen-bonding capabilities of their amides and carbonyl groups. Different workers have resolved mixtures of phenols, indoles, steroids, nucleic acid bases, nucleosides, dinitrosulfonyl (DNS), dinitrophenyl (DNP), and dimethylaminoazobenzene isothiocyanate (DABITC), derivatised amino acids and aromatic nitro compounds on polyamide.[31-34]

(*vii*) **Other stationary phases:** These are magnesium silicate, chitin, Sephadex[TM]. Magnesium silicate is suitable for separation of carbohydrates and their derivatives as reported.[35] Chitin, is a polysaccharide consisting of basically 2-acetamide-2- deoxy-D-glucan molecules linked through oxygen bridge with basic nature. It is used for amino acids separation and other applications like inorganic ions,nucleic acids, phenols and dyes.[36] Sephadex is a trade mark, in modified dextrin gels, which is hydrophilic and neutral in nature. It has been useful for separation of peptides and nucleic acids.[37]

(*viii*) **Mixed stationary phases:** Mixed phases of silica gel/kieselguhr, silica gel/alumina, cellulose /silica gel are sometimes suitable for very specific applications. Inorganic ions, herbicides and steroids[38-40] have been separated using mixed phase of silica gel/kieselguhr. Whereas cellulose/silica gel have applications in separation of food preservatives and antibiotics.[41, 42] Most mixed phase plates have to be freshly made in the lab.

(*ix*) **Dual phases:** Such plates are of recent innovation having two separate stationary phases on one plate. Two phases, in general, are normal and reversed phase, though other combinations are possible. Plates are developed in two different directions, one as a normal phase separation on silica gel and then after drying in a second direction as reversed phase separation on silica gel RP18. This finally results in two dimension "map" or "fingerprints" of components. Such technique allows resolution of large number of components as compared to normal plates. Separation of sulphonamides and bile acids[43] has been made by dual phase techniques. Only one sample can be analyzed on such a plate.

Table 6.3 specifies brief information about different types of sorbents and their utility.

Table 6.3: Different types of sorbent and their utility		
Sorbents	**Parameters/functions**	**Utility**
Aluminium oxide	Specific surface area: 150-200 m²/g Lower activity than silica gel	Unsaturated Al^3 + ions (Lewis acids) are selective for π-electron systems and other nucleophilic or basic partners. The O^2-ions are basic. The Al-OH bond acts as a dipole
Kieselguhr	Large pore volume (macrospores). Small specific surface area: 1-5 m²/g 90% silica acid (silicon dioxide), diatomaceous earth	Suitable for partition chromatography of polar substances and for use in concentrating zones
Silica 50,000	Pore size of 50,000 Å (5000 nm), and specific surface area 0.5 m²/g. Porous, extremely inert, synthetic silicon dioxide (comparable to kieselguhr)	Suitable for partition chromatography of polar substances and for use in concen-trating zones
Magnesium silicate (Florisil)	Polarity between that of silica and aluminium oxide. Composition 15% MgO and 85% SiO_2	No catalytic activity. Used to separate sugars, sugar derivatives, and other polyhydroxy compounds
Cellulose	Hydrogen bonding, strong swelling with water and alcohols derived from cotton, either fibrous or microcrystalline	Organic adsorbent suitable for the separation of hydrophilic substances such as amino acids, carbohydrates, nucleic acid derivatives, phosphates, etc. by partition chromatography

Contd...

Contd...

Sorbents	Parameters/functions	Utility
Acetylated cellulose	First reverse phase for chromatography	Used as chiral layer for separating optical isomers
PEI cellulose	Polyethyleneimine modified microcrystalline cellulose	Used as anion exchanger for the separation of mono and oligonucleotides, nucleosides
Polyamide (-CONH groups)	C6 hydrophilic, polycaprolactam.C11 hydrophobic, polyundecanamide	Selectivity for protic substances

6.4 RELATIONSHIP OF HUMIDITY WITH LAYERS

Various types of TLC sorbents, especially the widely used unmodified silica sorbent do adsorb water. Change in relative humidity in the lab or of the layer can affect important factors like R_f values, selectivity, solvent front migration and the position of multiple fronts. The relative humidity of the atmosphere is critical for reproducible work. If constant humidity cannot be ensured in a laboratory, it is desirable to pre-condition the plates for 30 minutes over saturated salt solutions or sulphuric acid solutions of particular concentrations. The relative humidity above selected salt solutions is given in Table 6.4. Special chambers like twin-trough, automated development chamber are now available which tackle the humidity problem to significant level, as these chambers accommodate sulphuric acid solutions or saturated salt solutions to control relative humidity. Heating the plates/sheets at 120° C for about half an hour has little effect because once the plate is exposed to ambient conditions during sample application or storage, it hardly takes few minutes to re-equilibrate water content to surrounding conditions. It is of little use to re-heat the plate.

Table 6.4: Saturated salt solution and relative humidity	
Saturated salt solution containing a large quantity of undissolved salt	**Relative humidity above the solution (20° C) (%)**
Di-sodium hydrogenphosphate $Na_2PO_4 \cdot 12\ H_2O$	95
Sodium carbonate $Na_2CO_3 \cdot 10\ H_2O$	92
Zinc sulfate $ZnSO_4 \cdot 7H_2O$	90
Potassium chloride KCl	86
Ammonium sulfate $(NH_4)_2SO_4$	80
Sodium chloride NaCl	76
Sodium chlorate $NaClO_3$	75
Sodium nitrite $NaNO_2$	65
Ammonium nitrate NH_4NO_3	63
Calcium nitrate $Ca(NO_3)_2 \cdot 4H_2O$	55
Sodium dichromate $Na_2Cr_2O_7 \cdot 2H_2O$	52
Potassium carbonate K_2CO_3	45
Zinc nitrate $Zn(NO_3)_2 \cdot 6H_2O$	42
Chromium trioxide CrO_3	35
Calcium chloride $CaCl_2 \cdot 6H_2O$	32
Potassium acetate $K(OOCCH_3)$	20
Lithium chloride $LiCl \cdot H_2O$	15

(ChromBook 06/07)

6.5 PRE-WASHING OF PLATES

It may be necessary in some cases, to remove impurities mostly coming from binder. A polar mobile phase can cause concentration of impurities at the solvent front. Impurities eluting with the front may not pose problems. Non-polar mobile phases do not cause any migration of impurities, hence no problem. Ideal way to pre-wash a plate is by blank development in a TLC chamber. Mobile phase for separation or methanol or mixtures of methanol and chloroform may be used. Direction of the blank development and the plate subsequently used for chromatography of the sample should be the same.

6.6 FLUORESCENT INDICATORS

Commercial pre-coated plates are available with inorganic fluorescence indicator, which shows green, yellow or blue fluorescence when excited at 254 nm by UV light. Indicators are uranyl acetate (green with yellowish tinge), manganese zinc silicate, zinc silicate, zinc cadmium sulphide (green), tin strontium phosphate and alkaline earth metal tungstate (blue). Detection depends upon the fluorescence quenched by sample component by absorption. Pink or violet spots/bands are seen for quenching on green fluorescence back ground, and grayish/black spots/bands on blue back grounds. Organic fluorescent indicators are also used. These are rhodamine dyes, fluorescein and hydroxypyrene sulphonate. Such indicators fluorescence at longer wavelength (366 nm) but rarely used. Plates are available with both types of indicators. Some indicators are phosphorescent but that too short lived.

6.7 CHANNELLED PLATES

Channelled plates are available with space of 1-2 mm between tracks. Self-made channelled plates can also be made. Each "channel" can behave as a separate plate, which is not desirable. They are also difficult to scan due to exposed glass beside the channel. Such plates offer no advantages.

6.8 CONCENTRATION ZONE PLATES

This offers a number of advantages for specialized separation. It consists of two different layer sections with a sharp border line, without any gap. Sample is applied on lower zone, which consists of kieselguhr or silicon dioxide of medium pore volume with a high pore diameter and very small internal surface area. Upper section of plate is coated with silica gel 60 sorbent. On elution, desired analytes migrate at the solvent front. On reaching interface, a concentrated band is formed that migrates into silica gel 60 layer and development continuous. They cannot be used for quantification.

6.9 IMPREGNATION OF SORBENTS

Such techniques improve resolution with buffers and complexing agents.

6.9.1 Impregnation before Layer Coatings

Silica gel slurry is pretreated with impregnation reagent. Buffer solutions, acids, bases, complexing reagents, salts or water soluble organic compounds are suitable for such techniques. Modified slurry is then coated on plate as usual. However, such techniques are not popular for quantification and are limited to water soluble agents.

6.9.2 Impregnation of Ready Plates

This can be achieved in following different ways.

(a) Plate is dipped in a solution of suitable concentration of the impregnating agent in an appropriate volatile solvent. The solvent is then evaporated at room temperature or at elevated temperature.

(b) Spray the impregnated solution and evaporate the solvent. Spraying is not uniform.

(c) Blank development of plate using impregnated solution and then removal of solvent. Separations of phenols, acids[44] by acidic plate and alkaloid and amine[45] by basic plate have been reported in literature. Mono-, di- and tri-saccharides[46-48] have been reported to be separated on silica gel 60 plates impregnated with 0.5 M sodium hydrogen orthophosphate. This technique is most reproducible.

6.9.3 Silver Nitrate Impregnation

Literature reveals the excellent separation of several types of lipids, fatty acids, steroids and terpenes, olefins[49-57] using silver nitrate impregnated silica gel 60 plates. The separation is based on the principle of interaction of silver ion with ethylenic π bonds present in solute molecules. The stronger interaction and then complexation lead to retention of compounds on plate. This technique allows *cis* and *trans* isomers of some organic compounds to resolve. Impregnation is done just before the plate is used.

6.9.4 Charge Transfer Plate

This technique relies on impregnating agent in layer which acts as a π-electron acceptor. These acceptors are mostly aromatic, unsaturated alicyclic or heterocyclic organic molecules with functional group of high electron affinity. 2, 4, 7-trinitrofluorenone, 2, 4, 6-trinitrophenol (picric acid), 1, 3, 5-trinitrobenzene, benzoquinone, tetramethyl uric acid, pyromellitic dianhydride, sodium desoxycholate, urea, nucleic acid bases, amino acids and caffeine[58-62] are used for impregnation.

Separation of PAH at very low level using caffeine impregnated plate is a significant work done in this direction. Here caffeine acts as a π-electron acceptor and PAH responds to varying degree of ability to π-electron donors. A charge transfer complex is formed.

6.10 PREPARATIVE LAYER CHROMATOGRAPHY (PLC)

This refers to the plate layers that are 0.5 mm or thicker up to 2 mm in general. The adsorbents are of the same specifications as analytical TLC. Samples can be loaded in mg quantities. Thicker layer adversely affects resolution. They are commercially available or can be made in the lab.

References

1. Unger KK. porous silica its properties and use as support in column liquid chromatography. *J Chromatography Library* **16**; 1-2, 1979
2. Lee KY, Nurok D, Zlatkis A. *J Chromatogr* **174**; 187, 1979
3. Touchstone JC, Levitt RE, Soloway RD, Levin SS. *J Chromatogr* **178**; 566, 1979
4. Scherz H, Stehlik G, Baucher E, Kaindl K. *Chromatogr Rev* **10**; 1, 1968
5. Ghebregzabher M, Rufini S, Monaldi B, Lato M. *J Chromatogr* **127**; 133, 1976
6. Jost W, Hauck HE. *Instrumental High Performance Thin-Layer Chromatography* (Interlaken 1982), Institute for Chromatography, 25-37, 1982
7. Jost W, Hauck HE. *J Chromatogr* **261**; 235-244, 1983
8. Jost W, Hauck HE. *Anal Biochem* **135**; 120-127, 1983
9. Klaus R, Fischer W, Hauck HE. *Chromatographia* **29**; 467-472, 1990
10. Jost W, Hauck HE. *Instrumental High Performance Thin-Layer Chromatography* (Wurzberg 1985), Institute for Chromatography, 83-91, 1985
11. Kang JS, Ebel S. *J Planar Chromatogr* **2**; 434-437, 1989
12. Hauck HE, Mack M, Reuke S, Herbert H. *J Planar Chromatogr* **2**; 268-275, 1989
13. Jost W, Hauck HE. *Instrumental High Performance Thin-Layer Chromatography* (Selvino/Bergamo 1987), Institute for Chromatography, Bad Durkheim, Germany, 241-253, 1987
14. Witherow K, Thorp RJ, Wilson ID, Warrander A. *J Planar Chromatogr* **3**; 169-172, 1990
15. Wainer IW, Brunner CA, Doyle TD. *J Chromatogr* **264**; 154, 1983
16. Gunther K, Martens J, Schickedanz M. *Angew Chem* **96**; 514, 1984
17. Mack M, Hauck HE, Herbert H. *J Planar Chromatogr* **1**; 304-308, 1988
18. Mack M, Hauck HE. *Chromatographia* **26**; 3-11, 1988
19. Th Brinkman UA, Kamminga D. *J Chromatogr* **330**; 375-378, 1985
20. Heathcote JG, Haworth C. *J Chromatogr* **43**; 84-92, 1969
21. Heathcote JG, Haworth C. *J Chromatogr* **41**; 380-385, 1969
22. McBridge RW, Jolly DW, Kadis BM, Nelson TE. *J Chromatogr* **168**; 290-291, 1979
23. Jones K, Heathcote JG. *J Chromatogr* **24**; 106, 1966
24. Heathcote JG, Davies DM, Haworth C, Oliver RW. *J Chromatogr* **55**; 377-384, 1971
25. Davies JM. *J Chromatogr* **69**; 333-339, 1972
26. Randerath K, Struck H. *J Chromatogr* **6**; 365, 1961
27. Mohammad A, Tiwari S, Chahar JP. *J Chromatogr Sci* **33**; 143-147, 1995
28. Randerath K. *Thin-Layer Chromatography* 195-197, 1963
29. Lepri L, Coas V, Desideri PG, Zocchi A. *J Planar Chromatogr* **7**; 376-381, 1994
30. Rossler H. *Thin-Layer Chromatography: A Laboratory Handbook*, E. Stahl (ed), 28, 1969
31. Bushan R. *J Chromatogr Sci* **55**; 353-387, 199
32. Soczewinski E, Szumilo H. *J Chromatogr* **81**; 99-107, 1973
33. Bhatia IS, Singh J, Bajaj KL. *J Chromatogr* **79**; 350-352, 1973
34. Bayliss RS, Knowles JR, Wybrandt GB. *J Biochem* **113**; 377-386, 1969
35. Rossler H. *Thin-Layer Chromatography: A Laboratory Handbook*, 29, 1969
36. Rozylo JK, Chomicz DG, Malinowska I. *Instrumental High Performance Thin-Layer Chromatography*, 173-187, 1985
37. Rossler H. *Thin-Layer Chromatography A Laboratory Handbook* E. Stahl (ed), 40-41, 1969
38. Egan H, Hammond EW, Thomson J. *Analyst* **89**; 480, 1964
39. Crepy O, Judas O, Lachese B. *J Chromatogr* **16**; 340-344, 1964
40. Gregorowicz Z, Kulicka J, Suwinka T. *Chem Anal* **16**; 169, 1971
41. Gossele JAW. *J Chromatogr* **63**; 433-437, 1971
42. Konig H, Schiiller M. *Z Anal Chem* **294**; 36, 1979
43. Sherma J. *Practice and Applications of Thin Layer Chromatography* on Whatman KC_{18} Reversed Phase Plates, TLC Technical Series, **1**; 10, 1981

44. Dallas MSJ. *Nature* **207**; 1388, 1965
45. Stahl E. *Arch Pharm* **292**; 411, 1959
46. Ghebregzabher M, Rufini S, Monalgi B, Lato M. *J Chromatogr* **127**; 133-162, 1976
47. Lee KY, Nurok D, Zlatkis A. *J Chromatogr* **174**; 187-193, 1979
48. Klaus R, Ripphahn J. *J Chromatogr* **244**; 99-124, 1982
49. Hammond EW. *Chromatography for the Analysis of Lipids*, 39-54, 1993
50. Gunstone FD. *Fatty Acid and Lipid Chemistry*, Blackie Academic & Professional, London, UK, 18-19, 1996
51. Touchstone JC. *J Chromatogr* B **671**; 169-195, 1995
52. Dobson G, Christie WW, Nikolova-Damyanova B. *J Chromatogr* B **671**; 197-222, 1995
53. Adlof RO. *J Chromatogr* A **741**; 135-138, 1996
54. Nikolova-Damyanova B, Christie WW, Herslof B. *J Planar Chromatogr* **7**; 382-385, 1994
55. Inomata M, Takaku F, Nagal Y, Saito M. *Anal Biochem* **125**; 197-202, 1982
56. Bhat HK, Ansari GAS. *J Chromatogr* **483**; 369-378, 1989
57. Aitzemuller K, Goncalves LAG. *J Chromatogr* **519**; 349-358, 1990
58. Funk W, Gluck V, Schuch B, Donnevert G. *J Planar Chromatogr* **2**; 28-32, 1989
59. Funk W, Donnevert G, Schuch B, Gluck V, Becker J. *J. Chromatogr* **2**; 317-320, 1989
60. Triska J, Vrchotova N, Safarik I, Ssfsrikova M. *J Chromatogr* A **793**; 403-408, 1998
61. Short GD, Young R. *Analyst* **94**; 259, 1969
62. Libickova V, Stuchlik M, Krasnec L. *J Chromatogr* **45**; 278, 1969

7

Mobile Phase in HPTLC

The mobile phase is the transporter of sample components, hence should be able to dissolve them. The sample should be soluble under equilibrium conditions. The mobile phase should be able to separate the desired components of the samples effectively by varying the strength of solvent(s). Mobile phase is also responsible for selectivity of the separation, besides the stationary phase and other variables in the chromatography system.

There is direct relationship between migration distance of mobile phase and time taken for development. However, solvent front velocity is not constant. As solvent migrates forward, the speed of migration lowers. Flow rates vary in accordance with the physical characteristics of the solvents present in mobile phase as well as shape of particles of stationary phase. Further, viscosity and surface tension also affect the migration rate. Hence velocity constant largely depends on the surface tension and viscosity of solvents. Therefore, solvents with high viscosity and surface tension will migrate at a slower rate than low viscosity and surface tension.

Table 7.1 gives details of important solvents with their velocity constant, surface tension and viscosity.

Table 7.1: Physical properties of solvents			
S. no. Solvent	Velocity constant (k)cm^{2}s^{-1}	Viscosity (cP)	Surface tension N m^{-1}
1. n-hexane	0.118	0.31	18.42×10^{-3}
2. Acetonitrile	0.114	0.35	19.10×10^{-3}
3. Acetone	0.112	0.36	23.32×10^{-3}
4. Tetrahydrofuran	0.103	0.55	27.31×10^{-3}
5. n-pentane	0.092	0.24	15.48×10^{-3} (25° C)
6. Water	0.082	1.00	80.10×10^{-3}
7. Ethyl acetate	0.080	0.45	23.75×10^{-3}
8. Chloroform	0.079	0.57	27.16×10^{-3}
9. Toluene	0.071	0.59	28.52×10^{-3}
10. Methanol	0.050	0.59	22.55×10^{-3}
11. 1, 4-dioxan	0.050	1.37	33.75×10^{-3}
12. Cyclohexane	0.047	0.94	24.98×10^{-3}
13. Ethanol	0.031	1.22	22.32×10^{-3}
14. Propan-2-ol	0.019	2.40	21.79×10^{-3} (15° C)

Solvent strength affects the migration of the sample components. Figure 7.1 shows the effect of solvent strength on migration of components.

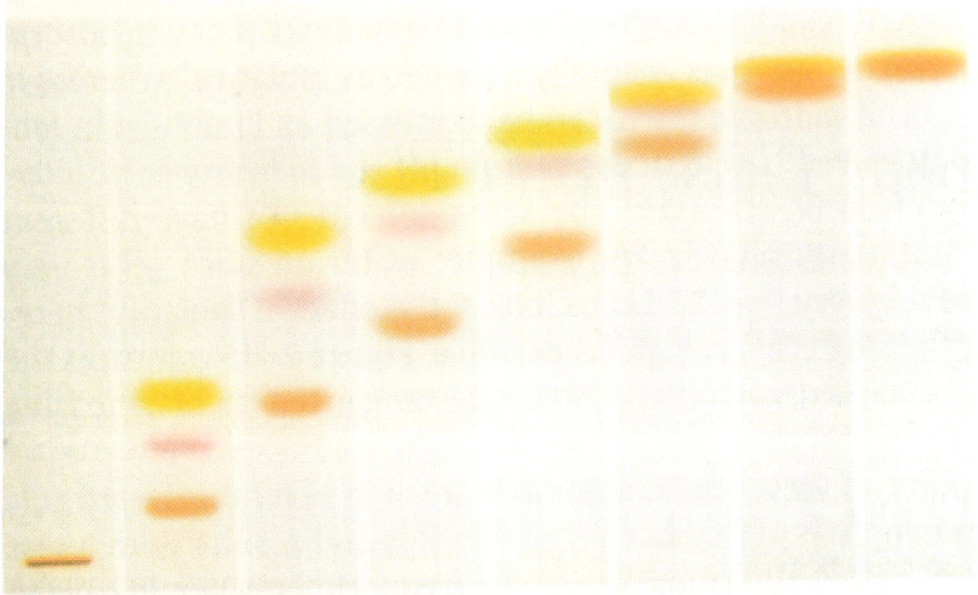

Fig. 7.1: Effect of solvent strength on migration of components

Note: *Effect of increasing solvent strength on the separation of a test dye mixture. Mobile phase– Hexane with (left to right) 0.5, 10, 15, 20, 30, 40, and 90% ethylacetate.*

Thus solvent strength generally describes the effect of mobile phase on the retention of sample components. Increasing solvent strength decreases retention, and thus increases R_f value. The same solvent will have different strength on different stationary phase. Solvent strength has different descriptions depending upon the retention mechanism.

Basically there are three main types of separation mechanisms. These are: (*i*) Adsorption, (*ii*) Partition, and (*iii*) Ion-exchange. However, separation mechanism cannot be attributed to one type of mechanism only. Other interactions may have important roles in separation mechanism. These may be ion-pairing, charge-transfer and π-π interaction. Therefore mechanism of separation may be complex in nature.

So far as solvent strength is concerned, it is regarded as: (*a*) "polarity" in adsorption chromatography and (*b*) "lipophily" in reversed phase partition chromatography. Solvent strength depends on pH in ion-exchange chromatography and ionic-strength in hydrophobic interaction.

An empirical way to express is to arrange several selected solvents in order of increasing strength which should be based on experimental values on a test sample. Experiments should be conducted with ultra pure solvents (LC grade) with a strong control on gas phase and its effect on layer pre-conditioning.

A more refined but empirical approach to solvent strength would be dimensionless solvent strength parameters ε^0 (defined as free energy of adsorption of the solvent molecules per unit area of adsorbent). Eluotropic series for common solvents for silica gel and aluminium oxide, as stated by Snyder[1–2] is described in Table 7.2.

S. no.	Solvent	Silica gel ε^0 value (expt.)	Alumina ε^0 value (expt.)	Silica gel ε^0 value (calc.)
		Table 7.2: Eluotropic Series for Common Solvents		
1.	n-pentane	0.00	0.00	0.00
2.	n-hexane		0.01	0.01
3.	iso-octane		0.01	0.01
4.	Cyclohexane		0.04	0.03
5.	Cyclopentane		0.05	0.04
6.	Carbon tetrachloride	0.11	0.18	0.14
7.	Isopropyl ether		0.28	0.22
8.	Toluene		0.29	0.22
9.	Chlorobenzene		0.30	0.23
10.	Benzene	0.25	0.32	0.25
11.	Chloroform	0.26	0.40	0.31
12.	Dichloromethane	0.32	0.42	0.32
13.	Methyl isobutyl ketone		0.43	0.33
14.	Tetrahydrofuran		0.45	0.35
15.	Diethyl ether	0.38	0.46	0.38
16.	1,2-dichloroethane		0.49	0.38
17.	Acetone	0.47	0.56	0.43
18.	1,4-dioxan	0.49	0.56	0.43
19.	Ethyl acetate	0.38	0.58	0.45
20.	Methyl acetate		0.60	0.46
21.	Amyl alcohol		0.61	0.47
22.	Aniline		0.62	0.48
23.	Acetonitrile	0.50	0.65	0.50
24.	Pyridine		0.71	0.55
25.	2-butoxyethanol		0.74	0.57
26.	1-propanol		0.82	0.63
27.	2-propanol		0.82	0.63
28.	Ethanol		0.88	0.68
29.	Methanol		0.95	0.73
30.	Ethanediol		1.11	0.85
31.	Acetic acid		<1	<1
32.	Water		>>1	>>1

Snyder et al.[3] in the year 1979 also suggested solvent strength table in reversed-phase chromatography for few solvents (Table 7.3). Thus solvent strengths for mixtures of solvents may be calculated from this table in approximation.

For adjusting R_f value of the desired component(s) in a proper way, solvent strength needs to be changed. For this purpose, solvent mixtures of varied composition should be used. But calculation of ε^0 for a mixture is a complicated task, as it is not linear with the proportions, e.g. if a "strong" solvent is added to a "weak" solvent, in the beginning, the strength increases drastically initially but later on slowly.

S. no.	Solvent	Solvent strength(S)
	Table 7.3: Solvent strength table	
1	Water	0
2	Methanol	3.0
3	Acetonitrile	3.1
4	Acetone	3.4
5	1, 4-dioxan	3.5
6	Ethanol	3.6
7	2-propanol	4.2
8	Tetrahydrofuran	4.4

It may be inferred that mixture of different solvents theoretically, will produce the same retention for the sample as long as ε^0 value of such mixtures are the same.

Table 7.4: Eluotropic series of solvents with quality parameters

Solvent	Polarity index according to Synder	Dielectricity constant DK (20bzw 25° C)	Molar Mass (g/mol)	Boiling point (°C)	Vapour pressure (20° C) (mbar)	MAK value (ml/m³) = (ppm)	Viscosity
n-heptane	–	1.9	100.21	98.4	48	500	0.40
n-hexane	0.0	1.9	86.18	68.9	160	50	0.31
Cyclohexane	0.0	2.0	84.16	80.7	104	300	0.94
Iso-octane	0.4	1.9	114.23	99.2	51	500	–
1, 1, 2-trichloro-trifluoroethane	–	2.4	187.38	47.7	368	500	–
Carbon tetrachloride	1.7	2.2	153.82	76.5	120	10	–
Toluene	2.3	2.4	92.14	110.6	29	100	0.58
tert-butyl methyl ether	2.9	–	88.15	55.2	417		–
Chloroform	4.4	4.8	119.38	61.7	210	10	0.56
Dichloroethane	3.7	10.6	98.97	83.4	87	5	–
Dichloromethane	3.4	9.1	84.93	40.0	453	100	0.43
1-butanol	3.9	17.8	74.12	117.2	6.7	100	2.95
Acetonitrile	6.2	37.5	41.05	81.6	97	40	0.39
2-propanol	4.3	18.3	60.10	82.4	43	400	2.27
Ethyl acetate	4.3	6.0	88.10	77.1	97	400	0.44
Acetone	5.4	20.7	58.08	56.2	233	1000	0.32
Ethanol	5.2	24.3	46.07	78.5	59	1000	1.20
1,4-dioxan	4.8	2.2	88.11	101.0	41	50	–
Tetrahydrofuran	4.2	7.4	72.11	66.0	200	200	0.47
Methanol	6.6	32.6	32.04	65.0	128	200	0.52
Water	9.0	80.2	18.01	100.0	23	–	0.95

(ChromBook)

The eluotropic series of solvents with their quality parameters in order of increasing elution power is given in Table 7.4. This may be helpful in selecting a suitable mobile phase for a particular separation problem. The table prescribes lists of eluotropic series for silica gel as stationary phase.

7.1 SELECTIVITY

It appears solvents of same strength should lead to similar R_f values. This is not true. There are different approaches to the theory of selectivity. Snyder[4] developed the basis of selectivity theory by classifying the solvents into groups according to their similarity in proton-acceptor (X_e), proton-donor (X_d) and dipolar attraction (X_n). The sum of selectivity parameter is always one as is seen from the Table 7.5 for a large number of solvents in a triangular coordinate system normally called as "selectivity triangle" and reported by Snyder[5] (Fig. 7.2).

Table 7.5: Polar index, selectivity parameters and selectivity groups of important solvents

Solvent	Pola-rity index (P)	Selectivity parameters			Total X_e+X_d $+X_n$	Selecti-vity group	Solvent stren-gth S_T on RP
		X_e	X_d	X_n			
Hexane	0.1	–	–	–	–	–	–
Isopropyl ether	2.4	0.48	0.14	0.38	1.00	I	
Diethyl ether	2.8	0.53	0.13	0.34	1.00	I	
2-propanol	3.9	0.55	0.19	0.27	1.01	II	
n-butanol	3.9	0.59	0.19	0.25	1.03	II	
1-propanol	4.0	0.54	0.19	0.27	1.00	II	
Ethanol	4.3	0.52	0.19	0.29	1.00	II	3.6
Methanol	5.1	0.48	0.22	0.31	1.01	II	3.0
Tetrahydrofuran	4.0	0.38	0.20	0.42	1.00	III	4.4
Acetic acid	6.0	0.39	0.31	0.30	1.00	IV	
Dichloromethane	3.1	0.29	0.18	0.53	1.00	V	
Ethyl acetate	4.4	0.34	0.23	0.43	1.01	VI	
Methylethyl ketone	4.7	0.35	0.22	0.43	1.00	VI	
Dioxan	4.8	0.36	0.24	0.40	1.00	VI	3.5
Acetone	5.1	0.35	0.23	0.42	1.00	VI	3.4
Acetonitrile	5.8	0.31	0.27	0.42	1.00	VI	3.1
Toluene	2.4	0.25	0.28	0.47	1.00	VII	
Benzene	2.7	0.23	0.32	0.45	1.00	VII	
p-xylene	2.5	0.27	0.28	0.45	1.00	VII	
Nitromethane	6.0	0.28	0.31	0.40	1.09	VII	
Chloroform	4.1	0.25	0.41	0.33	0.09	VIII	
Water	10.2	0.37	0.37	0.25	0.99	VIII	0.0

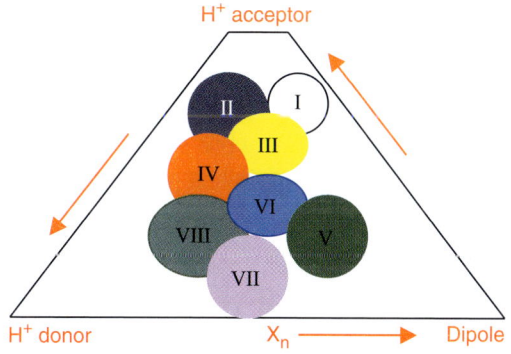

Fig. 7.2: Selectivity triangle

Majority of solvents in view of above could be divided into eight groups as mentioned in Table 7.6 according to Snyder et al.[6-7]

Group	Solvents
Table 7.6: Classification of solvent selectivity	
I	Aliphatic ethers
II	Aliphatic alcohols
III	Pyridine, tetrahydrofuran, ethers, amides (except formamide)
IV	Formamide, acetic acid, glycols
V	Dichloromethane, 1, 2-dichloroethane
VI	Aliphatic ketones and esters, 1, 4-dioxan, acetonitrile
VII	Aromatic hydrocarbons and ethers, aromatic halo- and nitro-compounds
VIII	Chloroform, water, nitro-methane, m-cresol

This triangle is normally applicable to normal and reversed phase separation. Solvents as mobile phase in adsorption chromatography can be classified as "localizing" and "non-localizing". Examples in this context are given in Figure 7.3.

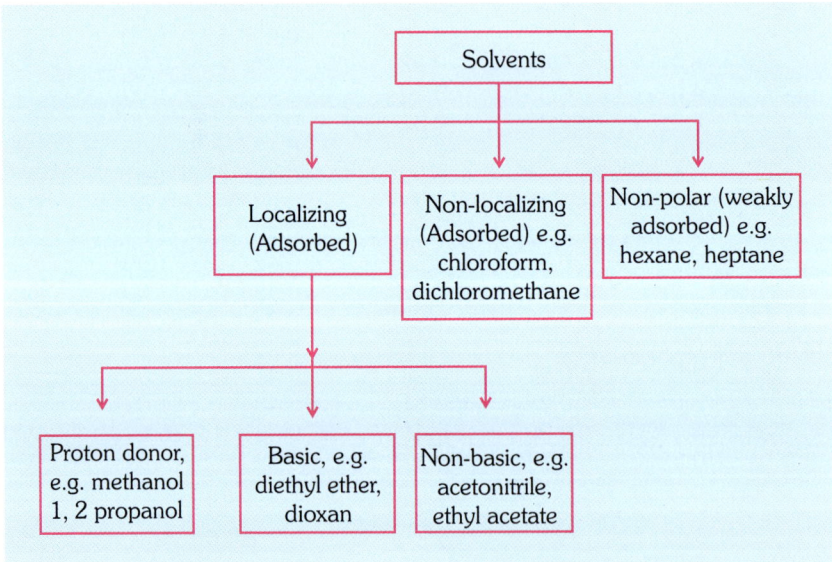

Fig. 7.3: Classification of solvents based on adsorption

Adsorption behaviour plays an important role while selecting a solvent. According to Snyder et al[8], adsorption behaviours of common solvents on silica gel is indicated in Table 7.7.

There are several other important factors while selecting a solvent. These are:

(i) Purity of the solvent is a must. Traces of polar impurities can change the solvent strength. Chromatographic grades of solvents are generally suitable. Presence of stabilizers, antioxidants may change the chemical behaviour of the solvent.

S. no.	Solvent	Localization	Basicity
Table 7.7: Adsorption behaviour of solvents			
1	Acetone		
2	Acetonitrile	Yes	No
3	n-butanol		
4	Chloroform	No	
5	Cyclohexane	No	
6	Dichloromethane	No	
7	Diethyl ether	Yes	Yes
8	Dioxan	Yes	Yes
9	Ethyl acetate	Yes	No
10	Heptane	No	
11	Hexane	No	
12	Isopropyl ether	Minor	
13	Methanol	Yes	Proton donor
14	1-propanol	Yes	Proton donor
15	2-propanol	Yes	Proton donor
16	Tetrahydrofuran	Yes	Yes
17	t-methylbutyl ether	Yes	Yes

(ii) Solvent bottles should be kept under proper storage condition.

(iii) Use of toxic or hazardous solvents, e.g. benzene must be avoided to possible extent.

(iv) Physical properties of solvent like boiling point/volatility, viscosity affects the chamber saturation, separation efficiency and drying of chromatographic plate. Hence, these are important factors while selecting a solvent.

(v) Mobile phase should be as simple as possible. Simpler the mobile phase composition, more rugged is the chromatography.

(vi) Small packing bottles should be purchased.

(vii) Date of opening the bottle and proposed date of expiry should be labelled.

7.2 PURITY OF SOLVENTS IN HPTLC

High quality solvents are necessary in HPTLC technique. Poor or analytical grade solvents are not suitable for quantification. Especially in quantitative analysis at ppm/ppb/ppt level and separation by using gradient methods, solvents should be isocratic and of gradient quality. Solvent consumption is very low in HPTLC and so the highest quality of solvents is affordable. The advantage with TLC is that any change caused by mobile phase quality affects standards and samples equally.

Quality specifications and purity of solvents as provided by E. Merck are given in Table 7.8, 7.9 and Figure 7.4.

Table 7.8: Specifications of LiChrosolv ® gradient grade products

S. no.	Product	Evap.Residue max.(mg/l)	Gradient at nm (max.mAU)			Fluorescence[1] at nm (max.ppb)	
			210	235	254	254	365
1.	Acetonitrile	4	2.0	–	0.5	1.0	0.5
2	Ethanol	5	–	5.0	2.0	–	–
3	Methanol	4	–	2.0	1.0	1.0	1.0
4	2-propanol	5	–	2.5	2.0	–	–
5	Water	5	5.0	–	0.5	1.0	0.5

[1]Calculated as Quinine in 0,05 mol/l H_2SO_4 (ChromBook 06/07).

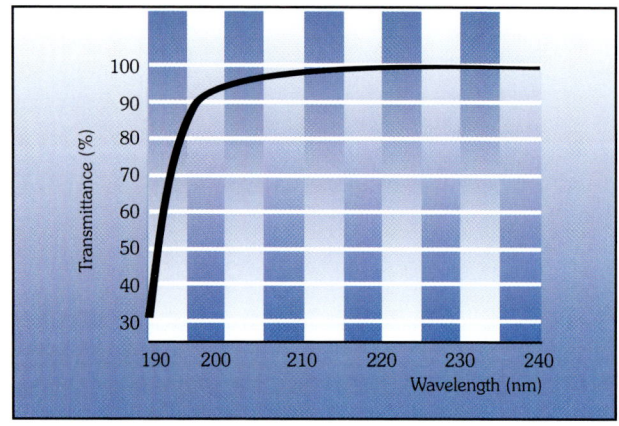

Reference: Water LiChrosolv®

Fig. 7.4: UV-spectrum acetonitrile LiChrosolv® hypergrade optical path: 1 cm

Table 7.9: Purity of different solvents

Solvent	Purity (GC) min. [%]	Evap. Residue max. [mg/l]	Water max. [%]	Acidity max. [meq/g]	Alkalinity max. [meq/g]	UV-transmission at nm [%]		
Acetone	99.8	2	0.05	0.0002	0.0002	335 (50%)	340 (80%)	350 (98%)
Acetonitrile hypergrade for LC/MS	99.9	1	0.01	0.0001	0.0002	191 (25%)	195 (85%)	200 (96%)
Acetonitrile gradient grade	99.9	2	0.02	0.0002	0.0002	193 (60%)	195 (80%)	230 (98%)
Acetonitrile isocratic grade	99.8	4	0.05	0.0005	0.0002	195 (70%)	200 (90%)	240 (98%)
Benzene	99.8	2	0.03	0.0002	0.0002	285 (70%)	290 (90%)	340(98%)
1-butanol	99.8	2	0.05	0.0002	0.0002	230 (75%)	240 (85%)	310 (99%)

contd…

contd…

Solvent	Purity (GC) min. [%]	Evap. Resi-due max. [mg/l]	Water max. [%]	Acidity max. [meq/g]	Alkalinity max. [meq/g]	UV-transmission at nm [%]		
Tert-butyl methyl ether	99.8	2	0.02	0.0002	0.0002	240 (60%)	255 (85%)	280 (98%)
Chloroform, stabilised	99.8	5	0.01	0.0002	0.0002	255 (70%)	260 (85%)	300 (98%)
1-chloro-butane	99.8	2	0.01	0.0002	0.0002	227 (60%)	232 (80%)	250 (98%)
Cyclohexane	99.9	2	0.01	0.0002	0.0002	230 (75%)	240 (90%)	260 (99%)
1,2-dichloro-ethane	99.8	2	0.02	0.0002	0.0002	240 (85%)	245 (90%)	270 (99%)
Dichloro-methane, stabilised	99.9	5	0.01	0.0002	0.0002	240 (70%)	245 (90%)	260 (99%)
1,4-dioxan	99.8	2	0.02	0.0002	0.0002	245 (50%)	270 (80%)	300 (98%)
Ethanol gradient grade	99.9	2	0.1	0.0002	0.0002	225 (60%)	240 (85%)	260 (98%)
Ethylacetate	99.8	2	0.05	0.0002	0.0002	260 (50%)	265 (80%)	270 (98%)
n-heptane	99.3	2	0.005	0.0002	0.0002	210 (50%)	220 (80%)	245 (98%)
n-hexane	98.0	1	0.01	0.0002	0.0002	210 (50%)	220 (80%)	245 (98%)
Isohexane	99.0	2	0.005	0.0002	0.0002	210 (60%)	220 (80%)	245(98%)
Iso-octane	99.0	2	0.01	0.0005	0.0002	210 (50%)	220 (80%)	270 (98%)
Methanol hypergrade for LC/MS	99.9	1	0.01	0.0002	0.0002	210 (35%)	220 (60%)	230 (75%)
Methanol gradient grade	99.9	2	0.02	0.0005	0.0002	220 (55%)	235 (83%)	260 (98%)
Methanol	99.8	3	0.03	0.0005	0.0002	225 (50%)	240 (80%)	265 (98%)
1-propanol	99.8	2	0.02	0.0005	0.0002	230 (70%)	240 (80%)	270 (98%)
2-propanol gradient grade	99.9	2	0.05	0.0005	0.0002	220 (80%)	230 (90%)	250 (98%)
Carbon tetra-chloride	99.9	5	0.01	0.0002	0.0002	270 (50%)	275 (80%)	290 (98%)
Tetrahydro-furane	99.9	1	0.02	0.0005	0.0002	260 (80%)	270 (90%)	310 (99%)
Toluene	99.9	2	0.05	0.0005	0.0002	300 (70%)	310 (80%)	350 (98%)
1,2,4-trichloro-benzene	99.0	2	0.01	0.002	0.0002	315 (50%)	320 (80%)	385 (98%)
Water gradient grade	–	5						

References

1. Snyder LR. *J Chromatogr* **16**; 55-88, 1964
2. Snyder LR. *J Chromatogr* **25**; 274-293, 1966
3. Snyder LR, Kirkland JJ. *Introduction to Modern Liquid Chromatography*, 2nd edn., Wiley-Interscience, New York, USA, 73. , 1979
4. Snyder LR. *J Chromatogr* **92**; 223-230, 1974
5. Snyder LR. *J Chromatogr Sci* **16**; 223-234, 1978
6. Snyder LR, Glajch JL. *J Chromatogr* **214**; 1-19, 1981
7. Snyder LR, Glajch JL. *J Chromatogr* **248**; 165-182, 1982
8. Snyder LR, Kirkland, JJ, Glajch JL. *Practical HPTLC method development*, 2nd edition, 1997

Chromatogram Development

D evelopment of chromatogram is the key element of HPTLC. Right selection of the chamber and proper handling of the process are equally important. Results may some what vary from chamber to chamber. However for detection purpose, if other conditions are same the difference may not matter. The glass development chamber is most widely used and so is ascending chromatography. In HPTLC, chromatographic development is an independent step unlike GC or HPLC, where the sample is introduced, separated and detected in an equilibrated chromatograph.

In HPTLC, the stationary, mobile and gas phases play important roles.

8.1 DEVELOPMENT CHAMBERS

There are different types of development chambers. These are designed with features to control the parameters of development reproducibility. The following factors have a bearing on separation.

  (i) Solvent vapour saturation of chamber
 (ii) Sorbent vapour adsorbed in layer
(iii) Solvent vapour demixing in chamber
 (iv) Humidity in the chamber

It is possible to eliminate unwanted effects by utilizing suitable features and improving resolution. Four prominent effects can be noticed in context of chambers.

(a) Unsaturated chambers give sharp bands unlike more diffused bands in saturated chambers.
(b) Secondary fronts are seen in unsaturated chambers and that too more pronounced in sandwich mode. These are on account of solvent demixing at the front.
(c) Lower R_f is observed in saturated chambers on pre-conditioned plate due to reduced layer activity.
(d) Saturation mode is far more reproducible.

Different types of chambers are deployed in development. These are:

(i) Flat bottom chamber (vapour unsaturated or vapour saturated).

(*ii*) Twin-trough chamber (bottom has two compartment) which has replaced flat bottom.
(*iii*) Sandwich chamber (unsaturated).
(*iv*) Horizontal chamber (double sided development, saturated or unsaturated).
(*v*) Automatic development chamber (ADC) (full environmental control and automation).
(*vi*) Forced flow development chamber (FFDC) (Development under pressure).
(*vii*) Vario chamber (saturated or unsaturated using six different mobile phases) for method development.
(*viii*) Automated multiple developments (AMD) for gradient high resolution chromatography.

8.2 FLAT-BOTTOM CHAMBERS

These are available in variety of sizes and economical too (Fig. 8.1).

Fig. 8.1: Flat-bottom all glass chamber with stainless steel lid

The side and rear walls lined with thick filter paper can make it a well-saturated chamber. The chamber is allowed to stand with mobile phase and covered tightly with lid for optimum time depending upon the dimensions (normally 20-30 minutes), so as to form equilibrium of developing solvent with its vapour. Putting "grease" on inner side of lid is not recommended for making it air tight, as this may contaminate the chamber and the plate. Care must be taken to place plate in saturated chamber as early as possible and sliding back the lid. The plate should be placed as vertical as possible with the layer facing the inside of the chamber, thus permitting proper exposure of the stationary phase to the solvent vapour. Only one plate should be used in one chamber.

In unsaturated chamber, amount of developing solvent should be least and walls are not lined with filter paper. Plate should be placed inside chamber immediately after pouring the required quantity of solvent, so that equilibrium between solvent and the vapour is partially established during development.

8.3 TWIN-TROUGH CHAMBERS (TTC)

Twin-trough chambers are not only economical but also a flexible substitute for flat-bottom chambers (especially in saturated mode). Twin-trough chambers require less solvent too as may be seen from Table 8.1.

S. no.	Chamber	Size of chamber (cm)	Unsaturated (ml)	Saturated (ml)
		Table 8.1: Volume of solvent required		
1.	Flat-bottom	10 × 10	15	20
2.	Flat-bottom	20 × 20	50	50
3.	Twin-trough	10 × 10	5	10
4.	Twin-trough	20 × 10	10	20
5.	Twin-trough	20 × 20	20	40
6.	Horizontal developing	10 × 10	2 /side trough	25 central trough+2/ side trough
7.	Horizontal developing	20 × 10	5 /side trough	50 central trough+5 in each side trough

It also offers better control on the development conditions, besides other advantages. For saturation, a filter paper is put on the rear side of the TTC. Entire mobile phase is poured over the filter paper for complete wetting. Solvent is then distributed evenly by tilting the TTC.

Proper preconditioning of the plate is done in a better way in TTC. The conditioning solvent (humidity or pH control agent or developing solvent) is put in the rear trough with filter paper lining, if required. The plate (layer facing inside of chamber) is placed in the front trough. After stipulated time, chromatogram is initiated by introducing developing solvent with a pipette in the front trough, after sliding open the lid. Reproducible results are possible in fully saturated mode. Figure 8.2 is a photo of TTC. Figure 8.3 shows advantages of TTC.

Fig. 8.2: Twin-trough chambers with stainless steel

Low solvent consumption   Reproducible pre-equilibration with solvent vapour   Start of development

Fig. 8.3: Advantages of TTC

8.4 SANDWICH MODE

A conventional flat bottom chamber is used as for "sandwich" chromatography. The layer is "sandwiched" between a glass plate and the plate support. A 1 mm thick Teflon liner is placed on 2 sides and to prevent the cover plate from touching the layer. The cover plate is also 2 cm short in length.

8.5 HORIZONTAL DEVELOPING CHAMBER (HDC)

This offers a good flexibility and is economical in operating costs. Horizontal chamber is illustrated in Figure 8.4.

Fig. 8.4: Horizontal developing chamber

Plate is placed face down into chamber. Mobile solvent is poured into solvent reservoir. Solvent is transferred through a capillary film into the layer to start chromatography. For unsaturated mode, counter plate is removed and central reservoir is left empty. But in case of saturated mode, mobile solvent is placed in central reservoir. Chromatography can be initiated when central reservoir is charged with conditioning solvent and plate is placed in its location without beginning chromatography. Preconditioning can be performed. The separation obtained in HDC is quite reproducible, especially in unsaturated mode, provided method is optimized and validated for this type of chamber. Partial saturation of layer or chamber does not take place as in a flat bottom or TTC. The biggest benefit is the scope of simultaneous development of a plate from two opposite sides. Development ceases when two solvents front meet in the middle of the plate. Using this technique the number of samples chromatographed can be doubled. It is especially useful for samples that exhibit little migration from the origin.

8.6 AUTOMATED DEVELOPING CHAMBER (ADC)

The main objective of using such chamber is to get reproducible results in addition to minimize the human error. Plate is inserted into the device and then layer is automatically brought into contact with the mobile phase. Development is controlled by measuring the migration distance of the solvent or by elapsed time. The plate is dried after completion of development process. Picture of an ADC is given in Figure 8.5.

Fig. 8.5: Automated developing chamber

The conditions for chromatogram development can be selected and data entered through a keyboard. Humidity of layer can be reproducibly controlled. Storage of data is possible for future reference. The progress of development is monitored using charged couple device sensor. The chromatogram can be dried with warm or cool air. In such chambers, activity of stationary phase is managed through humidity control and chromatography is independent of climatic changes. Another major advantage is that this machine uses the 20×10 cm. TTC for development and so it is easy to transfer methods from TTC to ADC.

8.7 FORCED-FLOW CHAMBERS

Forced-flow technique is also called as over-pressured layer chromatography (OPLC). The mobile phase is applied to the plate by constant volume pump along the plate length and solvent migration rate is fairly constant. Forced flow consists of upper and lower support blocks, which sandwich the TLC plate. The mobile phase is pumped under pressure through sorbent layer. This results in very short development time or a longer separation bed. Advantages of such technique are:

(i) The solvent velocity remains optimum over a longer adsorbent bed.
(ii) Higher efficiency is possible to achieve using fine sorbent particles.
(iii) A linear increase in efficiency with increasing migration can be obtained.
(iv) Analysis time is reduced drastically.
(v) Mobile phases having poor sorbent wetting quality may be used.
(vi) Solvent gradients techniques can be employed for best separation conditions.
(vii) Fractions can be eluted from the layer, similar to HPLC. OPLC requires special plates to contain the mobile phase on the plate.

This is a good concept but practical difficulties still exist that prevent OPLC from becoming a rugged and routine technique.

8.8 VARIO CHAMBER

The vario chamber is a multipurpose device, based on horizontal development. It is primarily developed for new method development. A scoring unit helps to remove the coated layer at certain fixed places converting a 10 × 10 cm glass plate into a 6 channeled plate. This enables 6 different mobile phases to be run independently in those six channels. Chromatogram development conditions can be optimized in 3 or 4 such plates in 2 to 3 hours.

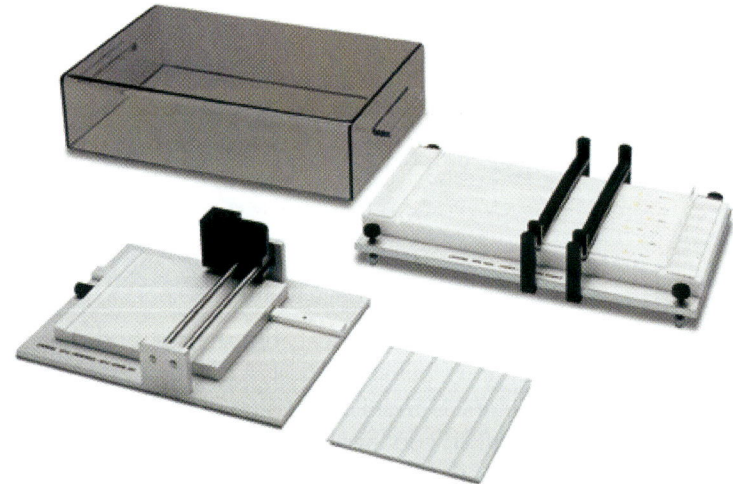

Fig. 8.6: Vario chamber

8.9 GRADIENT DEVELOPMENT

The separation of several components under isocratic conditions is often complicated by large difference in the polarity of the components.

Mobile phase of high strength is useful for separation of strongly retained components and mobile phase of low strength is suitable for separation of comparatively weaker components on silication limit. Gradient development in thin layer chromatography is becoming more popular due to its power of high resolution.

History of gradient elution in TLC started in 1962. Perry introduced a new technique known as Programmed Multiple Development. In this technique, the plate is developed several times in the same direction with the same solvent.

Niederwieser[1] defined gradient TLC as "a chromatographic technique using within separation area locally different separation conditions". Gradient techniques may be categorized keeping in view the variation in separation condition of mobile and stationary phases as: (*i*) mobile phase gradient technique (composition, pH and ionic strength); (*ii*) Stationary phase gradients (composition, impregnation and activity); and (*iii*) gradients connected with changes (temperature, flow rate). The biggest chances of achieving successful gradients are by changing the mobile phase strength. A continuous gradient is desirable for separation, but step gradients are easy to achieve. However, several steps [2, 3] used in step gradients may lead to comparable results with that of continuous gradient system[4]. Various techniques are used for gradient elution.

8.9.1 Mobile Phase Gradient

It may not be possible to separate multi-component mixtures with too much variation in R_f values by isocratic elution. Two or three component gradients consisting of weak solvent and strong solvent can be varied linearly. R_f depends on the strength of strong solvent in binary mobile phase. Hence it is possible to achieve gradient elution variations in sample retention exclusively by varying the mobile phase composition, but this is quite difficult to achieve manually for practical use.

8.9.2 Stationary Phase Gradient

It involves a continuous change or stepwise change in the composition/activity of the adsorbent layer along the plate.[5] A gradient in stationary phase can be made parallel to solvent flow direction or at its right angle (orthogonal gradients) with a special spreader. The latter one uses several different plates of varying adsorbent composition for evolving a best system. Such plates are not commercially available.

8.10 MULTIPLE DEVELOPMENTS

Here, the plate is repeatedly developed with the same or different mobile phases in same direction. This technique is most suitable for mixtures of large polarity variations such as pesticide residues[6], amino acid derivatives[7] and alkaloids[8], etc. Optimization of gradients can be obtained in following steps.

(*a*) By selection of base solvent (medium polarity) with minimum two modifiers (high polar and non-polar solvent).

(*b*) By developing a final gradient with suitable range of eluotropic strengths of solvent mixtures for improving the separation.

(*c*) By developing appropriate steps of a gradient.

It can be concluded that gradient development can be applied in following cases.

(*i*) For separation of complex samples.

(*ii*) To lower the detection limit by sharpening the chromatographic zones due to repeated development.

(*iii*) For speeding the search for better chromatographic system for difficult separations.

Only automation enables the reproducibility of mobile phase gradients.

8.11 AUTOMATED MULTIPLE DEVELOPMENT (AMD)

Burger[9] automated the process of multiple developments now known as AMD.

The principle of AMD is similar to manual multiple development. Use of different solvents is possible and is required with different development distances of each run that too by using two dimensions. Solvents are changed in respect to selectivity and/or solvent strength. The main feature is use of gradient elution, which allows repeated development of plates. It permits to obtain maximum attainable resolution without forced flow. Separation capacity of AMD is high, i.e. about 35-40. It can be tuned for each segment of chromatogram. This technique is to be used for complex samples and difficult separation, which justify the higher time required (45 min. to 2 hrs.). Some components may be carried away partly or totally by the vacuum.

The principle of AMD is as follows:

- Plate is developed repeatedly in same direction.
- In successive steps, a longer solvent migration distance is chosen than earlier one usually 3 mm longer.
- Between every run, solvent is completely removed by vacuum.
- Stepwise elution gradient in decreasing solvent strength is applied in up to 25 steps.
- Focusing effect in combination with gradient elution results in narrow bands. Peak width is about 1 mm; hence up to 40 components can be resolved with baseline separation within a separation distance of 80 mm.

An improved model of AMD is given in Figure 8.7

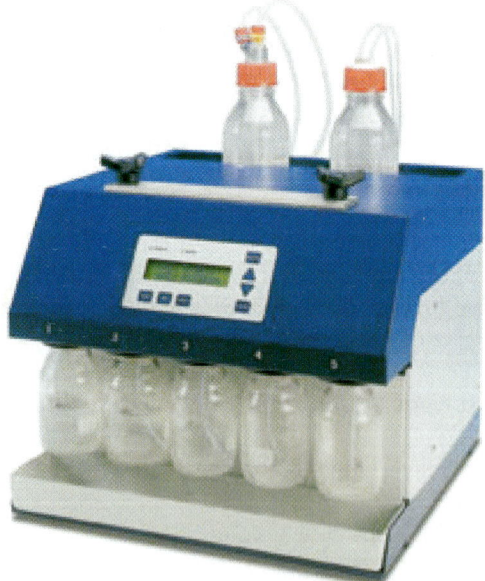

Fig. 8.7: Automated multiple development

AMD provides gradient elution with improved precision and reliability. Separation power is increased by a factor of 3 as compared to regular HPTLC. Gradient HPTLC requires ultra high purity solvents. 100 μm thin layers are recommended as they dry quickly, reducing the time required for analysis.

References

1. Niederwieser A. *Chromatographia* **2**; 23, 1969
2. Jandera P, Churacek J. *Advances in Chromatography*, Marcel Dekker, New York and Basel, 126, 1980
3. Jandera P, Churacek J. *Gradient Elution in Column Liquid Chromatography*, Elsevier, Amsterdam, 1985
4. Golkiewics W. *Chromatographia* **14**; 629, 1981
5. Snyder LR, Saunders DL. *J Chromatogr* **44**; 1, 1969
6. Lodi G, Betti A, Menziani E, Brandolini V, Tosi B. *J Planar Chromatogr* **4**; 106, 1991
7. Belay MT, Poole CF. *J Planar Chromatogr* **6**; 43, 1993
8. Queckenberg OR, Frahm AW. *J Planar Chromatogr* **6**; 55, 1993
9. Burger K, Fresenius Z. *Anal Chem* **318**; 228, 1984

Visual Detection

Since most compounds are not visibly coloured, visualization of chromatogram through some techniques is essential after development of chromatogram. The biggest advantage of HPTLC is that the sorbent is almost inert in nature, i.e. many reactions are possible on the plate without interference of the layer. A number of compounds absorb UV light or emit fluorescence when excited by visible or UV light. Therefore, HPTLC offers a unique possibility to visualize the result directly to the naked eye. Compounds absorbing UV light around 254 nm can be visualized with the help of fluorescence indicator embedded with the stationary phase (F_{254}). Plates with embedded indicator for 366 nm are available. Visualization is also possible by using suitable spraying reagent or dipping the plate in appropriate dipping reagent.

Identification of a component with a standard reference through comparison by visual examination may be done. Protective glasses and gloves should be used while making observation under UV light, as it is hazardous to eyes and skin. Comparisons of the results can also be made from other plates, its images, software library, etc. A permanent document of the chromatogram for independent verification can be made. It is possible to take photograph of the plate by digital camera. These images can be edited, archived, evaluated, and are GLP compliant with certified software. These can be used for quantitative determination using video densitometry.

9.1 EVALUATION BY NON-DESTRUCTIVE METHODS

Different non-destructive methods are available. These are:

9.1.1 Direct Visual Detection

Such methods are officially prescribed for evaluation of the chromatogram in day light by use of either reflected or transmitted light. This technique is possible in case of compounds adequately coloured, e.g. natural and synthetic dyes, nitrophenols, etc. Such compounds do not require any treatment for visualization. Dyes are legally permitted foodstuffs such as jam, jelly, food juices, wine, alcoholic drinks and margarine, etc. so this method is suitable for evaluating not only legally permitted dyes but also banned substances used as adulterants.

9.1.2 Evaluation Under Ultraviolet Light

This is a simple, clean, non-destructive and documentable method, which is extremely popular. A dual wavelength lamp assembly is used that emits either 254 nm or 366 nm.

At 254 nm, a plate is seen as green or blue background, if it contains a suitable fluorescent indicator. Fractions that absorb this wavelength appear as dark spots. The image here is universal, i.e. bright green or blue background and dark bands. At 366 nm, the reverse is observed. Fractions, that fluorescence appear as bright bands on a dark background. The bright bands can be of different colour. The image here is like a finger print with red, blue, green, yellow, etc. colours.

The image seen at 254 nm is actually "quenching of fluorescence" of the indicator while at 366 nm, it is emission of fluorescence.

Fluorescence is typically 10-1000 times more sensitive than absorbance because of the dark background and bright spots.

Fluorescence can often be enhanced by reagent treatment. This can have a stabilizing effect too, e.g. dipping a plate in a paraffin solution (33% v/v in hexane) increases the detection limits of aflatoxins 3 to 4 times.

The plate must be observed in total darkness and so a viewing box is a must, especially to see faint fluorescence. External white light interferes with fluorescence. Besides, the viewing box protects the eyes and skin of the analysts.

When observing a plate in the UV cabinet, always inspect at both short and long wavelengths. First observe at 366 nm and make the judgment only after 15-20 seconds, which is the time required for the eyes to adjust to the darkness.

The UV filter used in the lamp assembly must be replaced every 3 to 4 years, because it 'solarises', i.e. becomes opaque to UV light.

A colour photos at 254 nm and 366 nm before derivatization and after derivatization is given below as Figure 9.1(a–d).

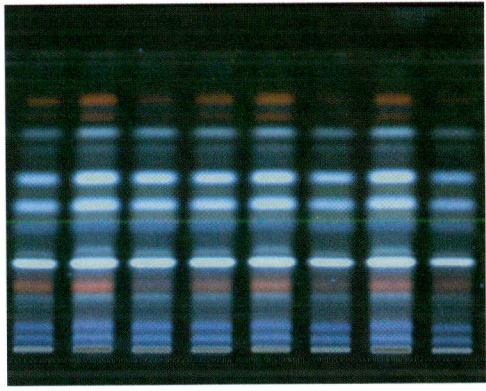

Image@366nm(B.D.)

Fig. 9.1 (a)

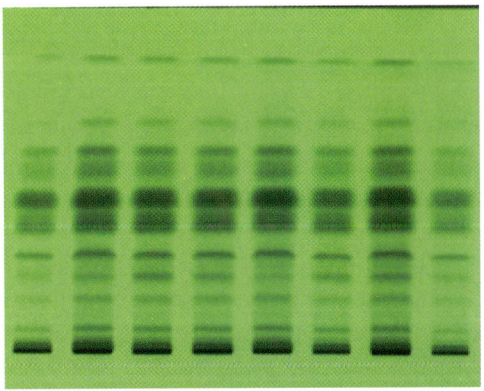

Image@254nm(B.D.)

Fig. 9.1 (b)

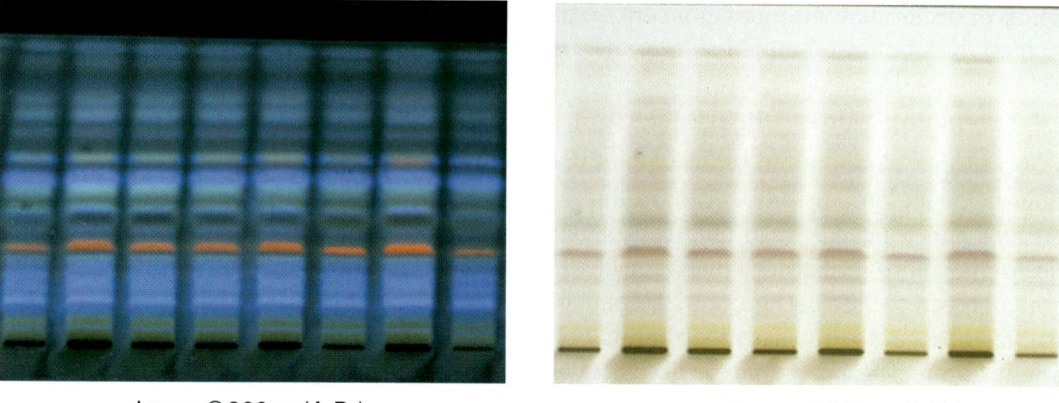

Image@366nm(A.D.)

Fig. 9.1 (c)

Image@366nm(A.D.)

Fig. 9.1 (d)

Note: *B. D. - Before derivatization; A. D. - After derivatization*

9.1.3 Chromatographic Zones

Chromatographic zones are colourless in visible light for most compounds. Exposure of UV light is normally done at shorter wavelength, i.e. 254 nm or at longer wavelength, i.e. 366 nm for fluorescing substances under observation. Picture of UV visualizer is shown in Figure 9.2.

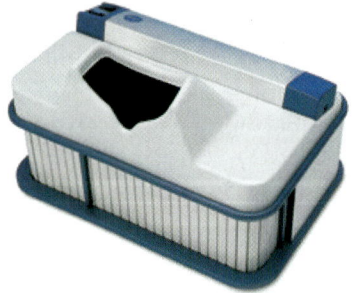

UV lamp in combination with viewing box

Fig. 9.2: UV cabinet

Chromatographic zones look dark on a lighter background in UV 254 nm or when fluorescence takes place, different visible spectrum colours are seen at 366 nm. Commercially pre-coated TLC plates are available which contains inorganic/organic fluorescence indicators which are an aid to visualization process. It is based on the fluorescence quenched by the compounds of the sample and such technique is very commonly known as "fluorescence quenching". Samples are best seen by continuous exposure to the UV light. It is based on the principle that energy is provided by electromagnetic radiation for bringing the electronic transition from ground state to an excited single state. But when the excited electron returns to ground state, they emit energy at a longer wavelength in visible range.

A dark chromatographic zone of analytes in these two backgrounds is often due to analytes which disturb the excitation process of the indicator in the layer exhibiting fluorescence through absorption of the excitation energy. This may be noted that layer showing fluorescence indicator may always be not mostly of advantage, because compounds fluorescence mostly at 366 nm and not at 254 nm. Hence plate containing indicator which fluorescence at 254 nm may be the advantage because both, i.e. fluorescence and fluorescence quenching can be observed depending upon the wavelength. When the analytes fluorescence at 254 nm or at both the wavelengths, then plate without any indicator should be used. Reagent treatments are also done for stabilizing and enhancing fluorescence. This can be very simple just by dipping the plate in reagent solution for few seconds thoroughly followed by treatment by detection reagent. The plate should be dried at room temperature. Suitable detection reagent should be applied to obtain colours chromatographic zones in visible light or in UV light, e.g. detection of aflatoxin can be enhanced 3-4-fold by dipping the plate in liquid paraffin (33% volume by volume in hexane).

9.2 REVERSIBLE REACTIONS

Out of important reversible reaction, following are non-destructive techniques.

9.2.1 Iodine Vapour

This is an almost universal reagent used for detecting organic compounds with double bonds. However, it should be noted that some reactions with iodine are irreversible. This method could be economical and rapid before ultimate characterization with a group specific reagent. Yellow brown zones on a lighter yellow/background are seen as iodine molecules attach themselves to the substance zones. Iodine crystals are put in a dry chromatography tank and the tank is allowed to be saturated with iodine vapours for few hours. Then, the plate is introduced into the chamber. After the chromatographic zones are recognized, the plate is removed for further examination and results are recorded. Iodine vapour's reversible reaction is suitable for a range of lipophilic organic molecules such as hydrocarbons, fats, waxes, few fatty acids, antioxidants and antibiotics, etc.

9.2.2 Ammonia Vapour

In combination with other reagents, ammonia vapour is suitable for improving the contrast between chromatographic zones and the plate background. Organic acids with pH indicators are most commonly used in visualization with this method. Bromocresol green and bromophenol blue as indicators can detect number of organic acids. But further treatment with ammonia vapour will sharpen the contrast between plate and components. However, like iodine, ammonia slowly evaporates and sensitivity comes back to that of prior to treatment. Ammonia vapour is exposed by holding the plate face down over a glass beaker containing the small ammonia solution. But twin-trough developing tank is also suitable for ammonia exposure.

9.3 NON-REVERSIBLE REACTIONS

(*i*) **Fluorescent dyes:** Fluorescein, dichlorofluorescein, eosin, rhodamine B and 6G, berberine, and pinacryptol yellow are used for detection of lipophilic substances. These dyes are dissolved in methanol or ethanol (0.01 to 0.10%), the plate is dipped and then air dried. The chromatographic zones look brightly fluorescent on a lighter background under

UV light at 254 nm. These dyes are not suitable for reverse phase plates. Ammonia vapour exposure after dye treatment may sometimes improve sensitivity. Fluorescein indicator is suitable for lipid, unsaturated compounds, and chlorinated hydrocarbon, etc. 2-7 di-chlorofluorescein is appropriate for saturated and unsaturated lipids. Rhodamine B is used as indicator for triglycerides, fatty acids and methyl esters, gangliosides, phenols, polyphenols, flavonoids, detergents. Rhodamine 6G is suitable for glycerides, sterols, phospholipids, fatty acid and their esters. Detection of sterols, lipids, saturated compounds is possible with berberine. Pinacryptol yellow is used for detection of anionic and non-ionic surfactants, sweetener and organic anions.

(*ii*) **pH indicators:** They are used for detection of acidic and basic substances. The indicators are sulfonthalein based, e.g. bromocresol green, bromothymol blue, bromophenol blue, and to a lesser extent bromocresol purple. Indicator is incorporated on the plate by dipping or spraying with approx.0.01 to 0.1% w/v ethanol/water solution and the pH is adjusted with the buffer salt or sodium hydroxide.

(*iii*) **Wetting/dipping:** Many reagents have been described for insitu derivatization.[1] Lipophilic substance on hydrophilic adsorbents like silica gel or aluminium oxide may be viewed and marked by spraying or dipping the plate in water. Lipophilic substances become visible as white zones clearly seen against light. It is possible to detect compounds such as herbicides, hydrocarbons, triterpenes, derivatives, etc. A dye solution like methylene blue or patent fast blue can also be used in place of water and the compounds appear pale on a transparent blue background. However, use of lipophilic dye solution, for visualization of lipophilic substance on a hydrophilic phase, gives dark zones on pale background.

9.4 DESTRUCTIVE TECHNIQUES

Reactions taking place on the plate between reagent and separated compounds which results in total change in the organic species is described as destructive technique. Visualized compounds are no more those which were applied in the sample.

(*i*) **Charring:** In this technique, developed chromatogram is treated with appropriate reagent and then the plate is heated at high temperature to degrade the organic compounds to carbon. Most common charring agent is sulfuric acid applied to the plate in dilute solution (5-15% v/v in methanol or water). Chromosulfuric and orthophosphoric acids are also useful in specific conditions. Time for heating will vary depending upon the compounds, i.e. from 5-20 minutes and the temperature between 100-180° C. However, binders present in the plate can be affected depending upon the temperature and time combination, leading to unwanted side effects.

(*ii*) **Thermochemical activation:** Zones on layer when heated to higher temperatures fluorescence under UV light. Sensitivity of detection by this technique is improved on moderately polar aminopropyl-bonded silica gel layers. Compounds with hetero-atoms such as nitrogen, oxygen, sulphur or phosphorus, as cited in the literature will readily respond to thermal activation than pure hydrocarbon.[2] Thermal activation process is effective for detection of few carbohydrates and fruit acids, etc.[3]

9.5 BIOCHEMICAL AND MICROBIOLOGICAL METHODS

These processes are useful for detection and quantification of antibiotics, alkaloids, insecticides, mycotoxins, and vitamins, etc. These are highly specific methods, thus reducing the interference of the matrix, require less sample clean up and enhance detection limit. Besides, information can also be assessed about possible degraded products and toxicology of the substance. Insitu bio-autographic methods with test organism are employed for detection.[4] Cell organelles in an agar or gelatin solutions are used as detector and applied directly on to the layer (this is known as bioautographic determination) or layer is placed on it. Active agents will diffuse from the layer to agar or gelatin suspension and react with test organism during incubation. Saponins can be detected by casting blood-gelatin suspension on the layer and then putting it for incubation. This results in transparent, colourless, hemolytic zones on a turbid red gelatin layer background. The principle is formation of enzymatic substrate reaction.[5] The developed plate is dipped in a substrate solution such as naphthylacetate/fast blue salt B, enabling the visualization of substance like organophosphate or carbamates through inhibition of enzymes. In these zones, reaction is inhibited, i.e. naphthol will not react to naphthylacetate, but reacts with fast blue salt B to form azo dyes. Bright zones appear on a red rose background.[6, 7]

9.6 SPRAYING TECHNIQUE

It is the most common technique for reagent transfer on to plate. It is rapid and simple. Equipment is easy to handle with small quantity of reagents. Spraying is flexible and so indispensable when reagents have to be applied in sequence. Spraying is the first choice while making a search for a suitable derivatization reagent. This technique has some drawbacks such as generation of obnoxious and hazardous fumes, which can be removed using a spray cabinet. One more problem is to achieve homogenous and defined derivatizations throughout the plate. Spraying equipments are illustrated in Figure 9.3.

TLC spray cabinet

Fig. 9.3: Spraying equipment (A)

For ensuring homogenous coverage of plate, selected sprayer should generate a fine mist and avoid any sputtering. Due care should be taken by regulating the function of correct gas pressure and distance of spray nozzle from plate. Spraying is preferred from up to down and left to right. It is possible to achieve consistent results.

Automatic TLC sprayer   Glass sprayer

Fig. 9.3: Spraying equipments (B)

9.7 DIPPING TECHNIQUE

A homogenous reagent transfer can be achieved by immersing the plate into derivatizing reagent. Dipping has to be done in a smoother way for avoiding tidemark. An immersion device (Fig. 9.4) can significantly improve the reproducibility. The concentration of the reagent is maintained and derivatization conditions can be standardized by optimizing the vertical speed and immersion time. In this process, no fumes are generated and exposures to hazardous chemicals are minimized, though little excess liquids cover the plate. After dipping, the plate must be dried very quickly.

Fig. 9.4: Immersion device

9.8 HEATING TECHNIQUES

Many chemical reactions that occur during derivatization need heating for completion. The plate heater with precise temperature control and uniform heating is the device of choice (Fig. 9.5). The heated surface must be resistant to reagents.

TLC plate heater

Fig. 9.5: Plate heating equipment

For quantitative analysis and for routine work, it is desirable that plate is heated evenly for reproducible results. Ovens have two shortcomings. Aggressive fumes produced during derivatizations may lead to cross-contamination and corrosion. The other problem is that derivatization process can, usually, not be seen. The oven should be dedicated to TLC work and thoroughly cleaned after each use. Plate heaters are another alternative. The equipment maintains uniform temperature across its surface. Glass plates may significantly bend on their edges due to low heat conductivity when a plate is put on the surface at temperatures more than 120° C. This leads to more efficient heating at the centre of the plate.

References

1. Fried B, Sherma J. *Thin-Layer Chromatography*, Marcel Dekker, New York, 39, 1986
2. Klaus R, Fischer W, Hauck HE. *LC-GC Int* **8(3)**; 151-156, 1995
3. Klaus R, Fischer W, Hauck HE. *Chromatographia* **29**; 496-471, 1990
4. Durackova Z, Betina V, Nemec P. *J Chromatogr* **116**; 155, 1976
5. Mendoza CE. *J Chromatogr* **78**; 29, 1973
6. Ackermann H. *J Chromatogr* **36**; 309, 1968
7. Weins C. Dissertation Saarbrucken 1996

10

Derivatization

In situ derivatization possibility is a strong point of HPTLC. Chemical reactions are possible in situ on the plate, before or after chromatography. Both the possibilities have their own advantages. However, the decision depends on sample matrix, level of detection/quantification and-interference present. Post-chromatographic derivatization is more popular technique for which several hundred references in literature are available as compared to a few for pre-chromatographic derivatization. The results are unique and specific when derivatization before chromatographic development has been recommended.

Reasons for derivatization can be:

– Transforming non-absorbing substances into detectable derivatives
– Lowering the detection limit
– To detect components selectively
– For visualization of all sample components
– To induce fluorescence
– To make separation easier
– To retain volatile compounds

10.1 PRE-CHROMATOGRAPHIC DERIVATIZATION

Pre-chromatographic derivatization is used to alter the properties of the sample wholly or partially before development. Such a process envisages derivatization of a particular class of compounds in which a group specific reagent has been used. A different mobile phase will normally have to be used to separate the sample, if it is pre-chromatography derivative. Derivatized compounds may have different colours or shades and can be easily visible.

Pre-chromatographic derivatization can be carried out on the layer. The sample is applied on the plate first. Then the reagent is superimposed or vice-versa. Reactions are faster and more complete if reagent is applied first. Derivatization can be extended to two-dimensional separations. First development is done with original sample as such. Before second development, the reagent is superimposed on the desired separated fraction. The second development is done perpendicular to the first.

Different types of derivatization reagents are used which cause: (*i*) oxidation and reduction; (*ii*) hydrolysis; (*iii*) halogenation; (*iv*) nitration and diazotization; (*v*) esterification and etherification; (*vi*) hydrozone formation; and (*vii*) dansylation, etc. Immense amount of work have been carried out in this direction and are available in literature.

The basic objectives of selecting a reagent are:

- Stable reaction products
- High yield at various concentrations
- Simple and rapid application
- Little or no interference due to excess reagent in chromatographic separation and analysis
- Reactions at room temperature or a little higher
- Environmentally safe reagents

Earlier work carried out for pre-chromatographic derivatization in food related material has been given in Table 10.1.

S. no.	Principle	Substances	Method, reagent and end products
			Table 10.1: Pre-chromatographic derivatization
1.	Oxidation	Alkaloids	(a) 10% chromic acid in glacial acetic acid superimposed. Development is done after a brief reaction period.[2] (b) Dehydration by heating the sample solution on silica gel layers.[3-4]
2.		Polyaromatic hydrocarbons (PAH)	(a) Sample solution with trifluoroacetic acid solution is applied. Heat to 100° C, cool and then develop. Trifluoroacetic acid catalyzes oxidation by atmospheric oxygen.[5] (b) Apply sample solution. Place in an iodine chamber for several hours. Then allow the iodine to evaporate. 3, 4-henzpyrene forms, e.g. bis-3, 4-benzpyrenyl.[6]
3.	Reduction	Alkaloids	Sodium borohydride solution is applied over the sample. The plate is then dried and developed.[2, 7]
4.		Fatty acids	Apply colloidal palladium solution to the start zone and dry at 80 to 90° C for 1 hr. Apply the sample solution, keep the plate for 1 hr. in a hydrogen-filled desiccator, then dry and develop.[8-9]
5.		Amino acids	The configuration is determined by reacting with a carbobenzyloxy-L-amino acid azide and reductively removing the protective group by hydrogen/palladium chloride solution.[10]
6.	Chlorination	Cholesterol, glycyrrhetic acid acetate	Sample is applied, wetted with anhydrous benzene, exposed for 4hr. to the vapors of thionyl chloride - benzene (1:1) in a desiccator. Dry and then develop. Chlorinated cholesterol or the chloride of 3-β-acetoxyglycyrrhetic acid is formed.[11]
7.	Bromination	Cholestanol, cholesterol	Apply sample, treat with a 2 to 3-fold excess of 0.1% bromine in chloroform. Only cholesterol is derivatized.
8.		Sorbic acid	Treated with bromine solution or bromine vapor; di-, tri- and tetrabromocaproic acids are produced.[13]

Contd…

Contd…

S. no.	Principle	Substances	Method, reagent and end products
9.		Capsaicinoids	Unsaturated capsaicinoids are completely brominated in bromine chamber.[14]
10.	Nitration	Polycyclic aromatic hydrocarbons (PAH)	Apply sample, dry. Place plate for 20 min. in a twin-trough chamber containing phosphorus pentoxide and 2 to 3 ml conc. nitric acid. PAH is nitrated by nitrous fumes.[15]
11.	Esterification and etherification	Aflatoxins	Apply sample extract with standard, then apply trifluoro-cetic acid, allow reacting at room temperature for 5 min, dry for 10 min. at 40° C and develop.[16-17]
12.		Aflatoxins, ochratoxin A, sterigmatocystine, penicillic acid, patulin	Apply sample, dry, apply trifluoroacetic anhydride, and keep at room temperature for 45 min., develop. The derivatives of patulin and penicillic acid possess appreciably different hR_f values.[18]
13.		Ochratoxin A, citrinin, penicillic acid, sterigmatocystine, zearalenone	Apply extracts of cereals or fungal cultures, apply 50 μl of pyridine-acetic anhydride (1:1) on top; remove excess reagent with stream of cold air and chromatograph.[19]
14.		Menthol, citronellol, linalool	Apply sample, and then acetylation mixture. Several applications of the reagent are necessary for complete reaction; heat to 100° C for 15 min; then chromatograph[20].
15.		Patulin	Apply the sample solution, then add acetic anhydride - pyridine (9:1), stand for 5 min. and dry for 15 min. in a stream of warm air[21-22].
16.		Alcohols	Apply sample, superimpose suspension of sodium acetate in acetic anhydride - glacial acetic acid (3:1), spray the plate with acetic anhydride, and allow for 3 hrs. at 50° C in a desiccator, evaporate and develop[23].
17.		Carboxylic acids, organophosphoric acids	Apply sample followed by ethereal diazomethane solution, dry and develop.[24-25]
18.		Triglycerides	Methylation is done with 0.5 N potassium methylate solution in methanol.[26]
19.		Peanut oil, glycerol phosphatides, cholesteryl esters etc.	Apply sample and then 2 mol/l of sodium methylate solution, dry for about 5 min. and develop. Only ester linkages react and not acid amide linkages.[27]
20.		Phospholipids, free fatty acids	Apply sample and 12% methanolic potassium hydroxide solution, moist with methanol for 5 min. Fatty acid methyl esters are produced, triglycerides do not react.[28]
21.		Fatty acids, n-hydroxy acids, ursolic acid	Apply sample followed by methanolic boron trifluoride solution, heat with a hot-air drier, cool and develop.[29]
22.		Sorbic acid, benzoic acid	Apply sample followed by 0.5% 4-bromophenacyl bromide in N, N-dimethylformamide. Heat to 80° C for 45 min. dry and chromatograph.[30]
23.		Phenols	Apply sample, then saturated sodium methylate solution, then treat with 4% 2, 4-dinitrofluorobenzene in acetone and heat to 190° C for 40 min. Chromatograph the dinitrophenyl ethers so produced.[31]

Contd…

Contd…

S. no.	Principle	Substances	Method, reagent and end products
24.	Dansylation	Carbamate and phenylurea herbicides	Treat sample for about 40 min with sodium hydroxide solution (c = 1 mol/l) at 80° C. Apply sample solution, then 0.2% dansyl chloride on it. Cover with a glass plate and allow in the dark at room temperature for 1 hr., then chromatograph.[32]
25.		Phenylurea herbicides	Sample is treated with KOH, apply sample solution spot with dansyl chloride. Stand for 30 min in the dark at room temperature, and develop the chromatogram.[33]
26.		Metoxurone and degradation products	Phenylurea herbicides are hydrolyzed to the corresponding aniline derivatives and reacted at the start with 4µl 0.25% dansyl chloride solution[34].
27.		Even-numbered and odd-numbered fatty acids (C_6-C_{24})	Apply dansyl semipiperazide or dansyl semicadaveride solution followed by samples then 1% N, N'-dicyclohexylcarbodiimide. Dry and develop.[35-36]
28.		Preservatives (benzoic, sorbic, propionic acids)	Apply sample followed by dansyl semipiperazide and N,N'-dicyclohexylcarbodiimide solution, dry and chromatograph.[37]
29.	Miscellaneous	Amino acids	Apply sample, treat with 2, 4-dinitrofluorobenzene solution. DNP-amino acids are formed.[38]
30.		Insecticides, e.g. eldrin, dieldrin, aldrin	Apply the sample solution, dry and add ethanolic zinc chloride solution, heat at 100° C for 10 min; cool, chromatograph the carbonyl compounds.[39,40]
31.		Alcohols	Put sample solution; add nitrophenyl isocyanate solution (10% in benzene). Dry and develop.[41]

10.2 POST-CHROMATOGRAPHIC DERIVATIZATION

This is an important tool in detection. Post-chromatographic detection can generate an additional set of data about the sample. There is no problem in detection of coloured components or those with intrinsic fluorescence in the chromatogram or components absorbing in UV at about 254 nm. There are certain components which do not exhibit such properties and thus are not detectable as such. Therefore, for detection/quantification purpose, it is necessary to transform such components into detectable derivatives for evaluation. Such reactions are carried out either as universal reactions or selectively on the functional groups.

Post-chromatographic derivatization can be done either as universal reactions or as functional groups selectivity. Universal reagents like water dyes, pH indicators and iodine vapour, nitric acid fumes, etc. reach with a larger variety of different compounds.

Derivatizations through gas phase provide rapid and simple transfer of reagent, but this has limited application. These include iodine, bromine and chlorine, besides volatile acids, bases, and gases like H_2S, NO. Twin-trough chambers are most suitable for such exposure to such modifier vapours.

10.2.1 Iodine Vapour/Solution

The reaction results in irreversible and probably in oxidative product. Iodine has fluorescence quenching properties, hence it looks like dark zone on plate containing F_{254} fluorescent indicator. Table 10.2 describes examples of iodine reaction.

S. no.	Substance	Reaction
	Table 10.2: Iodine reaction with common organic substances	
1	Polycyclic aromatic hydrocarbons, indole, and quinoline derivatives	Oxidation product is formed
2	Quinine alkaloids, barbiturates, unsaturated lipids, capsaicin, and calciferol	Iodine is added to the double bond
3	Opiates, brucine, ketazone, and trimethazone	Addition of iodine to the tertiary nitrogen for the opiates. Addition reaction with the- OCH_3 group of the brucine.
4	Thiols and thioethers	Oxidation of sulfur and addition on the double bond in the thiazone ring
5.	Alkaloids, phenthiazines, and sulfonamides	Complex is formed

It is possible to give starch treatment as in case of reversible iodine reactions. Iodine can also be used as a solution, which may be prepared in organic solvent such as petroleum ether, acetone, chloroform, methanol or ether. Sensitivity is increased in such cases and detection level goes to μg or even nanogram level.

10.2.2 Nitric Acid Vapour

Aromatic compounds, in general are nitrated with HNO_3 fumes. This is achieved by heating the plate at 160° C (approx.) for about 10 minutes and putting it immediately in an HNO_3 vapour filled chamber. Plate develops yellow or brown chromatographic zone. Besides, identification is possible under UV at 270 nm. Organic compounds such as sugar, xanthine derivatives fluoresce yellow/blue in long wavelength UV light (366 nm).

10.2.3 Other Reagents

(*a*) Many reagents produce their own characteristic colours, e.g.

 – **Phosphomolybdic acid**[42]: Exhibits bluish black zones in yellow back ground for a number of organic compounds.

 – **HCl vapour**[43]: Produces coloured product and then dark brown carbon with organic substances.

 – **Anisaldehyde-sulphuric acid**[44]: It is suitable for natural products, gives different colour zones.

 – **Antimony (III) or (V) chloride**[45]: Gives different colours on white background.

 – **Ammonium hydrogen carbonate vapour**[46]: On heating with several organic compounds, leads to fluorescence.

 – **Zirconium salts**[47]: Form yellow green to blue fluorescent zones.

(*b*) Some functional group selective reagents in post-chromatographic derivatization are given in Table 10.3

Table 10.3: Reaction with functional group

S. no.	Functional groups	Popular reagents
1	Acetylene compounds[48]	Dicobalt octacarbonyl
2	Aldehydes	2, 4-dinitrophenylhydrazine
3	Alcohols	Lead (IV) acetate dichlorofluorescein
4	Amines	Ninhydrin
5	Carboxylic groups	2, 6-dichlorophenylindophenyl (Tillmans' reagent)
6	Halogen-derivatives[49]	Ammonical silver nitrate (Dedonder's /Tollens' or Zaffaroni's reagent)
7	Ketones	2, 4-dinitrophenylhydrazine
8	Nitro-derivatives[50]	Benzocyanide, benzyltrimethyl ammoniumhydroxide
9	Peroxides	1-Naphthol/N^4-ethyl-N^4-(2-methyl-sulfonamidoethyl)-2-methyl, 1, 4-phenylendiamine
10	Phenols	7-Chloro-4-nitrobenzo-2-oxa-1, 3-diazole (NBD-chloride)
11	Thiols	7-Chloro-4-nitrobenzo-2-oxa-1, 3-diazole (NBD-chloride)

(c) Derivatization by inorganic reagents in the absorbents are mentioned in Table 10.4

Table 10.4: Derivatization by inorganic reagents

S. no.	Reagent	Substance detected	conditions, remarks
1	Ammonium sulfate[51]	Triglycerides, serum lipids	30-80 min at 150° C, gives fluorescent derivatives
2	Ammonium sulfate[52]	Lipids	Charring on heating
3	Aluminium oxide[53, 54]	Doxynivalenol in wheat	6 min at 120° C, gives a fluorescent derivative under UV light at 365 nm
4	Phosphomolybdic acid[55]	Essential oil components	Stabilization of silver nitrate impregnated adsorbent layer

(d) Few examples of fluorescence stabilization and intensity increase by treatment of chromatogram with viscous lipophilic compounds are given in Table 10.5.

Table 10.5: Lipophilic fluorescence intensifiers and area of application

S. no.	Fluorescence intensifier	Substances	Sensitivity increase/ stabilization	Remarks
1	Dodecane[56]	Polycyclic aromatic hydrocarbons	2 to 2.5 times	50% in n-hexane, significant time dependant zone enlargement
2	Palmitic and stearic acids[57]	Aflatoxins	15 to 30% B_1 and B_2	Unseparated fatty acids intensified the fluorescence
3	Liquid paraffin[58]	Aflatoxins	3 to 4 times	Dipping solution, 33% in n-hexane, investigation of fungal nutrients
4	Liquid paraffin[59]	Aflatoxin B_1	3 times approx.	Spray solution, 67% in n-hexane
5	Liquid paraffin[60]	Aflatoxins, sterigmatocystine	10 to 100 times	Foodstuff

Contd…

Contd...

S. no.	Fluorescence intensifier	Substances	Sensitivity increase/ stabilization	Remarks
6	Liquid paraffin[61]	Carbamate and urea herbicides	Stabilization & enhancement	Spray solution, 20% in toluene
7	Liquid paraffin[62]	Cholesterol, coprostanone, coprostanol, etc.	2 to 8 times	Dipping solution, 33% in n-hexane
8	Liquid paraffin[63]	Vitamin B$_1$ triethanolamine	2 times	Chloroform- liquid paraffin-triethanolamine (60:10:10)

(e) Other important visualization reagents have been indicated in Table 10.6.

Table 10.6: Other important visualization reagents

Visualization reagent	Reagent conditions	Detection of Compound groups
Ehrlich's reagent	4-dimethylaminobenzaldehyde (2% w/v) in 25% w/w hydrochloric acid/ ethanol (50:50 v/v). Then heat at 110° C for 2 min.	Amines and indoles
Gibb's reagent	2, 6-dibromoquinone-4-chloroimide (0.5% w/v) in methanol. After treatment, heat at 110° C for 5 min.	Phenols, indoles, thiols, barbiturates
Blue tetrazolium reagent	Blue tetrazolium (0.25% w/v) in sodium hydroxide soln. (6% w/v in water)/methanol (25:75 v/v)	Corticosteroids, and carbohydrates
Iron(III) chloride reagent	Iron (III) chloride (1% w/v) in ethanol/ water (95:5 v/v). heat at 110° C for 5 min. approx.	Phenols, ergot alkaloids, inorganic anions, enols, hydroxamic acids, cholesteryl esters
EP reagent	4-dimethylaminobenzaldehyde (0.2% w/v) and orthophosphoric acid (3% v/v) in acetic acid/water (50:50 v/v) then, heat at 80° C for 10 min.	Terpenes, sesquiterpene esters
N-bromosuccinimide reagent	0.5% w/v solution in acetone. heat at 120° C for 20 min.	Amino acids, Z-protected amino-acids, and hydroxyflavones, hydroxyquinones
o-phthaldehyde-sulphuric reagent	o-phthaldehyde (1% w/v) in methanol/ sulphuric acid (90:10 v/v), heat 80° C for 3 min.	Ergot alkaloids, β-blockers, indole derivatives, histidyl peptides

(f) In general, temperature, time and concentration of the reagent affect the result of the chemical derivatization. Hence due care should be taken to validate these factors depending upon the sample components. The derivatized zones are often not stable with time and temperature, which must be validated.

(g) Most popular derivatization reagents have been elucidated in Table 10.7. These must be available in a TLC laboratory.

Table 10.7: Most common derivatization reagents

Reagent	Preparation and use	Examination	Suitable for detection of -
Ammonia vapor	The plate is put in a chamber saturated with ammonia (25%) vapor. The plate may also be preheated if required.	White light, UV 366 nm	Opiates, mycotoxins, flavonoids, sennosides
Aniline-diphenylamine- phosphoric acid	4 g diphenylamine and 4 ml aniline are dissolved in 160 ml acetone. 30 ml o-phosphoric acid is added with care. The plate is then immersed in the reagent for one second, heated at 200° C for max. 10 min.	White light	Sugars, glycosides
Anisaldehydesulfuric acid	10 ml sulfuric acid is carefully added to an ice-cooled mixture of 170 ml methanol and 20 ml acetic acid. 1 ml anisaldehyde is added. The plate is immersed in the reagent for one second and then heated at 100° C for 2-5 min.	White light, UV 366 nm	Terpenoids, saponins, most lipophilic
2, 6-Dibromoquinone 4-chlorimide (Gibb's reagent)	50 mg 2,6-dibromoquinone (or dichloroquinone) 4-chlorimide is dissolved in 200 ml ethyl acetate. The plate is put in the reagent, dried in cold air, and then placed in a chamber saturated with ammonia until colors appear.	White light	Arbutin, vitamin B_6, phenols, cumarins, thiols, thiones, capsaicin, antioxidants, amines
Dragendorff's reagent	Solution A(0.85 g basic bismuth nitrate is dissolved in 10 ml acetic acid and 40 ml water under heating) Solution B(8 gm potassium iodide is dissolved in 30 ml water. Just before spraying, 1 ml of each solution is diluted with 4 ml acetic acid and 20 ml water).	White light	Alkaloids, heterocyclic nitrogen compounds, polyethylene glycol
Fast blue salt B	0.5 g fast blue salt B is dissolved in 100 ml water. The plate is sprayed/ or immersed, and then dried.	White light, UV 366 nm	Phenolic compounds, tannins, cannabinoids
Iodine (spraying solution) Iodine vapor	0.5 g iodine is dissolved in 100 ml ethanol. The plate is sprayed till background appears light yellow and is exposed to iodine vapor in the chromatographic chamber until zones appear	White light, UV 366 nm (after evaporation of non bonded iodine)	Conjugated double bonds, alkaloids, purine derivatives, lipids, carotenoids

Contd...

Contd…

Reagent	Preparation and use	Examination	Suitable for detection of -
Natural products/polyethylene-glycol	Solution A(1 g of diphenylbo--rinic acid aminoethylester is dissolved in 200 ml ethyl acetate). Solution B (10 g of polyethylene glycol 400 is dissolved in 200 ml dichloro-methane). The plate is heated at 100° C for 3 min, dipped while still hot in solution A, dried in cold air, then dipped in solution B.	UV 366 nm	Flavonoids, carbohydrates, anthocyanines, plant acids
Ninhydrin	0.6 g ninhydrin is dissolved in 190 ml isopropanol, and add 10 ml of acetic acid. Colors produced by the ninhydrin reagent can change with conce-ntration, solvent and modifiers.	White light	Amino acids, biogenic, amines, ephedrine
Phosphomoly-bdic acid	10 g phosphomolybdic acid is dissolved in 50 ml ethanol. The plate is sprayed and dried Sometimes heating of the plate is necessary. The color of zones can be optimized by exposing the plate to ammonia vapor.	White light	Fatty oils, phospholipids, reducing substances, esse-ntial oils compounds, morphine
Sulfuric acid	20 ml sulfuric acid is carefully added to 180 ml ice-cold met-hanol. The plate is dipped in reagent for one second, and heated at 100° C for 5 min.	White light, UV 366 nm	General reagent
Vanillin-inorganic acid	Diverse concentrations are used, mixed with acetic, hydro-chloric, phosphoric, sulfuric acid. The plate is sprayed, and heated at 100° C for 5 minutes.	White light, UV 366 nm	Terpenoids, sterols, salicin, ergot alkaloids, most of lipophilic compounds

Source: *Parameters of Planar Chromatography-useful Hint, Camag*

10.3 CONCLUSION

In practice, it is important to consider the following:
- Derivatization is done for visualization of fraction(s) or chromatogram. Reproducibility of the result may not be important depending on requirement of analysis and subsequent evaluation, e.g. in R and D or process monitoring.
- The color and/or fluorescence of the zones from the outcome of the chemical derivatization is affected by heating (time/temp.), reagent concentration. Besides, drying (time and

temperature) process of the developed plate prior to derivatization also affects the results. Therefore strict protocol should be followed for reproducible results.

- Derivatization is not needed if visualization is not essential. Better results are obtained without derivatization with the use of the scanner.
- If an HPTLC scanner is available, it is quite useful to scan the plate before derivatization. Thus two sets of data can be obtained without repeating chromatography.
- Since fluorescence detection is far more sensitive than absorbance, fluorescent derivatives are preferable.
- Dipping into the derivatization reagent is preferable to spraying.

References

1. Dunges W. GIT Fachz Lab. Supplement *Chromatographic*, 17-26, 1982
2. Kaess A, Mathis C. *Chromatogr Electrophor* Symp. Int 4th ed, 1966, Ann Arbor Science Publishers, Michigan 1968
3. Frijns JMG. *J Phaarm Weekbl* **106**; 605-623, CA 75, 121441 b, 1971
4. Polesuk J, Ma TS. *Mikrochim Acta (Vienna)* 677-682, 1970
5. Wilk M, Hoppe U, Taupp W, Rochlitz J. *J Chromatogr* **27**; 311-316, 1967
6. Wilk M, Bez W, Rochlitz J. *Tetrahedron* **22**; 2599-2608, 1966
7. Polesuk J, Ma TS. *J Chromatogr* **57**; 315-318, 1971
8. Kaufmann HP, Makus Z, Khoe TH. *Fette Seifen Anstrichm* **64**; 1-5, 1962
9. Kaufmann HP, Khoe TH. *Fette Seifen Anstrichm* **64**; 81-85, 1962·
10. Wieland T, Ottenheym H. *Pept Proc Eur Pept Symp* 8th ed 1966, 195, North-Holland, Amsterdam, 1967
11. Elgamal MHA, Fayez MBE. *Fresenius Z Anal Chem* **226**; 408-417, 1967
12. Cargill DI. *Analyst (London)* **87**; 865-869, 1962
13. Luck E, courtial W. *Dtsch Lebensm Rundsch* **61**; 78-79, 1965
14. Jork H. GDCh-*Workshop Nr 301, Dunnschichl-Chromatographic fur Fortgeschrittene, Universitat* des Saarlandes, Saarbrucken 1985
15. Wilk M, Hoppe U, Taupp W, Rochlitz J. *J Chromatogr* **27**; 311-316, 1967
16. *Bundesgesundheitsbl.* **18**; 231-233, 1975
17. Przybylski W. *J Assoc Off Anal Chem* **58**; 163-164, 1975
18. Gertz C. Boschemeyer, *Lebensm*, L. Z. *Unters Forsch* **171**; 335-340, 1980
19. Golinski F, Grabarkiewicz-Szczesna J. *J Assoc off Anal Chem* **67**; 1108-1110, 1984
20. Jork H, Kany E. GDCH-Workshop Nr. 301, *Dunnschichi-Chromatographie for Fortgeschrittene*, Saarbrucken 1986
21. Koch CE, Thurm V, Paul P. *Nahrung* **23**; 125-130, 1979
22. Leuenberger U, Gauch R, Baumgartner E. *J Chromatogr* **161**; 303-309, 1978
23. Elgamal MHA, Fayez MBE. *Fresenius Z Anal Chem* **226**; 408-417, 1967
24. Riess J. *J Chromatogr* **19**; 527-530, 1965
25. Maruyama Y. *Igaku to Seibutsugaku* **73**; 20, CA 69, 92757c, 1968
26. Shah S, Dutta J. *Lipids* **8**; 653-655, 1973
27. Oette K, Doss M. *J Chromatogr* **32**; 439-450, 1968
28. Kaufmann HP, Radwan SS, Ahmad AKS. *Fette Seifen Anstrichmittel* **68**; 261-268, 1966
29. Holloway PJ, Challen SB. *J Chromatogr* **25**; 336-346, 1966
30. Robbiani R, Buchi W. *Proc Euro Food Chem* III **2**; 216-223, Antwerpen, 1985
31. Cohen IC, Narcup J, Ruzicka JHA, Wheals BB. *J Chromatogr* **44**; 251-255, 1969
32. Frei RW, Lawrence JF, le Gay DS. *Analyst(London)* **98**; 9-18, 1973.
33. Lawrence JF, Laver GW. *J Assoc off Anal Chem* **57**; 1022-1025, 1974
34. Lantos J, Brinkman UAT, Frei RW. *J Chromatogr* **292**; 117-127, 1984

35. Junker-Buchheit A, Jork H, *Fresenius Z. Anal Chem* **331**; 387-393, 1988
36. Junker-Buchheit A, Jork H. *Spectrum (Darmstadt)* **2/88**; 22-25, 1988
37. Hansel W, Strommer R. *GIT Fachz Lab Supplement* **3**; Chromatographic, 21-26, 1987
38. Pataki G, Borko J, Curtius HC, Tancredi F. *Chromatographia* **1**; 406-417, 1968
39. Kurhekar MP, D'Souza FC, Meghal SK. *J Chromatogr* **147**; 432-434, 1978
40. Kurhekar MP, D'Souza FC, Pundlik MD, Meghal SK. *J Chromatogr* **209**; 101-102, 1981
41. Minyard JP, Tumlinson JH, Thompson AC, Hedin PA. *J Chromatogr* **29**; 88-93, 1967
42. Sherma J, Bennett S. *J Liq Chromatogr* **6**; 1193, 1983
43. Reh, E. and Jork, H., Fresenius Z. *Anal. Chem.*, **318**, 264, 1984
44. Martin PJ, Stahr HM, Hyde W, Domoto M. *J Liq Chromatogr* **9**; 1591, 1986
45. Wagner H, Seegert K, Sonnenbichler H, Ilyas M, Odenthal KP. *Planta Med* **53**; 444, 1987
46. Funk W, Fresenius Z. *Anal Chem* **318**; 206, 1984
47. Hagiwara T, Shigeoka S, Uehara S, Miyatake N, Akiyama K. *J High Resol Chromatogr* **7**; 161, 1984
48. Schulte KE, Ahrens F, Sprenger E. *Pharm Ztg* **108**; 1165, 1963
49. De Kruif N, Schouten A. *Parfumerie and Kosmetik* **72**; 386, 1991
50. Ebing W. *Chimia* **21**; 132, 1967
51. Mlekusch W, Truppe W, Paletta B. *J Chromatogr* **72**; 495-497, 1972; *Clin. Chim. Acta*, **49**; 73-77, 1973
52. Touchstone JC, Murawec T, Kasparow M, Wortman W. *J Chromatogr Sci* **10**; 490-493, 1972
53. Copius-Peereboom JW, Beekes HW. *J Chromatogr* **9**; 316-320, 1962
54. Trucksess M, Nesheim S, Eppley R. *J Assoc off Anal Chem* **67**; 40-43, 1984
55. Glass A. *J Chromatogr* **79**; 349, 1973
56. Shaun SJH, Butler HT, Poole CF. *J Chromatogr* **281**; 330-339, 1983
57. Zennie TM. *J Liq Chromatogr* **7**; 1383-1391, 1984
58. Chalela G, Schwantes HO, Funk W. *Fresenius Z Anal Chem* **319**; 527-532, 1984
59. Uchiyama S, Uchiyama M. *J Chromatogr* **153**; 135-142, 1978
60. Gertz C, Boschemeyer L. *Z Lebensm Unters Forsch* **171**; 335-340, 1980
61. Frei RW, Lawrence JF, LeGay DS. *Analyst (London)* **98**; 9-18, 1973
62. Schade M. Thesis, Fachhochschule Gießen, Fachbereich Technisches Gesundheitswesen, 1986
63. Derr P. Thesis, Fachhochschule Gießen, Fachbereich Technisches Gesundheitswesen, 1985

11

In situ Quantification

Densitometry is a solution to quantifying the components present in a sample in situ on the plate by absorbance in UV or visible light or emission in fluorescence. This was an important requirement for modern TLC to be accepted as an analytical tool comparable to HPLC, GC for reliable quantitative determination at micro- and sub-micro-level. The system is calibrated for determination of unknown by comparison with known-standard. However, in TLC this is validated only for a given plate. Therefore unknown and standards should be chromatographed on the same plate side by side. Plate performs as a background for evaluating the unknown substance with the known-quality of the standards. But it is necessary to adjust densitometry to optimize position with regards to noise, sensitivity, wavelength and other parameters for optimum results. Modern scanners provide reliable densitometric data. It is a fact that each TLC plates varies in different way though the other conditions of the analysis remain same. Therefore the obtained data may vary from the absolute values. Hence the same set may give different absolute values at the time of measurement on different plates. However, the calibrated results for the same unknown will be same on both the plates.

Modern scanners are used for quantitative evaluation, as well as to record absorbance spectra Digital imaging systems can also be used for quantification with in limited parameters.

11.1 SCANNING DENSITOMETRY

A TLC/HPTLC scanner is an advanced version of a traditional densitometer. A modern scanner has many facilities like multi-wavelength scanning, fluorescence scanning, and spectrum recording for identity as well as purity. The scanner is far more sensitive than a densitometer.

Scanning densitometry provides accurate and reliable evaluation in HPTLC for quantification of separated components. Spectrum data offers advantages over visualization and video densitometry. The reason being that monochromatic light in the wave range of 190-800 nm is tuned for absorption/fluorescence maximum of the separated compounds. The measurement is very sensitive, i.e. low nanogram level in case of absorbance and picogram range in case of fluorescence. Densitometry is normally done prior to derivatization but can be done subsequently, as well. Analytes without chromophoric groups should be derivatized to make them suitable for detection. Scanning densitometry is suitable for identification by comparison of profiles of analogue curves of every sample track which incorporate multi-wave scanning, e.g. sequential scanning of every chromatogram track with up to 30 wavelengths and evaluation of individual track at all wavelengths. UV spectra of separated components is recorded and used for identification purpose. A library of UV spectra, recorded from plates is commercially available. The chromatography, steps of HPTLC should be done with highest care and according to the validated methods to enable proper evaluation by densitometry.

11.1.1 Operation of the Spectrodensitometers

TLC/HPTLC scanners use monochromatic light in a single beam with a beam splitter as shown in Figure 11.1.

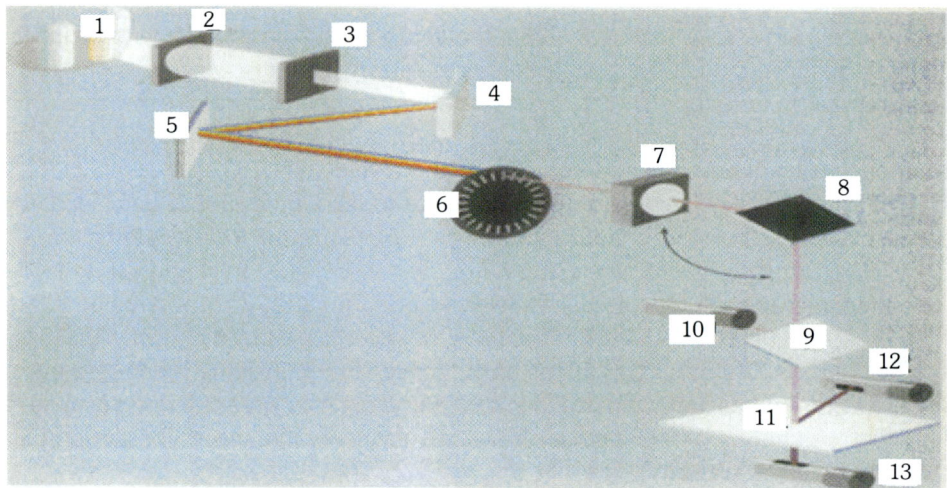

Fig. 11.1: Schematic set-up of a TLC scanner

Source: Camag, Switzerland

(*i*) lamp selector; (*ii*) entrance lens; (*iii*) monochromator entry slit; (*iv*) monochromator grating; (*v*) mirror; (*vi*) slit selector; (*vii*) micro-macro lens; (*viii*) mirror; (*ix*) beam splitter; (*x*) reference photomultiplier; (*xi*) TLC plate; (*xii*) measuring photomultiplier(remission); (*xiii*) measuring diode (transmission).

The principle is that beam of electromagnetic radiation hits layer and reflected from the surface, since the layer is opaque. The reflectance radiation is measured by photomultiplier. When the beam passes across a chromatographic zone, difference in optical response is seen because some radiation is absorbed and less is reflected from the surface of the plate if an absorbing fraction is present. This difference is the signal for detection and quantification of the substances on the plate. Commercially available spectrodensitometric scanner is shown in the Figure 11.2.

Through this process, a large amount of data is obtained by scanning of separate tracks at different wavelengths. Such data include height of the peak and area, position of the zone for all the separated components. Background on HPTLC plate may produce some noise. However, this can be measured using a software control and this noise value may be subtracted from the final chromatogram. Baseline adjustment is made to enable all the peaks to accurately integrate for quantification. Scanning direction should be from origin to solvent front on the TLC track. Radial scanning will be required for circular/anti-circular chromatograms.

Fig. 11.2: TLC scanner

Several criteria should be applied to get the best results for quantification by reflectance scanning. Whereas the type of the layer/plate has no significant effect on the reflectance scanning but accuracy and precision of the band application and selection of development solvent will have an effect, e.g. manual application of sample will give less reproducible result than automated band application. Precision of the band application is very important, because concentration of the sample should be uniform over all range of the zone. Sample overloading should be avoided. Short development distances lead to sharp peaks. Plate height is reduced with the decrease in particle size of sorbent and absence of voids. Therefore, precoated plates must be used for quantitative analysis. They improve resolution and reduce noise.

In case of spot application, a light beam of slit length longer than width of the largest development spot is selected, because the concentration of analyte will be different across the spot. Secondly, the spots become bigger in diameter as they migrate. Thirdly, the slit length should cover slight deviation which may occur in the direction of chromatographic development. Large slit length decreases the sensitivity. So the optimum slit length should be used to make possible light beam to cover all the fractions when it moves across the track. A too small slit may not cover the entire separated zone and also generate more noise. Due to greater diffusion of spots, sensitivity of detection decreases as R_f increases. At low R_f, the slit is too big compared to spot diameter which leads to poor sensitivity. Hence spots are not really suitable for quantification of most samples.

But in case of band application, the selection of slit length is easy and short slit length can be selected resulting in higher sensitivity. Normally, two-thirds of the slit length as compared to length of the chromatographic zone is selected. If these conditions are followed properly, the standard deviation and coefficient of variance is normally below 1-2%. Now the software facilitates large number of parameters and features to optimize quantitative analysis. Some practical advises are as below:

1. Chromatographic separation with sharp zones will benefit the evaluation and quantification.
2. Visual inspection prior to scanning will help to identify and remove any problem of the plate.
3. Spectral selectivity will provide more information by densitometer than by visual evaluation.
4. Width (slit dimension) should be kept between 70-80% of the band length application zone or approximately 120% of the diameter of the largest spot of the chromatogram.
5. Slit height should be about 0.4 mm to obtain optimum available light and resolution of the scan.

6. For qualitative measurement, the scanning speed may be high but it is advisable to keep the scanning speed medium for best result. High speed is required for multi-wavelength scans.
7. In many measurements, the settings selected by built-in automatic detector provide best adjustment.
8. The scan should be started below the application zone and end after the mobile front position.
9. The scan start and end positions should be adjusted to measure only components of interest, where possible.
10. A large bandwidth (20 nm) ensures low noise for quantitative measurements.
11. The scanning beam and sample tracks must be properly aligned. This must be checked with the help of compartment illumination.

11.1.2 Absorption Measurement

- Deuterium or tungsten lamp should normally be used for quantitative measurement at known absorption maximum of the analyte. The mercury lamp has very limited use due to its line spectrum and its strong emission line at 366 nm is used for fluorescence.
- When scanning a new sample, an automated multi-wavelength evaluation is done at 10 nm interval in UV range. This detects all peaks. Then spectra of all detected peaks are recorded to know the lambda max of each fraction.
- In the final method, a multi-wavelength scan can be chosen such that each fraction is measured at its lambda max for quantification.
- For finger printing or qualitative measurement, multi-wavelength scanning covering the entire UV range is useful to perform. This permits to locate the most suitable wavelength for general scanning for detection of all possible substances as well as for specific detection of individual components.
- The micro-position of lens in the scanner's optical system is a must for HPTLC. It condenses the large incident beam into a small one for HPTLC without losing energy.
- The concentrations of sample components must be so chosen that they fall in the linear range of response.
- Absorbance measurements are usually calculated by peak area.

11.1.3 Fluorescence Measurement

- Wherever possible, for quantitative evaluation fluorescence measurement should be used which is more sensitive than absorption technique and has less interference.
- Specific wavelength (usually 366 nm) within line spectrum of the mercury lamp should be used for excitation of the analyte to produce strong fluorescence signals.
- When optimum excitation wavelength is not known, fluorescence excitation spectrum should be recorded by use of deuterium lamp from 190-380 nm and K 400 universal cut off filter. A narrow band width, i.e. 5 nm should be chosen. To eliminate a false maximum at 254 nm, the plate without fluorescence indicator must be used.
- For samples containing heavy matrix or plate which has been derivatized with fluorescence inducing reagent will pose problem for evaluation. It is therefore useful that application position (showing a high sigma) is excluded from scan and the detector adjustment is manually done at a sample/matrix free position of the plate.

- Normally calibration data under fluorescence mode are linear over a wide range. However, crater effect is visible at high sample concentration because sample starts absorbing its own fluorescence. Reduction in the sample concentration solves such problem. Sometimes ghost peak on account of dust or lint during handling of the plate are seen, but such problem can be avoided. A check under UV 366 nm lamp will help to see that no dust particle interferes. To remove dust, a stream of nitrogen is used to blow it away.
- Peak height is usually used for quantitative calculations.

11.1.4 Theory of Spectrodensitometry

Beer's law is normally followed if there is a direct relationship between absorbance and concentration of the analyte in the solution. The absorbance is determined as a result of beam of electromagnetic radiation of fixed wavelength passing through a sample solution. The absorbed radiation by the solution is directly proportional to the concentration of the sample. But there is no linear relationship throughout the whole concentration range and this relies on the sample solution being transparent. Kubelka and Munk[1] described a different theory since TLC layer is opaque. They explained that what would happen if a beam of electromagnetic radiation impinged on an opaque layer. This led to a number of equations to be derived in order to establish a relationship between reflected radiation and concentration of the analyte. When a ray of light is impinged on the layer, some portion of the light is transmitted, some reflected in other direction and some are propagated within the sorbent in all the directions. It can be presumed that the transmitted and reflected portion of the incident light is made up only of the rays propagated within the sorbent in a perpendicular direction of the plain of the surface.

Since other light rays will go to longer path, they shall more strongly be absorbed and thus will less contribute to transmitted and reflected light. As soon as light leaves sorbent layer, scattering takes place in every direction at the layer air junction. A mathematical expression is derived for coefficient of light scatter(S) in terms of intensity of transmitted and reflected light, finite thickness of sorbent (l), and coefficient of absorption per unit thickness (K_A). The following formula has been proposed by Kubelka and Munk.

$$I_R = \frac{\sinh(b.S.I.)}{a.\sinh(b.S.I) + b.\cosh(b.S.I)} \qquad (Eq.\ 11.1)$$

$$I_T = \frac{b}{a.\sinh(b.S.I.) + b.\cosh(b.S.I)} \qquad (Eq.\ 11.2)$$

where

$$a = \frac{S.I + K_A.I}{S.I}$$

$$b = \sqrt{(a^2 - 1)}$$

I_R = Intensity of reflected light

I_T = Intensity of transmitted light

From these it is possible to infer that there is a non-linear relationship between intensity of reflected light and concentration of chromatographic zone. Practically, graphic displayed data may match a polynomial curve of general formula.

$$Y = a_0 + a_1 x + a_2 x^2 \qquad (Eq.\ 11.3)$$

Four to six standards should be developed along with sample on same plate for calibration curve over a wide range of concentration. It may be emphasized that relationship is almost linear over a small range of concentration. But when concentration of the sample is approximately known, then two standards near to the value can be taken for development. In practice, the chance of error is negligible[2].

Available software with the scanner enables to linearise the data graphically. The easy way is to convert the reflectance and concentration data into logarithms, reciprocals, or squared terms. Even if this approach is not successful to linearise the data (though it will be successful in many cases) then analyst must carry out non-linear regression analysis based on second order polynomials. The following equation gives authentic results.

$$\ln R_E = a_0 + a_1 . \ln c + a_2 . (\ln c)^2 \qquad \text{(Eq. 11.4)}$$

$$R_E = \text{reflectance signal}, c = \text{Sample concentration}$$

This should be noted that data fit to the equation is not compromised when three standards are used for the whole concentration range.

The data obtained in fluorescence mode could be used mathematically in a direct way. The fluorescence emission (F) can be expressed in the following way.

$$F = \emptyset . I_0 (1 - e^{-a} _m . l . c) \qquad \text{(Eq. 11.5)}$$

Where $\emptyset$ is the quantum yield and a_m is the molar absorptivity.
Whereas for low concentration, the formula can be simplified as:

$$F = \emptyset . I_o . a_m . l . c \qquad \text{(Eq. 11.6)}$$

It can, therefore, be established that fluorescence emission is linearly dependant upon the concentration of the sample, although the effects of absorption and scattering are not considered.[3,4]

11.2 EFFECT OF NITROGEN FLUSHING

A study was conducted to observe effect of nitrogen flushing for fluorescence compounds and for Aflatoxin determination. The following experiments will depict an idea of the effects which is not universally applicable for all fluorescence compounds. In each case, flushing should be tried wherever required for better separation.

11.2.1 Chromatographic Conditions

Stationary phase	:	HPTLC plate, silica gel 60F$_{254}$ (Merck) 10 × 10 cm or 20 × 20 cm.
Mobile phase	:	Toluene: Ethyl acetate: Methanol (7: 2: 2)
Sample application	:	Apply 5 µl and 10 µl of sample in duplicate.
Developing distance	:	80 mm
Tank saturation	:	Nil
Scan at sample	:	At 366 nm and >400 nm.

11.2.2 Sample as Such

As is sample (i.e. without flushing the scanner with nitrogen gas). Scan at 366 nm (Fig. 11. 3, Table 11.1)

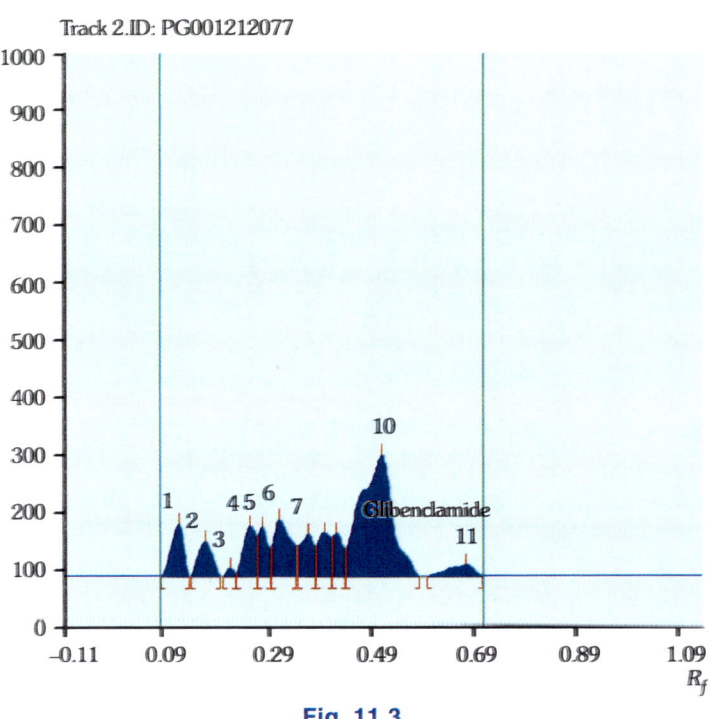

Fig. 11.3

Table 11.1

Track 2, ID: PG001212007

Peak	Start Position	Start Height	Max Position	Max Height	Max %	End Position	End Height	Area	Area %	Assigned substance
1	0.08 R_f	0.5 AU	0.12 R_f	90.8 AU	10.16 %	0.14 R_f	0.1 AU	1722.6 AU	6.82 %	unknown*
2	0.14 R_f	0.0 AU	0.17 R_f	60.2 AU	6.73 %	0.20 R_f	0.5 AU	1383.4 AU	5.47 %	unknown*
3	0.20 R_f	0.1 AU	0.22 R_f	15.1 AU	1.68 %	0.23 R_f	6.8 AU	171.0 AU	0.68 %	unknown*
4	0.23 R_f	7.2 AU	0.26 R_f	86.9 AU	9.72 %	0.27 R_f	39.8 AU	1734.0 AU	6.68 %	unknown*
5	0.27 R_f	70.5 AU	0.28 R_f	86.3 AU	9.65 %	0.29 R_f	50.2 AU	1322.9 AU	5.23 %	unknown*
6	0.30 R_f	50.7 AU	0.31 R_f	99.7 AU	11.15 %	0.34 R_f	55.0 AU	2672.7 AU	10.58 %	unknown*
7	0.35 R_f	55.1 AU	0.37 R_f	73.2 AU	8.19 %	0.38 R_f	53.2 AU	1571.1 AU	6.22 %	unknown*
8	0.38 R_f	53.8 AU	0.40 R_f	75.8 AU	8.48 %	0.41 R_f	38.0 AU	1587.7 AU	6.28 %	Glibenclamide
9	0.42 R_f	68.6 AU	0.42 R_f	74.6 AU	8.34 %	0.44 R_f	48.4 AU	1236.1 AU	4.89 %	unknown*
10	0.44 R_f	48.7 AU	0.51 R_f	210.4 AU	23.53 %	0.59 R_f	0.6 AU	11037.5 AU	43.67 %	unknown*
11	0.60 R_f	0.4 AU	0.68 R_f	21.2 AU	2.37 %	0.71 R_f	0.5 AU	834.4 AU	3.30 %	unknown*

11.2.3 Flushing with Nitrogen

After flushing the scanner with nitrogen gas for 10 min at pressure of 0.5 Bar. Scan at 366 nm (Figs. 11.4–11.6, Table 11.2)

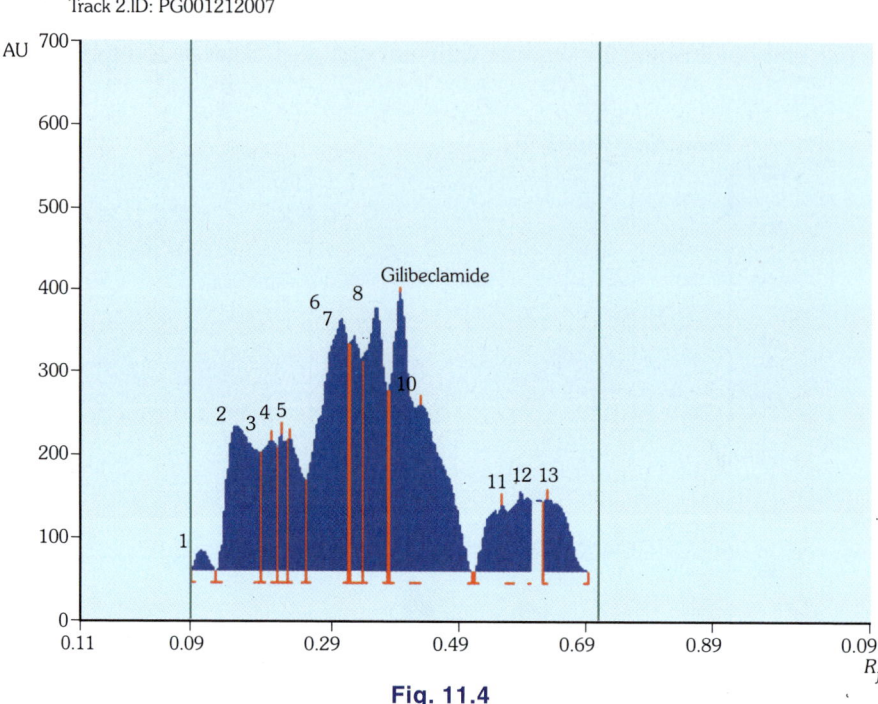

Track 2.ID: PG001212007

Fig. 11.4

Table 11.2

Peak	Start Position	Start Height	Max Position	Max Height	Max %	End Position	End Height	Area	Area %	Assigned substance
1	0.07 R_f	1.8 AU	0.08 R_f	24.1 AU	1.00 %	0.11 R_f	0.5 AU	391.5 AU	0.65 %	unknown*
2	0.11 R_f	0.2 AU	0.14 R_f	175.7 AU	7.33 %	0.18 R_f	44.0 AU	6440.8 AU	10.68 %	unknown*
3	0.18 R_f	144.2 AU	0.19 R_f	156.8 AU	6.54 %	0.20 R_f	50.2 AU	2874.3 AU	4.76 %	unknown*
4	0.20 R_f	148.7 AU	0.21 R_f	165.3 AU	6.89 %	0.22 R_f	56.8 AU	1874.8 AU	3.11 %	unknown*
5	0.22 R_f	158.2 AU	0.22 R_f	159.6 AU	6.65 %	0.25 R_f	09.9 AU	2883.8 AU	4.78 %	unknown*
6	0.25 R_f	110.2 AU	0.30 R_f	305.9 AU	12.76 %	0.32 R_f	73.9 AU	10739.7 AU	17.80 %	unknown*
7	0.32 R_f	274.3 AU	0.32 R_f	283.1 AU	11.81 %	0.34 R_f	54.6 AU	4089.2 AU	6.78 %	unknown*
8	0.34 R_f	255.2 AU	0.36 R_f	322.9 AU	13.46 %	0.38 R_f	17.7 AU	8159.8 AU	13.53 %	unknown*
9	0.38 R_f	218.2 AU	0.40 R_f	336.9 AU	14.05 %	0.42 R_f	37.2 AU	8097.7 AU	13.42 %	Glibenclamide
10	0.42 R_f	197.6 AU	0.43 R_f	201.2 AU	8.39 %	0.51 R_f	0.5 AU	7594.5 AU	12.59 %	unknown*
11	0.51 R_f	0.3 AU	0.56 R_f	81.6 AU	3.40 %	0.57 R_f	72.3 AU	2163.9 AU	3.59 %	unknown*
12	0.57 R_f	73.0 AU	0.59 R_f	97.9 AU	4.08 %	0.61 R_f	72.3 AU	2301.3 AU	3.81 %	unknown*
13	0.62 R_f	73.0 AU	0.63 R_f	87.1 AU	3.63 %	0.69 R_f	0.5 AU	2711.0 AU	4.49 %	unknown*

Track 2, ID: PG001212007

Fig. 11.5: Image of aflatoxin (B1, B2, G1, G2) at 366 nm

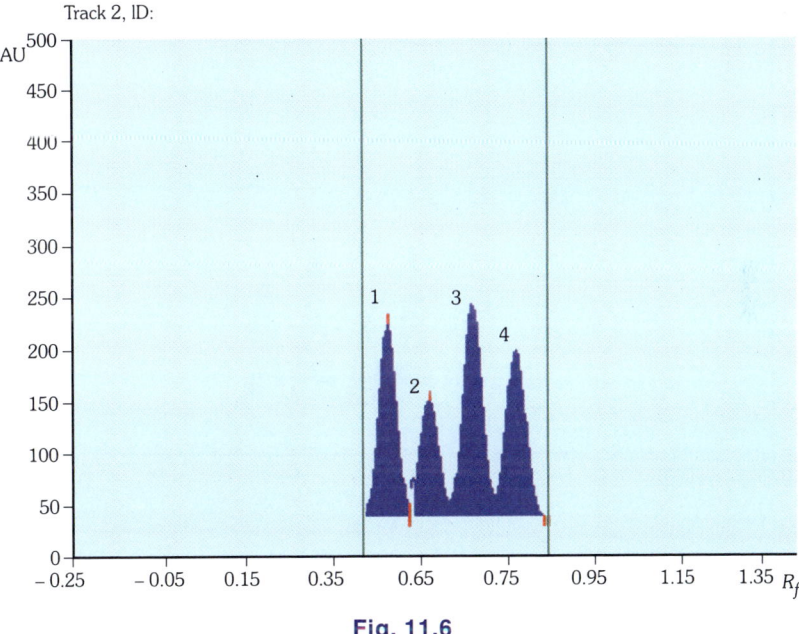

Fig. 11.6

11.2.4 As is sample (i.e. without flushing scanner nitrogen gas)

Scan at 366 nm (Table 11.3)

Table 11.3

Track 2, ID:

Peak	Start Position	Start Height	Max Position	Max Height	Max %	End Position	End Height	Area	Area %	Assigned substance
1	0.42 R_f	0.6 AU	0.47 R_f	183.9 AU	27.82 %	0.52 R_f	11.6 AU	5073.6 AU	26.19 %	unknown*
2	0.54 R_f	29.9 AU	0.57 R_f	112.9 AU	16.98 %	0.61 R_f	12.3 AU	3174.1 AU	16.39 %	unknown*
3	0.61 R_f	12.3 AU	0.67 R_f	204.8 AU	30.99 %	0.72 R_f	19.6 AU	6207.1 AU	32.04 %	unknown*
4	0.72 R_f	19.6 AU	0.77 R_f	160.0 AU	24.22 %	0.83 R_f	0.1 AU	4917.1 AU	25.38%	unknown*

11.2.5 Flushing the Scnner for 10 Mins

After flushing the scanner with nitrogen gas for 10 min at a pressure of 0.5 Bar: Scan at 366 nm (Fig. 11.7, Table 11.4)

Track 2, ID:

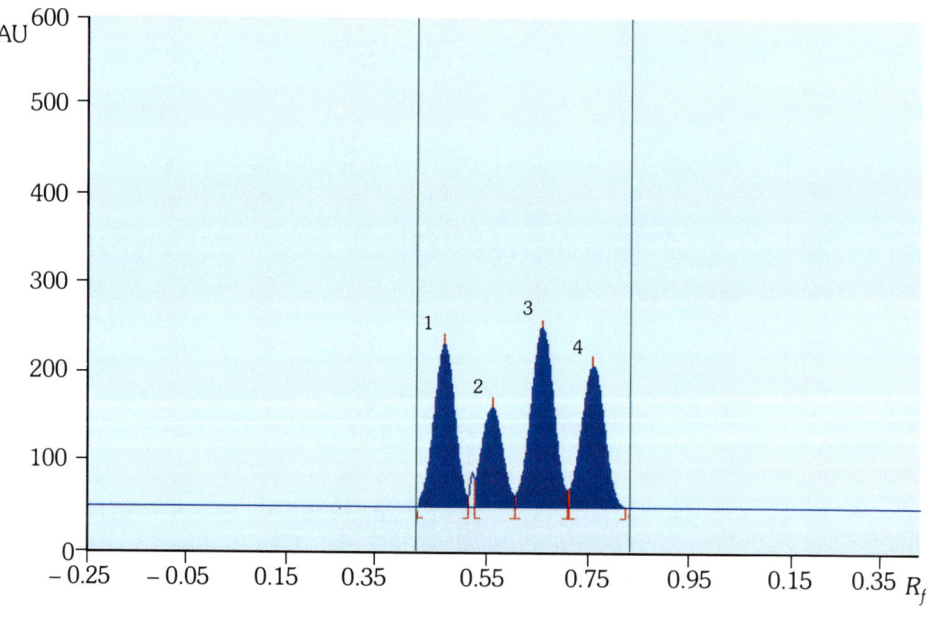

Fig. 11.7

Table 11.4

Track 2, ID:

Peak	Start Position	Start Height	Max Position	Max Height	Max %	End Position	End Height	Area	Area %	Assigned substance
1	0.42 R_f	0.6 AU	0.47 R_f	182.6 AU	27.86 %	0.52 R_f	11.5 AU	5048.1 AU	26.28 %	unknown*
2	0.54 R_f	29.6 AU	0.57 R_f	111.8 AU	17.05 %	0.61 R_f	12.6 AU	3154.9 AU	16.43 %	unknown*
3	0.61 R_f	13.0 AU	0.67 R_f	202.2 AU	30.85 %	0.72 R_f	19.3 AU	6158.1 AU	32.07 %	unknown*
4	0.72 R_f	19.6 AU	0.77 R_f	158.8 AU	24.23 %	0.83 R_f	0.0 AU	4844.7 AU	25.22%	unknown*

11.2.6 Flushing the Scanner for 20 Mins

After flushing the scanner, nitrogen gas for 20 min at pressure of 0.5 Bar: Scan at 366 nm (Fig. 11.8, Table 11.5)

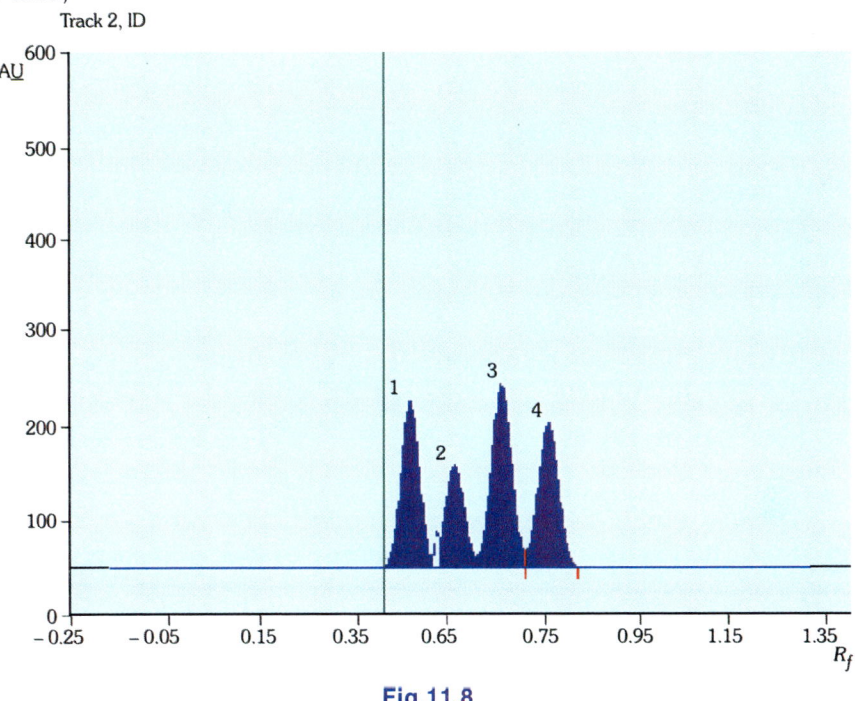

Fig.11.8

Table 11.5

Peak	Start Position	Start Height	Max Position	Max Height	Max %	End Position	End Height	Area	Area %	Assigned substance
1	0.42 R_f	0.5 AU	0.47 R_f	178.7 AU	28.02 %	0.52 R_f	11.1 AU	4936.7 AU	26.43 %	unknown*
2	0.54 R_f	28.8 AU	0.57 R_f	108.3 AU	16.98 %	0.61 R_f	11.6 AU	3064.7 AU	16.41 %	unknown*
3	0.61 R_f	11.8 AU	0.67 R_f	196.4 AU	30.79 %	0.72 R_f	18.4 AU	5964.7 AU	31.93 %	unknown*
4	0.72 R_f	18.6 AU	0.77 R_f	154.4 AU	24.20 %	0.83 R_f	0.2 AU	4712.0 AU	25.23 %	unknown*

Track 2, ID:

11.2.7 Flushing the Scanner for 30 Mins

After flushing the scanner, nitrogen gas for 30 min at a pressure of 0.5 Bar: Scan at 366 nm (Fig. 11.9, Table 11.6)

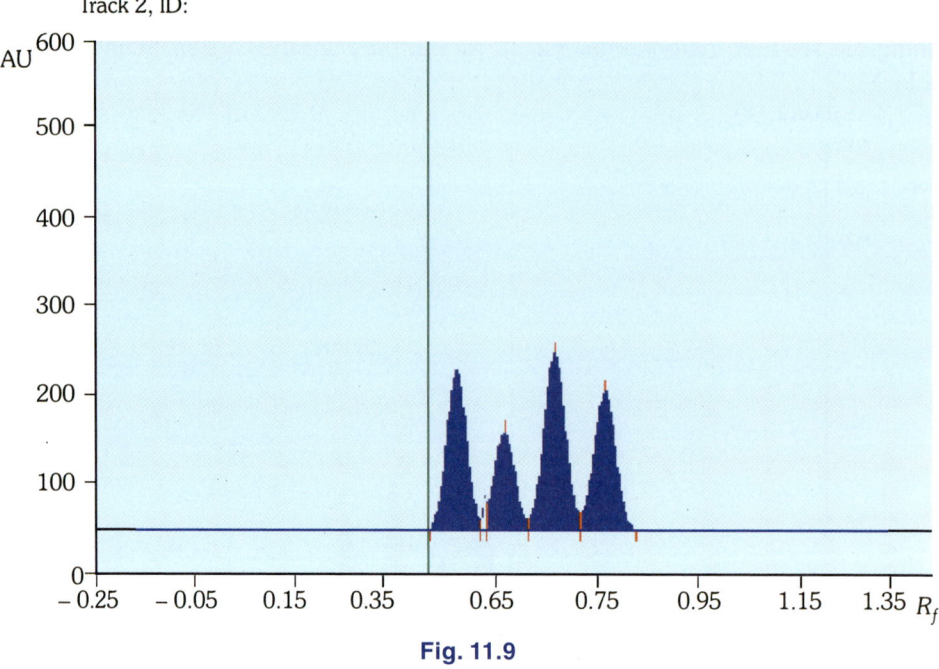

Fig. 11.9

Table 11.6

Peak	Start Position	Start Height	Max Position	Max Height	Max %	End Position	End Height	Area	Area %	Assigned substance
1	0.42 R_f	0.7 AU	0.47 R_f	182.0 AU	27.94 %	0.52 R_f	11.3 AU	5039.3 AU	26.41 %	unknown*
2	0.54 R_f	29.7 AU	0.57 R_f	110.9 AU	17.03 %	0.61 R_f	12.3 AU	3133.6 AU	16.42 %	unknown*
3	0.61 R_f	12.4 AU	0.67 R_f	201.0 AU	30.86 %	0.72 R_f	19.0 AU	6110.9 AU	32.03 %	unknown*
4	0.72 R_f	19.1 AU	0.77 R_f	157.5 AU	24.17 %	0.83 R_f	0.2 AU	4796.0 AU	25.14 %	unknown*

Track 2, ID:

11.2.8 Conclusion

Flushing with nitrogen eliminates oxygen from the light path. From the scan data, it is observed that some fluorescent compounds show increase in area, while some show decrease in area after flushing, nitrogen gas. It is not applicable to Aflatoxin B_1, B_2, G_1, G_2, as there is no change in area after flushing nitrogen. Flushing the scanner may not be advantageous every time.

11.3 SCANNERS

Scanners can automatically record absorbance spectra of all the desired or selected, separated fractions on a plate. Automation here means that positions of the fractions on a plate must be automatically available for spectrum measurements, e.g. on a plate with say 18 samples tracks and say 5 fractions per track would need to record $18 \times 5 = 90$ spectra for identification. Spectra for identification are recorded at the peak max position.

For purity checking of fractions, it is essential to record 3 spectra on each fraction and compare them. In such a case, the number of spectra recorded will be $18 \times 5 \times 3 = 270$. However, time take to obtain such a large amount of data is only a few minutes in automated scanners.

Creating a file manually describing the X and Y position of each fraction on each track is tedious, error prone and non-GLP.

Spectra obtained from plates can be used to create a library. UV spectra recorded in a UV spectrophotometer will not be the same as on a plate because one is a solution the other is a solid, one is transmission the other reflectance and thirdly the solvent can shift the lambda max and so can silica gel.

In order to know the correct lambda max, UV spectra in HPTLC are recorded at a narrow bandwidth of 5 nm.

11.4 VIDEO IMAGING AND IMAGE DENSITOMETRY

Video densitometry is done using a digital image of chromatogram and evaluation with appropriate software. Scanning densitometer identifies the representation of chromatogram as a number of peaks, which can be quantified. Video imaging results in storage of all the chromatograms on the plate as a computer file. The data is subsequently quantified. Video densitometry works on grouping of pixels of the image as per tracks and then evaluating on a gray scale the saturated pixels. Pixels of fractions having same R_f are averaged and after that plotted as a function of distance in the direction of chromatography. The result is analog curve of the chromatogram. It is possible to compare the curves of individual tracks or individual peaks. If chromatography and imaging have been done under similar conditions, comparisons of different plates are also possible while adopting this method for identification purpose. The number, position and related area/height of peaks on the tracks are compared and evaluated. Quantitative determination in absorbance (dark zone on bright background) or fluorescence (bright zone on dark backgrounds) can be done if the reference standards have been chromatographed along with the unknown sample. But on downside, spectra selectivity in the UV-Vis range, i.e. 180 to 800 nm is not available for evaluation. Only 254 nm, 366 nm in UV and white light can be used. The other drawback related to capture of settings during imaging selected manually to obtain a good picture is not valid on modern devices. Today's truly GLP compliant systems do not allow access to raw data of the images and therefore are very secure.

HPTLC photo documentation systems illuminate the plate with white light or 254 nm short wave UV and 366 nm long wave UV. 12 bit high resolution 3 CCD colour industrial camera with fixed focus is used to ensure very high reproducibility (Fig. 11.10). Consumer cameras can be used but they are mass produced and so not reproducible. Their images can be edited since they are non-GLP. While the scanner offers monochromatic wavelengths from 190 to 800 nm, the imaging devices offer only 3 options - white light, UV 254 nm and UV 366 nm.

It is possible to control the brightness, colour, contrast and intensity with TLC software and this will be stored in the file for reference and future use. The raw data of the image is not accessible. The quantification of components is done by comparing with standards on the same plate. It is possible to quantify the results for a number of analytes with a good accuracy and reliability.[5-10] Video densitometry is an acceptable method, but spectrodensitometry is more accurate, reproducible and versatile.

Documentation system with high resolution 12-bit industrial camera

Fig.11.10: CCD camera

Video imaging is a quick and easy method with the possibility of having a hard copy photographic image of the developed plate. It serves as a permanent visual record of the analysis. But data acquisition with CCD camera is simpler and convenient in two-dimensional chromatography though quantification takes more time.[11]

It is to be noted that in HPTLC documentation, resolution is far more important than number of pixels. A 12 bit camera has 4096 levels of sensitivity, i.e. grey levels, while a 8 bit camera has only 256 levels, even if both cameras have the same number of pixels.

A fixed focus camera is a must for reproducibility. Another important aspect is that the colours of substances with reference to their intensity, shade and tint will appear different to the human eye, the camera screen the PC screen and on the print out. Each of these four visualisers have their own variables, e.g. in a printout—different papers, different inks, age of paper, intensity and shade of incident white light, etc. Thus it is obvious that a coloured HPTLC plate image is printed on a PC monitor. It is strongly recommended to compare two images on any PC monitor in order to eliminate all variables.

11.5 REPORTING AND RECORD KEEPING

To compare different samples visually, it is necessary that the different samples treacle from different plates be displayed side by side. Modern software for image documentation from some manufacturers allows such type of image comparison. The same image looks a little different when seen on the camera's LCD screen and a PC screen and on different types of printers and different types of papers. The best way, therefore, is to compare the images on the PC screen. Image comparison software is therefore essential for high quality work.

As per GLP requirement, it is necessary that all the records are permanently maintained for a certain time period depending upon the individual requirement. The information should be kept in a well-organized manner with due traceability. The raw data should be inaccessible.

All records should be systematically maintained and the analytical work should be in accordance with the standard operating procedures (SOP) specified. Good record keeping is a requirement for accreditation of the laboratories and is useful for conducting research activities and method development. Since all kinds of imaging systems are available in the market, it is imperative to use one for HPTLC work that meets the above reporting and record keeping requirements.

References

1. Pollak V, Touchstone JC, Sherma J, Wiley J (eds). *Densitometry in Thin Layer Chromatography, Practice and Applications*, Chichester, UK, 11-45, 1979
2. Ebel S, Hocke J. *J High Res Chromatogr Chromatogr Commun* 156-160, 1978
3. Butler HT, Poole CF. *J High Res Chromatogr Chromatogr Commun* 77-81, 1983
4. Berkhan EW. *J Planar Chromatrogr* 81-87, 1988
5. Prosek M, Pukl M. *Handbook of Thin Layer Chromatography*, F. Sherma and B. Fried (eds), Marcel Dekker, New York, USA, 279, 1996
6. Vovk I, Prosek M. *J Chromatogr A* **768**; 329-333, 1997
7. Vovk I, Prosek M. *J Chromatogr A* **779**; 329-336, 1997
8. Vovk I, Golc-Wondra A, Prosek M. *J Planar Chromatogr* **10**; 416-419, 1997
9. Petrovic M, Kastelan-Macan M, Babic S. *J Planar Chromatogr* **11**; 353-357, 1998
10. Vovk I, Franko M, Gibkes J, Prosek M, Bicanic D. *J Planar Chromatogr* **11**; 379-382, 1998
11. Gibkes J, Vovk I, Bolte J, Bicanic D, Bein B, Franko M. *J Chromatogr A* **786**; 163-170, 1997

12

Coupling Techniques with HPTLC

C oupling techniques of other instruments with HPTLC enhance the capabilities of analysts depending upon their aims of analysis. Literature has revealed that HPLC, MS (mass spectrometry), NMR, FTIR (Fourier transform infrared spectroscopy), and Raman spectroscopy have been successfully coupled with HPTLC. However, many instruments are very expensive. Therefore, nature of the analysis has to be kept into the view with the cost of instruments.

HPTLC has advantage of coupling with other analytical techniques because it is an offline process and separation, development and detection are not continuous. Further advantage is that a number of plates can be run simultaneously and the separation can be evaluated qualitatively before detection. Such facilities enable the analysts to concentrate on the fractions of interest. Unknowns can be detected and their structure can be elucidated by coupling techniques. The separation is "stored" as the separated components of the sample remain on the sorbent layer after development of chromatogram. Chromatogram can be stored as long as the components are stable. Moreover, evaluation of the TLC chromatogram by coupling methods can be done at different location where such special equipments are available.

Several references are available on HPTLC coupling.[1–3] Specific detection using the coupling techniques are especially useful when sample matrix is complex. Important coupling techniques have been discussed here in brief.

12.1 HPTLC WITH HPLC

Linking of HPTLC with HPLC is a powerful combination since sorbent used on the plate and column for different selectivity. For initial separation, reversed phase HPLC column is used. The retention times of the interested peak (from start to end of the peak) will decide as to which effluent cut should be transferred to normal phase HPTLC plate through band spraying.[4] This process can be online. AMD technique can be used for obtaining further enhanced resolution of analytes in HPLC peak.[5]

12.2 HPTLC AND MS

Several reviews of TLC/MS[6–8] which highlight on different aspects of design of instruments and applications, are available. There is a greater advantage in analysis of mixtures for presence/absence of particular compounds. Large information can be obtained for each spot.

Hyphenated technique with mass spectrometry has proved one of the most useful tools for this technique. After separation, the fraction(s) is eluted from the sorbent and fed to the ion source of the mass spectrometer. This procedure is simple and fast. A suitable interface is essential. Mass spectra are obtained directly from the layer. Probes or plate scanners have also been designed suitably for planar chromatography.

TLC in tandem mass spectrometry (MS/MS) is an effective hyphenated technique. Applications are limited by cost. However, the main problem is the insignificant contribution and relatively limited fragment ion data available from a mass spectrum. Solution to this problem is tandem mass spectrometry (MS/MS) wherein two spectrometers are linked and the molecular ion of the compound of interest is selected in the first mass spectrometer and is separated. Fragmentation of the molecular ion then induced and fragment ion is focused and determined by scanning in the second mass spectrometer. Since the fragments correspond to ion selected in the first mass spectrometer, the spectrum enables to analyze the compound to be characterized very efficiently and without interference.

Instruments designs may involve different mass analyzers. Quadruple mass filters and ion traps provide advantages of comparatively small size. The successful adoption of HPTLC/MS is a simple interface device which transforms distribution of sample on xy plane into a sequence of sample molecules in a gas or liquid phase. The method must be simple and robust with varied options and advantages.

12.3 HPTLC AND FTIR

This is a very powerful technique and makes use of standard reference spectra from FTIR reference library. Positive identification of structural isomers is possible. When spectra are not available, data related to molecular structure of separated compounds can be interpreted. Since most chemical compounds have IR absorption, the method can be made applicable to many substances including those which is non-UV absorbing. In situ HPTLC-FTIR approach is well established by several workers.[9–11]

Off-line approach is simpler and the sample can be taken from the spot or zone of the plate, then dissolved in a suitable solvent and transferred to FTIR spectrometer. Although the transfer is time consuming, it is possible to measure full IR spectrum. Now the sample preparation has been included and equipments are available to elute analytes from the plate without scraping of any sorbent with the help of solvent like methanol[12] and then it is transferred to small quantity of potassium bromide powder, which is allowed to evaporate. The powder is ground and potassium bromide pellet is made for transfer to FTIR spectrometer. The application of HPTLC with FTIR is useful in number of compounds like phthalates[13], polyaromatic hydrocarbons, amino acids[14], corticosteroids[15], phospholipids[16], dyes[17] and phenols.[18]

12.4 HPTLC AND RAMAN SPECTROSCOPY

Raman spectroscopy depends on vibration modes of molecule for examination as in case of infrared spectroscopy. Therefore, unique spectra are produced that can be applied for identification of compounds. The main advantage in Raman spectroscopy is that sorbent gives weak Raman spectra which results in low background interference. A special purified HPTLC silica gel. Aluminium sheet is available for doing such experiments which is designed to minimum background interference with accessible spectra range of 80-3500 cm^{-1}. There is a ten-fold increase in signal/noise ratio as compared with conventional HPTLC plates. The surface enhanced Raman spectroscopy is able to detect analytes at nanogram level and has been useful for identification of dyes[19] and DNA bases.[20-22]

References

1. Somsen GW, Modern W, Wilson ID. *J Chromatogr A* **703**; 613-665, 1995
2. Wilson ID, Morden W. *LC GC International* **12(2)**; 72-80, 1999
3. Busch KL. *Handbook of Thin-Layer Chromatography*, Sherma J, Fried B (eds). Marcel Dekker, New York, USA, 183-209, 1991
4. Jaenchen DE, Traitler H, Studer A, Kaiser RE (eds). *Instrumental Thin-Layer Chromatography*, Institute for Chromatography, Bad Durkheim, Germany, 185-192, 1987
5. Burger K. *Instrumental Thin-Layer Chromatography*, Kaiser RE (ed). Institute for Chromatography, Bad Durkheim, Germany, 33-34, 1989
6. Morden W. In: Wilson ID, Poole CF, Adland TR, Cooke M (eds). *Encyclopedia of Separation Science*, New York, Academic Press, 2000.
7. Busch KL, Mullis JO, Carlson RE. *J Liq Chromatogr* **16**; 1713, 1993
8. Busch KL. *J Planar Chromatogr Mod TLC* **5**; 72, 1992
9. Stahlmann SA. *J Planar Chromatogr* **12**; 5-12, 1999
10. Percival CJ, Griffiths PR. *Anal Chem* **47**; 154, 1975
11. Fuller MP, Griffiths PR. *Anal Chem* **50**; 1906, 1978
12. Issaq HJ. *J Liq Chromatogr* **6**; 1213, 1983
13. Bush SG, Breaux A. *J Mokrochim Acta* **1**; 17, 1988
14. Tajima T, Wada K, Ichimura K. *Vibr Spectrosc* **3**; 211, 1992
15. Herman JA, Safer KH. *Planar Chromatography in the Life Sciences*, JC Touchstone (ed), J. Wiley, Chichester, UK, 157-166, 1990
16. Herman JA, Shafer KH. *Planar Chromatography in the Life Sciences*, JC Touchstone (ed), J.Wiley, Chichester, UK, 157-166, 1990
17. Fuller MP, Griffiths PR. *Appl Spectrosc* **34**; 533, 1980
18. Bode U, Heise HM. *Mikrochim Acta* **1**; 143, 1988
19. Rau A. *J Raman Spectrosc* **24**; 251, 1993
20. Koglin E. *J Planar Chromatogr* **2**; 194-197, 1989
21. Koglin E. *J Planar Chromatogr* **3**; 117-120, 1990
22. Koglin E. *J Planar Chromatogr* **6**; 88-92, 1993

13

Good Practices Compliance Operations in HPTLC

There has been continual need for reliable methods in determining the compliance with national as well as international requirements in the areas of food analysis. It is vital to determine compliance with food safety standards in conformity with the provision of agreement on the application of sanitary and phytosanitary measures (SPS agreement) and thus aiding in resolution of food safety disputes with the World Trade Organization.

The rules of Good Manufacturing Practices (GMP) for instrument and accessories and Good Laboratory Practices (GLP) are by and large, equivalent, but have to be implemented in their own areas. GLP Hand Book for Laboratory Technicians[1] describes in detail the basic rules for following such practices. Nowadays analytical chemists have tremendous advantages over a few decades ago because the methods have been automated thus reducing the chances of manual error. Enormous documentation are now available on GLP/GMP and other good practices related to sampling, analysis and interpretation of results.

Documenting the good practices requires five Ws; who has done, what, when, and where, with what. How this has been done and why this has been done?

The American Food and Drug administration published first draft on GLP as early as in 1976 which was finalized in 1979. These practices were adopted by Organization for Economic Cooperation and Development (OECD) in 1981 for mutual acceptance of test results for evaluation of chemicals. While following good practices, it has to be emphasized that no analytical results are falsified, as these can lead to fatal consequences.

Food practices are followed by sequences of Standard operating procedures (SOPs). SOP is defined as per ISO/IEC as "a set of instructions having the force of a directive, covering those features of operations that level themselves to a definite or standardized procedure without loss of effectiveness. According to ISO 9000, SOP is a process document that describes in detail the way that an operator should perform a given operation.

For achieving good practices in laboratory, necessary documents such as policy, objectives of the laboratory, quality manual, standard operating procedures, test procedures, working procedures, instructions, orders, and standards to be followed are used. SOPs have been described by a Pharmaceutical-Technical Committee of the German Association of Pharmaceuticals Manufacturers (BAH)[2] as "Standard Operating Procedures are documents that give information and instructions for the performance of operations that do not necessarily refer to one particular product or one particular task, e.g. the cleaning of equipment and plant, clothing, or climate control. They are essentially general instructions of an organization, administrative or of technical nature and not specific instructions referring to specific products manufacturing, methods, cleaning instructions, etc. The product-specific requirements and instructions are rather to be found in documents such as manufacturing methods for the particular products, operating instructions for particular machines and plant, cleaning and disinfection instructions for particular equipment and plant, etc. Only number of written documents on good practices cannot produce the best results, because poor training of the laboratory personnel or non-availability of technical competent personnel can lead to abnormal results. W. Gunther remarked[3] "GLP gives the laboratory legal but not technical security." The written documents are important for training. It is desirable for each laboratory to formulate their own SOPs for the test methods depending upon the food commodities also. The following points should be kept in mind for observance of the good practices.

13.1 GOOD PRACTICES

Good practices to be followed are:

13.1.1 Traceability

Put an id. no. of each plate. Readymade, pre-coded plates are available. Coded HPTLC plates are useful for simplification of the quality assurance, documentation and archiving of test results. Manual writing on the plates should be avoided. The name of plate manufacturer, item no., serial number of the plate, batch number should be visible thus avoiding any possibility of mix-up. These can be digitally documented along with the results.[4]

13.1.2 Use of Official and Validated Methods in Routine Work

Only official/validated method should be used since there are enormous numbers of food commodities and they are very complex in nature. Therefore their extraction procedure, method of sampling, etc. also vary depending upon the nature of the commodities. Whereas the official methods are preferred in general, but it would be desirable in case of official method also to standardize according to the nature of the food commodities and then methods are laid down for a particular food commodity and for testing of quality factor or adulterant.

13.1.3 Qualified and Calibrated HPTLC Equipment and their Accessories, Calibrated Glass Apparatus, etc.

HPTLC should be periodically qualified and be demonstrated for its suitability for their intended uses. Calibration is a part of qualification. Similarly the accessories and glass apparatus used in measurement should be calibrated to National/International traceability.

13.1.4 Certified Reference Material

Reference materials being used for quantitative determination must be certified with National/International traceability by competent and authorized institution.

13.1.5 Test Procedures and Documentation

Test procedure enables the analytical work to be done in a reproducible manner and should be written in detail with each step taking into consideration the possibility of the error. Analysis should be conducted exactly as per prescribed and standardized procedures. Necessary parameters should be documented.

The documents of the good practices should be open to improvement depending upon the observations in the analytical reports and difficulties being faced during analysis with the objective to achieve best reproducible results. However, the change in documents should be made by authorized person. Due to advancement in use of laboratory management software and electronic data processing, this is becoming more important that the software being used in laboratories are validated. Access to computers should be only for authorized personnel.

13.1.6 Validation of Methods

There is continual need for reliable and accurate analytical methods for use in laboratory in assuring the compliance with National/International regulations in the area of food quality and safety. The reliability of the method is determined by an established system for validation procedure.

Validation can be defined as a formal procedure evaluation of the suitability of a given analytical method for its intended use and also by determining reliability of the result.[5] General principle of the validation as given in the guidance document and validation to test method by National Accreditation Board and Collaborative Laboratories in India are as follows:

The objective is to determine repeatability and reproducibility of the test method by adequate lab data regarding suitability of the method for intended analytical activities. The standard official test methods or the methods given by the regulatory bodies should be considered as validated for their intended use in the testing laboratory. But it should be verified that the methods are well tried for their repeatability, reproducibility as per their requirement in the light of regulatory standards for a particular food commodity. In case of modified test method or the in-house developed tests, the level of repeatability and reproducibility must be ensured with adequate documentary evidences to meet the intended requirement and to specify the customer needs for the suitability of the method along with the demonstration of technical capability of the test method in context of the specified range. Following factors should be taken into account as they affect the results.

1. Nature of sample
2. Homogeneity
3. Test method
4. Equipments
5. Personal and their technical competency
6. Test environment

Only validity of the analytical method is not enough. The validation process should consist of the following steps.

1. Instruments/equipments and accessories validation through their qualification.
2. Software validation
3. Method validation
4. System suitability

13.2 USE OF QUALIFIED/CALIBRATED EQUIPMENTS, ACCESSORIES AND OTHER GLASS APPARATUS, ETC.

For obtaining a good and reproducible result, it is necessary that all the equipments and glass apparatus for measurement are properly calibrated with their certified counter parts. For this purpose, the certification is a must and becoming more important in the field of analysis. It is considered most vital that the user should have an interaction with the manufacturer of the instrument and should be specific about his exact requirement of the analysis and the level to which detection/quantification have to be made. The instrument/software has to be decided, developed and produced in a validated environment according to the requirement of GMP and ISO 9000. The instrument should meet the requirements of "design qualification" (DQ), "installation qualification" (IQ), "operation qualification" (OQ), "performance qualification" (PQ) and "maintenance qualification (MQ)". Design qualification is done in the factory where instrument is manufactured, whereas IQ and OQ are done before use at the user side and after that calibration maintenance system suitability is done. PQ is done by the customer as it is dependent on sample. Each sample will have its own PQ while IQ and OQ are dependent on the instrument. When using the instruments, a user record is maintained.

Each laboratory should have an SOP for every instrument and the procedure for placing the instruments. A record should also be maintained to keep the instruments mentioning the spare parts inventories. A guideline for maintenance of the instrument is given below.

(a) The identity of the item of equipment and its software.
(b) The manufacturer's name, type identification, and serial number or other unique identification.
(c) Check that equipment complies with the specification during installation.
(d) The current location, where appropriate.
(e) The manufacturer's instructions, if available or reference to their location.
(f) Dates, results and copies of reports and certificates of all calibrations, adjustments, acceptance criteria, and the due date of next calibration.
(g) The maintenance plan where appropriate, and maintenance carried out to date.
(h) Any damage, malfunction, modification or repair to the equipment.
    The following documents should be examined with different qualifications.

A. Design Qualification(DQ)

1. Technical requirement of the users in detail and its suitability for the particular model and design of the equipments along with required accessories.
2. Detail inspection of the equipment.
3. Technical evaluation of the equipments in context of the specification as claimed by the manufacturer.
4. Other requirements to be complied by the manufacturer.
    This is entirely managed by the manufacturer before they start production.

B. Installation Qualification(IQ)

1. Verification of the consignment for completeness along with the documents for site requirement.

2. Installation by the supplier along with the documents how the instrument is installed, who performed the installation and other miscellaneous details.
3. Acceptance of the installation by the customer.

C. Operational Qualification(OQ)

1. To ensure that the instruments are operative according to the defined and claimed specification for accuracy, precision, etc. by conducting the specified tests.
2. Calibration of the equipment.
3. Validation of the hardware and software.
4. Exhaustive training to the end users along with handing over of the documents related to installation.

   IQ and OQ must be done at site during installation. IQ must be repeated if the instrument is shifted. OQ can be repeated at a fixed interval, e.g. 6 months or one year and also after every repair, change of parts, etc. IQ and OQ are the responsibility of the expert who is authorized by the manufacturer to IQ and OQ.

D. Performance Qualification (PQ)

1. Testing of the instruments under actual running condition for the desired working range.
2. Checking the reproducibility of the results during continuous performance of the instruments.

   PQ is the responsibility of the buyer.

E. Maintenance Qualification (MQ)

1. To maintain a continuous record regarding maintenance of the instruments.
2. To ensure the calibration at prefixed interval of time.
3. To carry out preventive maintenance steps as required for proper maintenance of the instruments.
4. To lay down SOP for maintaining of operation.

13.3 MAINTENANCE OF COMPREHENSIVE DOCUMENTS

The following documents should be maintained.

1. Documents for DQ, IQ, OQ and MQ.
2. SOP
3. SOP for cleaning and routine maintenance of theinstruments by customer.
4. SOP for annual maintenance of the instruments by the supplier.
5. User manual and other documents supplied by the manufacturer regarding use of instruments.
6. Log book.
7. Operating instructions.
8. Relevant services like maintenance, upgrade, breakdown for software, hardware, validation, etc.

References

1. Christ GA, Harston SJ, Hembeck HW. *GLP-Handbuch for Praktiker, GIT Darmstadt,* ISBN 3-928865-03-X, 1992

2. Bundesfachverband der Arzneimittel-Hersteller e.V. (Hrsg): "Standardverfahrensanweisungen (SOPs) der fiktiven Firma "Muster" fur die Arzneimittel-Herstellung einschl. Verwandter Produkte, Teil I", Übierstr. 71-73, 53173 Bonn (no ISBN Number)

3. Gunther W. GLP = Gute Laborpraxix = Richtige Analytik, *CLB Chemie in Labor and Biotechnik* **43**(10); 536-540, 1992

4. Wieland G, Lasercodierte DC. HPTLC Fertigplatten erobern die GLP-Laboratorien, circular from the LPRO/CHROM 1 department of Merck KGaA (Darmstadt) 26. 3. 06

5. Renger B, Jehle H, Fischer M, Funk W. Validierung von Analysenverfahren in der pharmazeutischen Analytik, Beispiel: Gehaltsbestimmung von Thoephyllin in einer Brausetablette mittels HPTLC. *Pharm Ind* **56**(11); 993-1000, 1994

14

Method Development and Validation of Analytical Method

etailed study of existing literature is used for development of new analysis methods. There can be many reasons for development of new methods of analysis. These may be due to following reasons.

- Lack of suitable existing method for a particular analyte in specific sample matrix.
- Existing methods may not be suitable, convenient, economical and fast.
- Existing methods may not be able to provide required sensitivity or analyte selectivity in a sample of interest.
- HPTLC techniques may prove better, suitable, cheaper, convenient and easy to determine, and meet legal and scientific requirements.

It is further necessary to translate the goals into a method development design. Then optimization process may remove variables. While optimizing conditions, the first stage of development should be improvement in terms of resolution, peak shape, time, detection limits, quantification limits and above all, quantification of specific analyte(s) of interest. Results obtained should be evaluated against the goal set for method development. The following general criteria should be kept in mind.

- There should be adequate and clear chromatographic resolution.
- The limits of detection should, by and large, be lower by at least one order of magnitude than needed.
- Calibration curve should be linear as much as possible.
- Sample preparation and sample throughput is optimized.
- Interference is identified and minimized.
- Data is stored and properly recorded.
- Reproducibility of analytical figures should be exhibited within acceptable accuracy and precision.
- Cost of analysis should be acceptable.

The following steps should be taken for method development, optimization and validation.

- Analyte standard characterization.
- Detailed method of analysis.
- Study of literatures and existing methodology available.
- Selection of technique.

– Setting up the instruments and conducting the initial studies.
– Optimization of the method.
– To establish analytical figures of merit with standards.
– To evaluate the new method with real samples and to derive figures of merit.
– To validate figures of merit.
– To standardize percentage recovery of real samples and to establish quantitative analysis parameters.
– To validate the method.
– To document protocols and procedures in detail stepwise.
– To conduct inter-laboratory collaborative studies for establishing the reliability of the method developed within analytical protocol.
– Comparison of inter-laboratory collaborative studies for improvement in method if required and to make summary report.
– Final validated method, validation procedures, results and publication of the method.

14.1 METHOD VALIDATION

Method validation is necessarily to be performed after method development for its legal acceptance. This is done to ensure that an analytical procedure is accurate, specific, reproducible and rugged over a specific range suitable to analyte for determination. This provides an assurance that method will produce reliable results what it is intended for. For example, the Codex Alimentarious Commission (CAC) requires method performances information should be available in order to include the method of analysis in a codex commodity. This basically incorporates precision (repeatability, reproducibility), accuracy, specificity, limit of detection, sensitivity, limit of quantification, applicability and practicability as may be appropriate.

The ideal way to validate a method is to ensure that method has done well through a collaborative study in accordance with established internationally harmonized protocols for design, conduct and interpretation of method performance studies. This normally asks for a study design involving a minimum 5 test samples, participation of 8 laboratories with valid data and should invariably include blind replicates or split levels to assess within laboratory the repeatability parameters.

Procedural manual of the CAC[1] has described guidelines for acceptance procedures for Codex standards and inclusions of specific provisions in the Codex standard and related texts. Study regarding determinative method to be conducted according to internationally harmonized ISO/IUPAC/AOAC protocol[2] could require a minimum of up to 5 samples including blind replicates or split level samples with 8 participating laboratories.[3]

The CAC has established the required information for all methods.[4] The Codex Committee on Methods of Analysis and Sampling (CCMAS)[5] has in principle accepted alternative approach for assessment of analytical methods, which is criteria-based approach, i.e. defined set of selection criteria, to which methods must comply, is established without endorsing particular methods in specific for adoption.

Further, procedures prescribed amongst various codex committees also differ for identifying methods. Codex Committee on Pesticide Residues (CCPR) used the following criteria for selection of methods recommended for analysis in CAC.

– Publication of the methods.
– To make available results of collaborative studies or validation in a number of laboratories.

- Capability of determining more than one residue, i.e. multi-residue analysis.
- Suitability for testing at or below MRL levels for many commodities.
- Applicability in laboratories for routine analysis.

Multi-laboratory model adopted for analysis is preferred option for validation of method of analysis. If it is not possible to conduct full collaborative study, a three-laboratory model or equivalent may be applied, which can include a 2 laboratories "peer review" system or also internal validation within a single laboratory. However, the rationale behind such model should be justified which will permit users to evaluate recommended method accordingly.

Normally, method of analysis for validation should be subjected to collaborative studies in conformity with internationally accepted guidelines. If it is not possible to carry out collaborative study according to internationally accepted guidelines, then the three-laboratory model or equivalent should be applied. However, the other validation protocol should be followed. If one or two laboratory approach is adopted, then rationale for selecting this validity model should be described.

14.2 VALIDITY CHARACTERISTICS

Validity characteristics comprise of the following analytical performance parameters.

1. **Specificity:** It is the ability to measure accurately and specifically the analyte of interest in presence of other components accepted in the sample matrix. The details related to specificity should relate at least to such substances which are expected to give rise to interfering signal when measuring principle is used. For example, in residual analysis, interfering substances may give a response similar to the residue under measurement. Random interference must be performed by the analysis of a set of representative blank samples. For identification, specificity is normally demonstrated by its ability to discriminate between components of closely related structure or by comparison with a non-reference material. Whereas for assay and impurities test, specificity is exhibited by the resolution of two closest eluting components. These components are normally the major or active components and an impurity. Impurity must be demonstrated to ensure that the assay is not affected by the presence of spiked material.

2. **Accuracy and recovery:** Accuracy is the measurement of closeness of agreement between the true value of the analyte concentration and the mean result that is obtained during procedure a large number of times to a set of homogenous samples. It is very closely related to systematic error and recovery of the analyte. Normally accuracy at or below the MRL or required level must be equal to or greater than the accuracy above the MRL or level of interest. The percentage recovery of an analyte which is added to a blank test sample is a related measurement that compares the amount found by the analysis to the amount added to the sample. For interpretation of recovery, it is important to recognize that added analyte to a sample may not behave in the same manner as the same biologically active analyte. In case where the methods involve a number of steps including extraction, isolation, purification and concentration, the recovery may be low at lower concentration. However, this is regardless of what average recoveries are observed. 100% recovery with low variability is always desirable.

3. **Precision:** Precision of method relates to closeness of agreement between independent test results derived from homogenous test materials analyzed within stipulated conditions. Repeatability and reproducibility can be best estimated when the validation is carried out in a collaborative manner. Wherever it is not possible to carry out collaborative studies, the

laboratories should obtain suitability of methods repeatability and within laboratory reproducibility from data obtained in the laboratory. The precision of a method can be estimated within laboratory using measurement reliability, procedures, and estimation of measurement of uncertainty.[6]

4. **Limit of detection(LOD):** It is defined as a lowest concentration of an analyte present in the sample which can be detected, though not necessarily quantified. It is a limit test which specifies as to whether an analyte is above or below a certain value. It is also expressed as a concentration at a specified signal to noise ratio (usually 2- or 3- to 1 ratio). LOD can also be calculated on the basis of standard deviation (SD) of the response and the slopes(s) of the calibrated curve at the levels approaching LOD according to formula LOD = 3. 3(SD/s). The standard deviation of the response can be determined on the basis of standard deviation of the blank on the residual standard deviation of the regression line or the standard deviation of y intercepts of the regression lines. The method used to determine LOD must be documented and supported by analyzing appropriate number of samples at the limit to validate it.

5. **Limit of quantitation (LOQ):** It is the smallest quantity above which the determination of analyte is possible with specified degree of accuracy and repeatability (within laboratory reproducibility). LOQ is expressed as concentration with the precision and accuracy of the measurement. Sometimes signal to noise ratio of 10-1 is used for determining LOQ. This is a good rule of thumb. But it must be noted that determination of LOQ is a compromise between concentration and the required precision and accuracy. As LOQ concentration level decreases, the precision also decreases. However, if greater precision is necessary, then higher concentration must be reported for LOQ, e.g. signal to noise ratio 20 to 1.

6. **Sensitivity:** Sensitivity is the change in analytical response divided by the corresponding change in the concentration of the standard (calibration curve), i.e. the slope of analytical calibration. A method can be considered sensitive if a small change in concentration of the analyte causes a large change in the analytical measurement. The analytical response may vary with the magnitude of analyte concentration. However, it is normally constant over a reasonable range of concentration.[7] In ideal situation, calibration should be a straight line expressing a direct linear relationship between analytical response and concentration of standard.

7. **Linearity and range:** It is the ability of the method to demonstrate quantitative results which are directly proportionate to analyte concentration within an acceptable range. It is normally reported as variance of the slope of the regression line. Range is the upper and lower level of analyte that have been demonstrated to be determined with precision, accuracy and linearity under a specified method.

8. **Ruggedness:** Ruggedness indicates reproducibility of result obtained under different variety of conditions expressed as percentage relative standard deviation (RSD). These conditions may include difference in the laboratories analysts, instruments, reagents and experimental periods.

9. **Robustness:** It is the capacity of the method to remain unaffected by small deliberate variations in the method parameters. The robustness may be evaluated by change in the method parameters such as organic solvent, pH, ionic strength or temperature and then determining the effects on the results on the method. Robustness should be considered in the beginning at the time of development of method. In case the results are affected by variations in method parameters, then these parameters should be well controlled and precautionary statements included in the method documentation.

10. System suitability: System suitability itself is an integral part of chromatographic methods and should be verified in respect of the adequacy of the resolution and reproducibility of the system for the analysis to be performed. System suitability is based on the concepts that equipments, electronics, analytical operations and the sample constitute an integral part that can be evaluated as a whole. System suitability is a checking of the system to ensure system performance before or during the analysis of the unknowns.

14.3 METHOD VALIDATION IN HPTLC

In case of HPTLC, method validation can be of two kinds:

1. **Qualitative methods:** These methods are focused on R_f values, sequences and colours of zones. It is preferred to do the densitometric evaluation of the chromatogram based on image of the plate. For the purpose of reference, a sample on same or a different plate should be used. An electronic image can also serve the purpose as a reference.

2. **Quantitative methods:** In this method, the areas/heights of the analytes are determined against the calibrated curves obtained from calibrated standards on the same plate for determining the amount of separated substances.

International Conference on Harmonization (ICH) provides the framework for validation. Table 14.1 gives a brief summary of different validation parameters for different type of methods.

However, method validation as adopted by Association of Analytical Communities (AOAC) International describes several levels of validation which ultimately result in declaring official method. In such method, collaborative trials include 8-10 independent laboratories. Table 14.2 briefs the parameters considered to be most useful for different types of method. However, one can always decide about the parameters for the purpose of validation.

Table 14.1: Parameters for validation as per International Conference on Harmonization

S. no.	Analytical procedure characteristics	Identification	Testing for impurities/ active ingredients		Assay * Dissolution (only measurement) *Content/potency
			Quantitative	Limit	
1.	Accuracy	–	+	–	+
2.	Precision	–	+	–	+
3.	Repeatability	–	+	–.	+
4.	Intermediate precision	–	+*	–	+*
5.	Specificity†	+	+	+	–
6.	Detection limit	–	–‡	+	–
7.	Quantitation limit	–	+	–	+
8.	Linearity	–	+	–	+
9.	Range	–	+	–	+

- This characteristic is not normally evaluated.
+ This characteristic is normally evaluated.
* In cases where reproducibility has been performed, intermediate precision is not required.
† Lack of specificity of one analytical procedure should be compensated by other supporting analytical procedure(s).
‡ May be necessary in some cases.

Reference-International Conference on Harmonization, Note for Guidance on Validation of Analytical Methods. Text and Methodology (CPMP/ICH/381/95; ICH Topic Q2 [R1]).

S. no.	Analytical Procedure	Qualitative methods *	Quantitative methods †
	Table 14.2: Suggested validation parameters for HPTLC analysis		
1.	Specificity	+	+
2.	Precision(on the plate)	+	+
3.	Repeatability	+	+
4.	Intermediate precision	+	+
5.	Reproducibility ‡	+	+
6.	Robustness §	+	+
7.	Accuracy	–	+
8.	Detection limit	–	+¶
9.	Quantitation limit	–	+
10.	Linearity	–	+
11.	Range	–	+

  *   Identification, process control, stability tests, batch-to-batch consistency, mix-up.
  †   Assay of marker substances
  ‡   Only for multi-laboratory validated methods
  §   Optional, but highly recommended
  ¶   Detection of impurities

14.3.1 Validation of Qualitative HPTLC Methods

Qualitative methods are basically more useful for detecting the adulterants in food. It is considered that validation of specificity, precision and robustness should be done. Stability test can be done in the shelf life study of a product. The methods should be able to detect the degradation in the fingerprint and changes in the quantity of marker compounds. Following parameters can be used for validation.

1. **Specificity:** This gives an idea whether samples of same identity will give similar results which can be differentiated from those obtained by sample of different identity. For the purpose of reference, fingerprints are authenticated with reference material and chemical reference substances can also be used. Fingerprints of the test sample may be compared in respect of number, position, colour and intensity zones visually. Specificity should confirm that excipients do not interfere in the analysis.

2. **Precision:** In qualitative analysis, precision refers to the R_f value of the separated compounds. The sequence of fingerprint, of background colour, and of separated zones should be compared for validation. It is adequate that zones presenting same compounds form horizontal lines parallel to each other. The precision on a plate can be obtained as a standard deviation from measured R_f value of different zones of fingerprints seen in the parallel analysis of replicates of the same reference material samples. R_f value should not differ more than 0.01. In case of qualitative analysis, repeatability should be established visually by observing the individual plates. On the basis of R_f values, precision of the mean of each compound per plate can be calculated and acceptable criteria should be that R_f of same substance should not vary more than 0.02 from plate to plate. However, the

intermediate precision should be accepted if R_f values of the same substance should not vary more than 0.05 between plates from different experiments on different dates. Several factors affect the intermediate precision even if the intention to keep everything is constant. For example, temperature and relative humidity may vary. Different plates may show different behaviour. However, the intermediate precision gives good idea of long-term stability of a method. This is especially important if method is intended for routine analysis. Though reproducibility is not a validation parameter as described in ICH guidelines, but it is considered most important for establishing official methods and for legal requirements. It gives an idea of expected range of results when a method is performed in different laboratories. Reproducibility can be determined in collaborative studies duly coordinated by a technically competent personal and all the participating laboratories should be supplied with methods which have already been validated in the primary laboratory. It is expected that participating laboratories should use aliquot portion of the same samples. Each laboratory should have a different checklist with a provision to record the work. Deviation from original method and other observations, problems or suggestions for clarifications is seen across while performing the test. Thereafter the results obtained from all the laboratories are collected, examined and reproducibility is determined. Reproducibility can be accepted if R_f value of the same compound does not deviate more than 0.05 as compared to those obtained in the primary laboratory. However, while deriving such values this must be seen that the plates belong to the same manufacturer in all the different laboratories so that there is no variation in the quality of the plate.

14.3.2 Validation of Quantitative HPTLC Method

1. **Specificity:** This is a very important factor for quantitative determination and the baseline resolution (R_s above 1.25) of the targeted compound must be validated. Specificity can be enhanced by densitometric evaluation at a wavelength unique for the absorption of the compound. Absorption spectrum is very useful too. Derivatization can also be done for increasing specificity.

2. **Linearity, range of working, LOD, LOQ:** The linear working range can be determined by diluting sample and varying application volume. For assay, the target value should be near to middle of the working range by diluting the sample accordingly. It is necessary that working range should cover 75-125% of the target value. There are several methods for validating linearity. It is common to apply 3 or 4 concentration levels of standards for calibration.

3. **Precision:** Precision in a quantitative method affects results of determination. This normally covers two aspects. The first is variation on account of TLC process and the second is variation due to handling of the samples. TLC related precision can be obtained from several application of the sample across the entire plate. These can be calculated from calibrated or non-calibrated data. In the other case, sample is applied and the calibration data may require at least one standard on the same plate. The calibrated data can give repeatability.

4. **Accuracy and recovery:** Accuracy can be correlated with the recovery at the upper, lower and in the middle of the working range. A mixture of excipients can be spiked with a

known amount of analyte. Then the mixture is put to analysis including the sample preparation. Recovery can be calculated as quotient of measured amount and added amount of analyte. The other approach is with the sample of low and known analyte content. The recovered amount of the spiked sample should be corrected by the original content. Recovery should be consistent and high for acceptance.

14.4 METHOD VALIDATION PROTOCOL

Before outlining an experimental design on protocol for method validation, it is necessary to view basic assumptions as enumerated below. These may vary from lab to lab or method to method.

1. Selectivity has been demonstrated, measured and documented during the course of validation.
2. The method has been developed under optimized conditions and the robustness must be the basic parameter to be investigated.
3. As soon as the practical data is obtained, statistically valid approach should be used to confirm the values on which to take the decisions.

After the method has been validated, it should be transferred to other users worldwide for use. There should a constant communication between users so that the required goals of method development, optimization and validation have been achieved in an efficient manner. Once, it has been agreed upon, the method can be applied by one and all. Documentation of the method must include a detailed routine procedure, the method validity, report system suitability, and plan for method implementation. The end users must plan verification of the method performance before documenting SOP. At some point of time, it may be necessary to revalidate a method which can be done in a reactive or proactive way. Reactive validation should be done in response to changes in nature of the commodities or other changes such as dilution, sample preparation in the method. But, proactive revalidation can be done by taking advantage of new technology or to automate the procedure.

As validation is a constant evolving procedure, a well defined and documented process brings in transparency with evidence that system and method is suitable for its intended use. By doing method development, optimization and validation in a logical manner, the laboratory resources can be used in more efficient way.

A typical example of protocol for validation of qualitative method by HPTLC is given by E Reich et al.[8]

14.5 PROCEDURE FOR VALIDATION OF METHOD

Validation of method in food should normally be done with the following protocol.

Protocol

Validation of method number (identification number to be given) for identification/qualitative analysis (commodity name to be mentioned) by HPTLC.

1. Purpose for developing the method

2. Acceptance criteria
3. Details of the personnel involved in the study
   (a) Program in-charge
   (b) Reviewer
   (c) Analyst of the main laboratory
   (d) Analyst of the substantiating laboratory
   (e) Other personnel of additional laboratories
4. Details of the method
   (a) Sample preparation
   (b) Reference material preparation
   (c) Derivatization preparation
   (d) Stationary phase detail
   (e) Method of sample application
   (f) Humidity and temperature record
   (g) Types of chromatography
       - Chamber type and other configuration
       - Developing solvent
       - Development distance
       - Drying
   (h) Derivatization
   (i) Documentation
   (j) Images of chromatogram
   (k) Evaluation of results and system suitability
5. Validation
   (a) Materials
       (i) Chemicals and solvent, e.g. name of the manufacturer, quality, quantity and traceability
       (ii) Reference materials with traceability certification
       (iii) Samples with source, batch no. authenticity certificate no.
       (iv) Details of the instruments
       (v) Details of the plate material
   (b) Stability
       (i) Stability of analyte in solution and on plate*
       (ii) Stability of the analyte during chromatography*
       (iii) Stability of derivatization product*
   (c) Specificity
       (i) Image comparison with reference material*
       (ii) Detection of contaminants/adulteration/ingredients*
       (iii) Analysis of finished products along with interference of matrix*
   (d) Precision and repeatability*
   (e) Intermediate precision*

(f) Reproducibility
(g) Robustness

  (i)   Type of chamber*
  (ii)   Details of the chamber saturation*
  (iii) Developing distance*
  (iv)   Waiting times*
  (v)   Relative humidity*
  *This should be done for each section. Description of the experiment and acceptance criteria

(h) Results with photodocumentation
(i) Whether method is acceptable or not?

6. Conclusions, approval with signatures

(a) Primary laboratory conclusion along with date and signature
(b) Substantiating laboratories conclusion along with date and signature
(c) Comments of the reviewer with conclusion along with date and signature
(d) Final approval by the competent authority

14.6 EXAMPLE OF VALIDATION

An example of validation of quantitative analysis of caffeine in Indian black tea samples[9, 10], is given below.

HPTLC method has been established for determination of caffeine in black tea samples. The analysis was performed on silica gel, with toluene–acetone (4:6), as mobile phase. Caffeine was quantified densitometrically at λ = 276 nm. Recovery of caffeine was 100%, 98.54%, 98.99%, for pure and spiked samples respectively. Method precision and specificity were validated. The method is simple, reproducible, precise and accurate.

14.6.1 Introduction

Caffeine

Caffeine: 3, 7-dihydro-1, 3, 7-trimethyl-1H-purine-2, 6-dione; 1, 3, 7-trimethylxanthine; 1, 3, 7-trimethyl-2, 6-dioxopurine; caffeine; thein; guaranine; methyltheobromine; No-Doz.
$C_8 H_{10} N_4 O_2$; mol wt 194.19. C 49.48%, H 5.19%, N 28.85%, O 16.48%.
It is present in tea, coffee, mate leaves; also in guarana paste and cola nuts.

14.6.2 Materials, Reagents and Standard Solutions

Analytically pure caffeine from Loba cheme, Mumbai, India was used. All chemicals including methanol, toluene and acetone were of analytical reagent grade (Merck, Mumbai, India) and were used without further purification. Tea samples were obtained commercially.

A stock solution containing 1 mg/ml caffeine was prepared in methanol. Calibration solutions (0.01 mg/ml) were prepared by diluting the stock solution.

14.6.3 Sample Preparation

The commercially obtained black tea samples were pulverized and 75 mg of finely powdered tea samples after passing through Mesh No. 300 was transferred to 50 ml volumetric flask, extracted with methanol for 20 min by sonication. Dilute to the volume with the same solvent. A sample solution of (3 µl, equivalent 3000 ng) was applied for assay of caffeine.

14.6.4 Chromatography

Chromatography was performed on 20 × 10 cm aluminum backed silica gel 60 F_{254} HPTLC plates (Merck, Darmstadt, Germany). Before use, the plates were dried in an oven at 50° C for 5 min. Samples and standards were applied as 8 mm bands by spraying at a rate of 150 nl/s by means of Camag Linomat 5 sample applicator equipped with a 100 µl syringe. The distance between bands was 11.3 mm; distance from the side was 12 mm.

The developing solvent was allowed to ascend to 80 mm, with toluene- acetone (4: 6), as a mobile phase in a Camag Automatic Developing Chamber-2, saturated for 20 min by lining with thick Whatman filter paper. The room temperature was 25 ± 2° C. Humidity was controlled at 47% RH throughout with saturated potassium thiocyanate.

The average development time was 13 min. Densitometric scanning at λ = 276 nm was then performed with a Camag TLC scanner-3 equipped with WIN CATS software, version 1.4.2, using a deuterium light source, the slit dimensions were 6.00 × 0.45 mm The UV spectrum was recorded between 190-400 nm to confirm match with standard.

14.6.5 Validation of the Method

(*a*) **Linearity:** Standard solutions equivalent to 40, 80, 120, 160, 200 ng per band caffeine were applied to an HPTLC plate. The plate was developed, dried and scanned as described above. A calibration plot was constructed by plotting peak areas against amount of caffeine. The linearity of the response for caffeine was assessed in the concentration range 40 to 200 ng per band. The slope, intercept, and correlation coefficient were also determined. Over the concentration range studied, the correlation coefficient for the calibration plot was r = 0.99831 and the slope was 15.34 (n = 6). The linear calibration range was, therefore, found to be between 40 to 200 ng per band.

(*b*) **Sensitivity:** The sensitivity of measurement of caffeine by use of the proposed method was estimated in terms of limit of quantification (LOQ) and the lowest concentration detected under the chromatographic conditions as the limit of detection (LOD). The LOD and LOQ were calculated by the use of the equations LOD = 3 × N/B and LOQ = 10 ×

N/B where N is the standard deviation of the peak areas of the caffeine (n = 5), taken as the measure of the noise, and B is the slope of the corresponding calibration plot. The limit of detection (LOD) was 37.3 ng and the limit of quantification (LOQ) was 111.92 ng for this method.

(c) Precision: Precision was measured by using standard solutions containing caffeine at concentrations covering the entire calibration range. The precision of the method in terms of intra-day variation of R_f was determined by analyzing caffeine standard solutions in the range (40-200 ng per band) three times on the same day. Inter-day precision of R_f was assessed by analyzing these solutions (40-200 ng per spot) on three different days over a period of one week. The results of the precision studies are shown in Table 14.3.

Table 14.3: Results from determination of caffeine			
Concentration[ng/band]	**Intr-day precision R_f**		
ΔR_f	P48_070720_01	P48_070720_02	P48_070720_03
40	0.36	0.36	0.36	0.00
80	0.36	0.36	0.36	0.00
120	0.36	0.36	0.36	0.00
160	0.36	0.36	0.36	0.00
200	0.37	0.36	0.36	0.01
Concentration[ng/band]	**Intr-day precision R_f**		
ΔR_f	P48_070718_00	P48_070719_00	P48_070720_01
40	0.37	0.37	0.36	0.01
80	0.37	0.37	0.36	0.01
120	0.37	0.38	0.36	0.02
160	0.37	0.38	0.36	0.02
200	0.37	0.38	0.37	0.01

(d) Accuracy: The accuracy of the method was determined by use of standard additions at three different levels, i.e. multiple-level recovery studies. Sample stock solution of the tea sample at a concentration of 1500 ng/ml was prepared. 50%, 100% and 150% of the standard caffeine solutions were added to the solution, and the recovery (%) was determined. Values were found to be within the acceptable limits given in Table 14.4.

Table 14.4: Results from recovery studies				
Sample	Initial amount [ng]	Amount added [%]	Amount recovered [ng]	Recovery [%]	RSD [%]
Sample A	75000	0	78.64		0.161
Sample B	75000	50	118.91	100	3.534
Sample C	75000	100	156.58	98.54	0.716
Sample D	75000	150	196.99	98.99	2.073

(e) Specificity: The mobile phase designed for the method resolved the sample components very efficiently, as shown in Figure 14.1. The R_f value of caffeine was 0.37. A typical absorption spectrum of caffeine on a plate is shown in Figure 14.2. The wavelength 276 nm was selected for detection because it resulted in better detection sensitivity of caffeine. The band of caffeine from the tea sample was identified by comparing its R_f value and its absorbance/

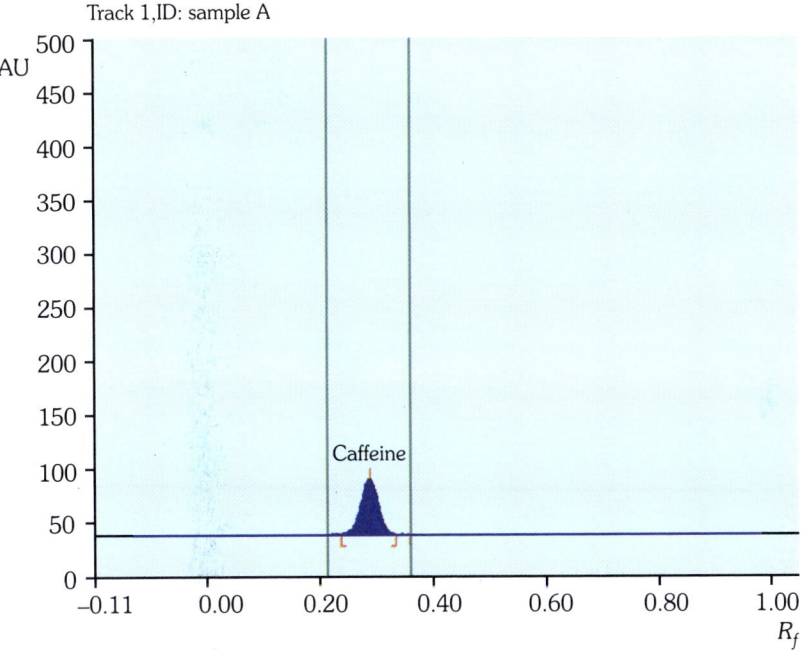

Fig. 14.1 (a): Typical HPTLC chromatogram obtained from a solution of black tea sample

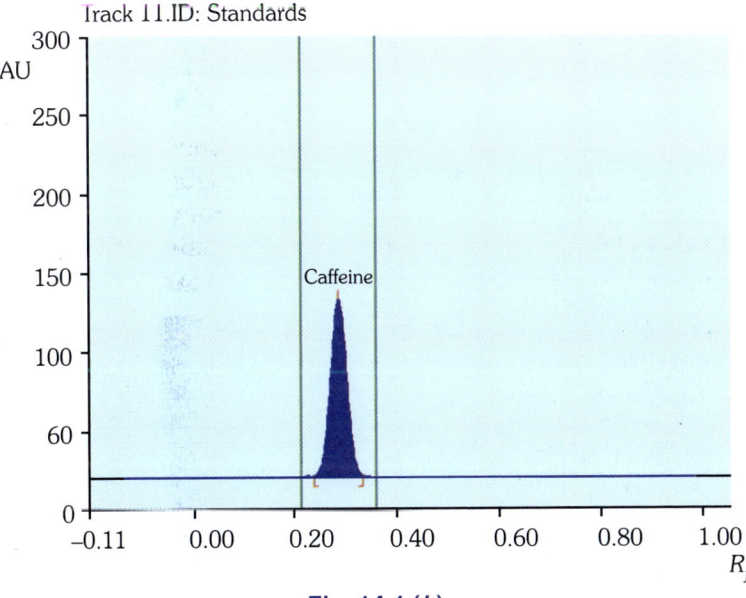

Fig. 14.1 (b)

reflectance spectrum with those of standard caffeine. The peak purity of caffeine was tested by comparison of spectra acquired at the peak-start (S), and both peak flanks. The correlation between these spectra were indicative of the purity of the caffeine peak (correlation r(S, M) = 1.0000, r (M, E) = 1.0000 as shown in Figure 14.3.

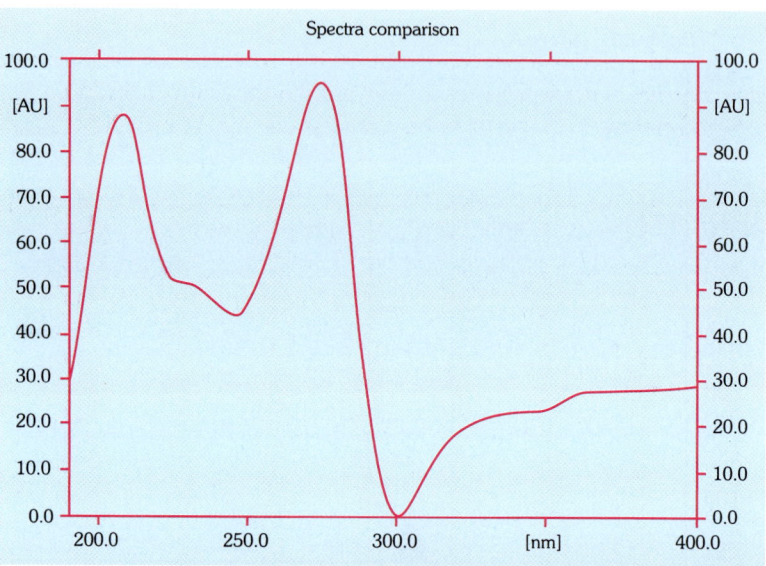

Fig. 14.2: Peak of standard solution of caffeine

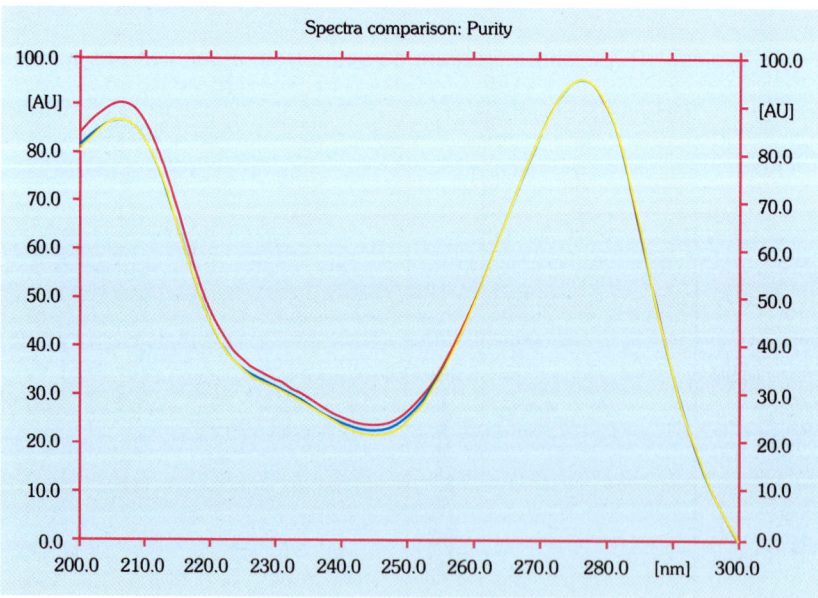

Peak purity spectra of caffeine from tea sample, scanned at the peakapex, and on both peak flanks positions of the band (r > 0.99).

Fig. 14.3: Absorption spectrum of caffeine on plate

(*f*) **System suitability:** According to USP 23, section 621, system-suitability tests are an integral part of a chromatographic analysis and should be used to verify that the resolution and reproducibility of the chromatographic system are adequate for the analysis. To ascertain the effectiveness of the method developed in this study, system-suitability tests were performed on a freshly prepared standard stock solution of caffeine. This method of analysis met the entire requirement.

(*g*) **Ruggedness and robustness:** Ruggedness is a measure of the reproducibility of a test result under normal variation in operating conditions from instrument to instrument and from analyst to analyst. The results of ruggedness testing are reported in the Table 14.4 (Fig. 14.4).

Table: 14.4: Results from ruggedness testing	
	Tea sample
Analyst I	99.17%
Analyst II	102.18%

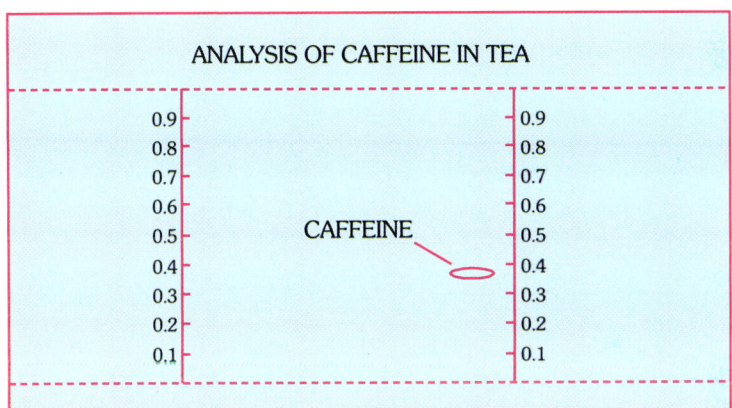

Fig. 14.4: Results from ruggedness testing.

Robustness is a measure of the capacity of a method to remain unaffected by small but deliberate variations in the method conditions, and is an indication of the reliability of the method. The method is acceptable for the parameter as given in Table 14.5.

Table 14.5: Results from robustness testing		
Chamber type	:	Automatic developing chamber -2
Chamber saturation	:	20 min.
Development distance	:	80 mm.
Development time	:	11.10–13.50 mm:ss
Relative humidity	:	44–48%
Temp.	:	26°–30° C.

(*h*) **Repeatability:** The repeatability of the sample application was assessed by applying standard caffeine solution (4 µl) six times on an HPTLC plate, then development of the plate and recording the peak height and area. The %RSD for peak height and peak area values of caffeine was found to be 1.85% and 1.63%, respectively. Repeatability of measurement of peak height and area was determined by spotting 4 μl standard caffeine solution on the HPTLC plate and developing on the plate. The separated band was scanned six times without changing the position of the plate. %RSD for measurement of peak height and peak areas of caffeine was 0.358% and 0.541%, respectively.

(*i*) **Stability studies:** To test the stability of caffeine on the HPTLC plates, analyte was tested against freshly prepared solutions. No decomposition of the drug was observed during chromatogram development. No decrease in the concentration of caffeine on the plate was observed within 3h (Fig. 14.5).

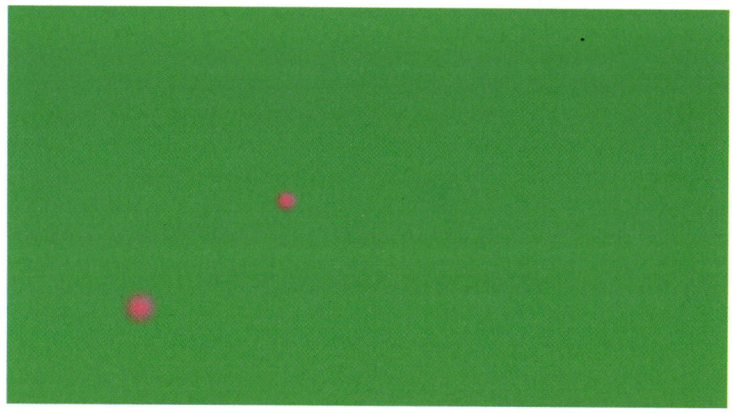

Fig.14.5: Black tea sample, 2D chromatogram

References

1. FAO/WHO. 1997. *Codex Alimentarius Commission Procedural Manual*, 10th ed. 35-41, FAO, Rome
2. FAO/WHO. 1995. *Codex Alimentarius*, Volume 3, *Residues of Veterinary Drugs in Foods*, 2nd ed., Section 4, FAO, Rome
3. Horwitz W. Protocol for the design, conduct and interpretation of method performance studies. *Pure and Applied Chemistry* **67**; 331-343, 1995
4. Thompson M, Wood R. Harmonized guidelines for internal quality control in analytical chemistry laboratories. *Pure and Applied Chemistry* **67**; 649-666, 1995
5. Thompson M, Wood R. 993. International harmonized protocol for proficiency testing of (chemical) analytical laboratories. *Pure and Applied Chemistry* **65**; 2132-2144
6. ISO/IEC. 1995. Guide to the Expression of Uncertainty in Measurement. International Organization for Standardization, Geneva.
7. Freiser H, Nancollas G. *Compendium of Analytical Nomenclature. Definitive Rules.* 2nd ed. Blackwell Scientific Publications, Oxford, UK 1987
8. Reich E, Schibli A. *High-Performance Thin-Layer Chromatography for the Analysis of Medicinal Plants*, 2006
9. Ranganna-*Handbook of Analysis and Quality Control for Fruit and Vegetable Products.* 2nd ed. The Merck index-10th edition

Guidelines for Standard Operating Procedure of Food Analysis by HPTLC

15.1 PURPOSE

To lay down SOP for obtaining acceptable reproducible results in food analysis using HPTLC

15.2 PROCEDURE COVERED

Sampling to the final results

15.3 PROCEDURE

(a) Sampling

Sampling is beyond doubt, the utmost important factor responsible for variations. A typical example of sampling plan in grains for aflatoxin is given below:

Robertson et.al[1] demonstrated the problems in sampling grains for mycotoxins. The variations in peanuts[2], cottonseed[3] and corn[4] in non-processed grains have been shown by different authors. The several difficulties while sampling shelled peanuts for analysis of aflatoxin have been extensively discussed by Whitaker et al.[5–10] Practical problems can be big when lots or consignments are very large especially in case of rail/road wagons, large stacks. Sample drawn in such cases are, by and large, superficial and cannot achieve scientific norms. To arrive at a consensus opinion FAO, UN in the "technical consultation on sampling plans for aflatoxin analysis in peanuts and corn" proposed a sampling plan that could produce consistent results. The size of the sample was enhanced to 5.0 kg in place of 4.0 kg to diminish the coefficient of variation of the results. Besides sampling, care should be taken for sealing, packing with proper traceability, in addition to transportation under ideal conditions. A typical example of sampling plan for analysis of aflatoxin in peanuts and corn are given below.[11]

(b) Collection of Samples

(a) It is desirable and convenient to take samples when the lots are in movement. Aflatoxin content should be determined from representative samples during building or dismantling the stacks, loading or unloading trucks, wagons, etc. When it is not possible to do automatic and mobile sample collection, sample collector can manually pass a cup through the stream at periodical intervals to collect sample through incremental technique.

(b) When large lots are considered, it is preferred to utilize automatic sampling technique.

(c) In case of unprocessed materials, each sample should comprise of minimum 100 incremental samples drawn in a systematic random manner.

Incremental sample: Collect the small portion incremental sample taken to compose a bulk sample.

- Take a portion resulting from the division of bulk sample known as working sample.
- Take a portion that effectively will be analyzed and is known as analytical portion.

The minimum sample weigh taken for in shell peanuts is 7.0 kg and for shelled peanuts and corn is 5.0 kg.

An example of peanuts and corn sampling is given below.[11]

(a) *In-shell peanuts*:

$$NS = 6\sqrt{SL} \qquad \text{(Eq. 15.1)}$$

where:

NS = minimum number of sacks to be sampled

SL = number of sacks of the lot

0.2 kg is taken from each sack as incremental sample to produce bulk of 7 kg. This is a minimum bulk sample. If the numbers of sacks are to be sampled according to the formula not enough, more sacks be sampled.

(b) *Shelled peanuts*:

$$NS = 4\sqrt{SL} \qquad \text{(Eq. 15.2)}$$

When NS is equal to or higher than SL, all sacks must be sampled and some of them, twice.

0.2 kg should be taken from each sack to total of 5.0 kg.

This is a minimum size of sample. If lots have fewer number of sacks available for sampling according to the formula, more sacks should be sampled, if necessary even twice to collect minimum of 5 kg.

Peanuts in bulk

(a) *In-shell peanuts*:

$$NP = 24\sqrt{TL} \qquad \text{(Eq. 15.3)}$$

NP = minimum number of points to be sampled

TL = number of tons of the lot

(b) *Shelled peanuts*:

$$NS = \sqrt{TL} \qquad \text{(Eq. 15.4)}$$

The same recommendations apply as for peanuts in sacks

Shelled corn in sacks

$$NP = 4\sqrt{SL} \qquad\qquad \text{(Eq. 15.5)}$$

The same recommendations apply as for peanuts in sacks

Shelled corn in bulk

$$NP = 4\sqrt{SL} \qquad\qquad \text{(Eq. 15.6)}$$

The same recommendations apply as for peanuts in sacks.

Required quantity of the sample is needed for analysis in laboratory. Quantity depends upon the factor to be identified/determined and the nature of food commodity along with the available lot size from where the sample is drawn, in addition to location, e.g. manufacturer's premises, wholesaler, storage area, retailer, etc. Proper sampling equipments need to be used depending upon the physical status of the commodity, lot size and package in which it is packed. Official methods should be followed for sampling plan for different food commodities.[12]

(c) **Quantity of representative sample for analysis:** Quantity of representative samples may vary from food to food and the parameters to be tested. However, quantity should be ideal to obtain best results. Legal requirements for quality of samples should also be taken into consideration while drawl of samples and quantity of representative samples being sent to laboratory.

(d) **Sample preparation (mixing, grinding, etc.) and extraction, concentration:** The factor in mind to be kept while performing the sampling and sample preparation of mycotoxins contamination of particular products such as grains and nuts is that contamination occurs in pockets of high concentration and may not be randomly distributed. Therefore, total laboratory sample must be included in the sample preparation. The aim should be the maximum size reduction and thoroughness of mixing to achieve effective distribution of contaminated portions. For example, one contaminated peanut (approx. 0. 5 gm) can contain enough aflatoxin to result in significant level when mixed with 10, 000 peanuts say 5 kg or 10 lbs. To obtain more than 1 piece of contaminated nut in each 50 gm portion, the single bad nut must be reduced to 100 pieces and these 100 pieces must be uniformly blended through entire mass. To achieve this degree of size reduction, nut must be ground to pass no. 20 sieve. Although further size reduction may not be needed with flours, liquids or paste, thorough mixing is still needed before removal of analytical sample. Adsorption of aflatoxin on sediments in liquid commodities is possible.

Batch type size reduction equipment like Hobart vertical cutter/mixer (VCM), blender and food cutter reduce particle size and mix in one operation.

With other type of size reduction equipment and when product is in finally ground state, mixing is needed. Free flowing dry materials can be mixed in double cone or twin shell blender.

Grains are most easily ground in disk mill such as Bauer or Romer grinding/sub-sampling mill. Soft materials are best handled with meat chopper. Pastes and powders can be mixed in food cutter or with flat beater in planetary mixer.

Greater homogeneity of nut meals is archived by reducing to paste with disk mill, liquefying the paste with n-heptane and mixing and further grinding slurry with blender. Practical homogeneity of hard, in shell nuts is achieved by size reduction in hammer mill, followed by mixing in planetary mixer or by simultaneous size reduction and mixing in Hobart VCM.

Nut meals can be handled in Hobart-VCM in same manner as in-shell nuts, if mixed with equal weight of grinding aid such as coarse-ground oyster shells.

Bread: When total solids of original loaf are not desired. Cut loaf or ½ loaf of bread in to thick slices, spread slices on paper and let it dry in warm room until sufficiently crisp and brittle to grind well in mill. Grind entire test sample to pass no. 20 sieve, mix well, and keep in air tight container.[13]

Butter: Soften sample in test sample container, by warming in water bath kept at as low temp. as practicable $\leq 39°$ C. Avoid overheating which causes visible separation of curd. Shake frequently during softening process to reincorporate any separated fat, and observe fluidity of sample. Optimum consistency is attained when emulsion is still intact but fluid enough to reveal test sample level almost immediately. Remove from bath and frequently shake vigorously or place test sample container in mechanical shaking machine that simulates hand shaking with arm 23 cm long, set to oscillate at 425 ± 25 times/min. Continue shaking until test sample cools to thick, creamy consistency and test sample level can no longer readily be seen. Promptly weigh test portion for analysis.[14]

When handling large samples, coarse-grind and mix entire sample, then remove approx 1/20th portion and again regrind to finer size for drawing analytical sample.

Draw the sample with same precautions as in case of lot sample wherever practical; divide by riffling or similar random dividing procedure until sub-division is close to desired analytical sample weight. Where such sub-division is not practical, composite no. of small randomly taken portions. With liquids, suspend any particulate matter before drawing analytical sample.[15–19]

(e) **Plate material:** Standard readymade plates should be used for much of the work. In a number of cases, plates are used without pretreatment unless impurities fronts are produced during chromatography on account of contamination of plates. It is recommended to pre wash the plates for reproducibility and quantitative analysis. Pre-washing should be done as follows.

(a) Upper side of the plate is marked with a pencil for direction of development.
(b) Plate is developed with 20 ml methanol per trough in a 20×10 cm twin-trough chamber to the upper edge.
(c) The plate is dried for approx. 15–20 minutes in an oven at 120° C.
(d) The plate is equilibrated in laboratory atmosphere in a suitable container protecting it from dust and fumes.

Plates can be handled on both side edges or on the top edge.

(f) **Sample application:** Spray-on technique is used and sample is applied as bands preferably with automated sample applicator. The speed of application can be set matching the properties of the solvent selected. For eliminating volumes error, use of high volatile solvents should be avoided for sample as well as standards for quantitative analysis.

Following parameters are guidelines for sample application (Table 15.1).

Few instruments do not provide facilities for programming of distance between bands. In such cases, distance between tracks (centre to centre) and band length must be selected for keeping the minimum distance requirement.

Table 15.1: Parameters for sample application	
Parameters	**mm**
Space between bands (mini.)	2
Band length	8
Distance from lower edge plate	8
Distance between left and right edge of plate (mini.)	10

(g) **Preparation and storage of developing solvents:** Developing solvents comprising of more than one components should be made by measuring the required volume (respectively weight) of each solvent separately and then it is transformed to a solvent bottle of appropriate size. Mixture is shaken thoroughly by closing the bottle air tight and shaking it. Micro-pipette should be used for measurement of small volume (near about 1-2 ml). Graduated pipette may be used for measuring 20-25 ml. Graduated cylinder is used for measuring volume beyond 25 ml or so. For eliminating error, developing solvent should be used in a volume which should be sufficient for one day.

(h) **Development:** Plates should be developed in a saturated twin-trough chambers by following procedures.

1. Appropriate volume, i.e. 10 ml approx. for 10×10 cm plate and near about 20 ml for 20×20 cm plate is proposed.
2. Chamber is opened and piece of filter paper of correct size (10×10 or 20×20 cm) is placed in the rear trough.
3. Solvent is put into the chamber to enable filter paper thoroughly wetted and adheres to rear wall of the chamber.
4. The chamber is tilted to about 45° C to one side to equalize the volume of solvent in both the troughs.
5. The chamber is placed on table and allowed to stand for about 15-20 minutes.
6. The desired developing distance is marked with pencil (approx. 70 mm from lower side of the plate) on the right edge of the plate.
7. Slide the lid of the chamber to a side and plate is put into front trough. The layer should face the filter paper and back of the plate should be against the front wall of the TTC.
8. Put the lid and develop plate to the mark.
9. Lid is opened and the plate is taken out.
10. The plate is dried in a vertical position in the direction of chromatogram for 5 minutes in cold air stream.
11. The left mobile phase and filter paper should be discarded after each development.
12. The chamber should be cleared and dried every time before the next run.

(i) **Derivatization:** Spraying or dipping technique may be used for derivatization of the sample on plate. Dipping is more preferred technique over spraying. Spraying should be done in a spray cabinet/fume hood. Plate heater is used for heating if required. Details of the derivatization procedure need to be observed.

1. *Dipping technique:*
   (i) Tank is charged with adequate reagent for complete immersion of the chromatogram.

    (ii) Plate is fixed in the holder of the immersion device and the parameter is fixed in accordance with the method and press start button.

   (iii) Excess reagent is allowed to drip off the plate and wipe off back side of the plate with paper towel. Then the plate is removed from plate holder.

   (iv) The plate is dried with cold air vertically in the direction of chromatograph or the plate may be heated.

2. *Spraying technique:*

    (i) The spray bottle is filled up to approx. 50 ml of the reagent.

   (ii) The plate is put in spray cabinet or fume hood upright against a filter paper or paper towel.

   (iii) The plate is sprayed in both the direction i.e. horizontal and vertical to cover it homogenously.

   (iv) The plate is dried with cold air or is heated as the method prescribes.

3. *Heating of the plate:*

    (i) Desired temperature is selected and put on the plate heater.

   (ii) Stabilize the required temperature.

   (iii) The plate is put on the heater.

   (iv) The plate is removed after heating it for required time.

(j) Documentation, labelling and imaging:

    (i) Electronic documentation is done on the developed plate under UV at 254 nm, 366 nm and white light. In case, if particular light does not respond well, this must be documented. Images are labelled and recorded in worksheet.

   (ii) Individual mark with traceability is given to each plate and written with the pencil on the top corner.

   (iii) Electronic images are given file number matching the plate ID, with description of light used for capturing. A proper coding system with traceability should be used.

(k) Quantitative determination: Scanning densitometer is used for quantitative determination. The analysis file is labelled to depict all the other activities and additional information is given in case multiple evaluations under various conditions are done.

(l) Documentation of works: All the works done are properly documented with traceability.

15.4 WORKSHEET

Standard proforma of the worksheet is given below.

    (i) Name and address of the laboratory with phone no., fax no., e-mail, etc. along with legal identity if any

   (ii) Nature and quantity of the sample received with lot no. if any, packing, sealing, etc.

   (iii) Code no./project no. given

   (iv) Physical appearance of the sample

(v) Mode of homogenization of the sample for analysis
(vi) Date of receipt of the sample
(vii) Source of receipt of sample
(viii) Laboratory temperature and humidity
(ix) Name of the analyst
(x) Status of the sample (Whether legal or research, survey)
(xi) Details of the standards used with description and lot no.

Chemicals used:

S. No.	Name, purity or quality of chemicals	Name of manufacturer	Batch no.	Date of purchase with expiry if any

Plates used:

S.No.	Plate, nature, size, etc.	Name of manufacturer	Batch no.	I.D.mark given

Materials used:

S.No.	Name of article	Quality with ID mark	Name of the manufacturer

   Test no.1. plate no:
(i) Objective
(ii) Experiments:

    (a) Extraction and sample preparation
    (b) Standard preparation
    (c) Sample application
    (d) Instrument details:
       Band length
       Distance between tracks
       Application position x: y:
       Scheme of application

Track	Volume (μl)	Nature
1		
2		
3		

Development:

(i) Chamber
(ii) Saturation
(iii) Developing distance from application point/lower end of plate
(iv) Developing solvent
(v) Time of development
(vi) Drying of plate

Derivatization reagent:

(i) Name of Reagent
(ii) Reagent preparation with date of preparation
(iii) Reagent use (spraying, dipping time, heating, temp., etc.)

Reagents:

(i) Image under UV 254 nm
(ii) Image under 366 nm UV
(iii) Image of derivatized plate white light
(iv) Image of derivatized plate UV 366 nm

Conclusions/Opinion

Finding of the result of each plate be given.
Name and signature of authorized signatory.

References

1. Robertson JA, Lee LS, Cucullu AF, Goldblatt LA. Assay of aflatoxin in peanut peanut products using acetone-hexane-water for extraction. *J Am Oil Chem Soc* **42**; 1965
2. Cucullu AF, Lee IS, Mayne RY, Goldblatt LA. Determination of aflatoxins in peanuts and peanut sections. *J Am Oil Chem Soc* **43**; 89-92, 1966
3. Koltun SP, Gardner HK, Jr Dollear FG, Rayner ET. Physical properties and content of individual cateye fluorescent cottonseeds. *J Am Oil Chem Soc* **51**; 178-1, 1974
4. Shotwell OL, Goulden ML, Hesseltine CW. Aflatoxin distribution in contaminated Cereal, *J Am Oil Chem Soc* **51**; 492-499, 1974
5. Whitaker T, Wiser E. Theoretical investigations into the accuracy of sampling shell peanuts for aflatoxins, *J Am Oil Chem Soc* **46**; 377-379, 1969
6. Whitaker T. Dickens JW, Wiser E. Design and analysis of sampling plans to establish aflatoxin concentrations in shelled peanuts. *J Am Oil Chem Soc* **47**; 501-504, 1970
7. Whitaker T, Dickens JW, Monroe R. Comparison of the observed distribution of in shelled peanuts to the negative binomial distribution. *J Am Oil Chem Soc* **49**; 59, 1972
8. Whitaker T, Dickens JW, Monroe R. Variability of aflatoxin test results. *J Am Oil Chem Soc* **51**; 214-218, 1974
9. Whitaker T, Dickens JW, Wiser E, Monroe R. Development of the method to estimate sampling plans used to estimate aflatoxin concentration in lots of shelled peanuts. IUP, Technical Reports, No.10, 1974
10. Whitaker T. Sampling granular foodstuffs for aflatoxin. *Pure App Chem* **49**; 1709, 1977
11. Homero Fonseca, Sampling Plan for the Analysis of Aflatoxin in Peanuts and Corn: An Update, *Brazilian Journal of Microbiology* **33**; 2, Sao Paulo, April/June 2002

12. Indian Standard IS:1548-1981-Manual on Basic Principles of Lot sampling (second revision) Indian Standards Institution, Manak Bhawan, 9, Bahadur Shah, Zafar Marg, New Delhi 110 002

13. *J. AOAC* **9**; 42, 1967; **15**; 72, 1932; **17**; 65, 1934

14. *J. AOAC* **21**; 361, 1938; **35**; 194, 1952; **42**; 36, 1959

15. AOAC official Methods of Analysis of AOAC Interaction 18th Edition, 2005

16. Indian Standard IS: 1548-1981- Manual on Basic Principles of Lot sampling (second revision)

17. *Cereal Foods World* **29**; 771, 1984

18. *Pure Appl Chem* **49**; 1709, 1977, **58**; 305, 1986

19. *J. AOAC* **62**; 1182, 1979, **63**, 95, 1980

20. *J. Am Oil. Chem Soc* **58**; 852, 1981

HPTLC in Food Analysis

Food is usually defined as "any substance, whether processed, partially processed or unprocessed, which is intended for human consumption and includes primary food to the extent defined, genetically modified or engineered food or food containing such ingredients, infant food, packaged drinking water, alcoholic drink, chewing gum, and any substance, including water used into the food during its manufacture, preparation or treatment but does not include any animal feed, live animals unless they are prepared or processed for placing on the market for human consumption, plants prior to harvesting, drugs and medicinal products, cosmetics, narcotic or psychotropic substances".

Food adulterant means any material which is or could be employed for making the food unsafe or sub-standard or mis-branded or containing extraneous matter.

Food additives is any substance not normally consumed as a food by itself or used as a typical ingredient of the food, whether or not it has nutritive value, the intentional addition of which to food for a technological (including organoleptic) purpose in the manufacture, processing, preparation, treatment, packing, packaging, transport or holding of such food results, or may be reasonably expected to result (directly or indirectly), in it or its by-products becoming a component or otherwise affecting the characteristics of such food but does not include "contaminants" or substances added to food for maintaining or improving nutritional qualities.

Food is a complex composition comprising of number of organic and inorganic ingredients. This is not possible to lay down quality and purity standards of all the food materials available in the market. However, specifications for significant number of general foods are available in various food laws.

Primary objectives of the food laws implementing authorities are to ensure that safe, fit for human consumption and wholesome food, in addition to claim of food contents are provided to the consumers. For this, comprehensive tests are necessary in a laboratory not only for checking the compliance of quality standards as per requirements, but also for ensuring the compliances of declaration on food package, nutritional value, contaminants limits, and additives besides checking adulterants. It is also the motive and lawful aim of the food producer, manufacturer, processor and any agency involved in sale to ascertain their contribution in the maintenance of food quality and safety to enable supply of food of desired quality and safety to the consumer. Therefore, testing at different levels, i.e. at the time of production, raw materials, in-process, final product and testing for controlling the hazards, etc. are of vital importance for overall maintenance of quality and safety of food.

To establish the food quality and safety, it is necessary that a simple, rapid, accurate and inexpensive method is available in the food laboratory. Thin-layer chromatography is widely used in food analysis and quality assessment throughout the world. Due to advancement in instrumentation and automation, HPTLC has become one of the important tool which gives high resolution and better separation with accurate and quantitative results. It has its own advantages over other techniques.

16.1 SAMPLING

Sampling is an important part of an analytical procedure. The basic objective is to obtain a sample for analysis which should be representative of the lot from which it is drawn. There can be two types of sampling procedures. One is manual and other is continuous. Manual sampling procedure is done with different equipments namely triers, probes, or sampling tubes, augers or drill-type samplers, syringe-type samplers, etc. depending upon the nature of the food commodities. In continuous sampling procedure, a tiny fraction of material is mechanically diverted with the help of sampler or sample boxes at fixed time intervals. In case of solid sample, riffle cutter may be used, whereas in case of liquid sample, mainstream line can be applied. The samples to laboratory should be sent in a proper sealed condition in a suitable packaging material or container depending upon the nature of the food and stored properly during the transportation. Relevant information/label declaration in context of tests must be provided to the laboratory. Further, it should be received in the laboratory at a prescribed temperature depending upon the nature of the food with the basic objective that the quality of the food does not deteriorate during transportation and storage period.

The laboratory has no control on the field sampling of the food products. But it must be assumed that portion received for analysis is representative of the whole lot. Variation in results of the sample taken by different persons from the same lot must be viewed from scientific angles. There can be different reasons for variations like non-representative sample or different procedure for analysis of chemical and safety parameters.

Liquid foods are reasonably homogeneous and can be made representative by suitable means while solid and semisolid foods are always heterogeneous. Therefore mechanical grinding, mixing, rolling, agitation, stirring or any other means to make the sample homogeneous prior to such sample is often necessary.[1,2,70] While making the sample representative, care should be taken to ensure that any mechanical or manual process used for this process does not change the quality of the food, e.g. generation of self heat, etc. Freezing is an important technique to prevent changes in quality of food before analysis. Frozen food must be thawed before analysis without affecting the food composition. Therefore, slow thawing without heat and in a closed container is preferred. It prevents condensation of moisture on the sample. If any separated liquid is noticed in the sample, it must be made homogeneous before analysis.

16.2 SAMPLE PREPARATION

Sample preparation consists of following steps:
1. Extraction of desired components from the sample.
2. To remove interfering matters by suitable process such as purification, clean up procedure, etc. thereby concentrating the sample.

Since food is a complex matter, the ingredients of food must be taken into account while following these steps, e.g. fatty foods should be handled differently from the non-fatty foods and desired chemical components of the food should be extracted.[144]

16.2.1 Extraction

For extraction, care has to be taken to separate the components of interest from bulk matrix in a manageable way. Liquid solid extraction between solvent and solid substrate can be done in a short period, i.e. 2-4 minutes in a high speed blender or for a longer period, i.e. about 30 minutes by shaking in a flask. Presoaking treatment of the sample by keeping overnight prior to extraction can give improved result. Liquid samples may be extracted in a separating funnel. Mini-columns can be used to absorb fractions and then selectively eluted by appropriate solvents. A number of instruments are available for extraction purposes. Pesticides residue extractor, an instrument, is suitable for better, economical and consistent extraction from fatty food. Accelerated solvent extraction is an automated extraction method, which takes advantage of the temperature dependence of extraction kinetics and requires small volume of solvent, in addition to reduction in extraction time from hours to few minutes. This is based on operation at higher temperatures and pressures than traditional solvent based extraction techniques. This instrument streamlines sample preparation and is suitable for extraction of food samples. These instruments can be utilized depending upon the nature of the food commodities. However, there is a need to validate the method and develop standard operating procedures. Steam distillation method can be used for extraction of components of interest especially in case of volatiles like benzoic acid and sorbic acid. All types of extractions like liquid-liquid, liquid-solid involve partitioning of a material between two phases.[72] In case of liquid-liquid extraction, partitioning is basically governed by relative solubility of the extractant in the two immiscible liquids.[72] Liquid-solid extraction is completed by physical occlusion of extractant in an inert solid material. Partitioning of a substance is rarely completed in one extraction. Therefore multiple extractions are recommended for quantitative determination. The total volume for extraction is used in several aliquots for the extraction rather than all solvent for one extraction. Reliable quantitative results are, thus obtained. Mixture of solvents as compared to single solvent is seen to be more effective especially for lipids. Lipids are not suitable for extracting from food. Solvents of low polarity will extract glycerides, sterols and small amounts of complex lipids. However, mixtures of low polarity with polar solvents will extract the majority of complex lipids. In case food is naturally acidic or alkaline, then this property can be utilized to affect the degree of partitioning between water solution by adding small amounts of acid or alkali to the extraction solvents. Sometimes components present in the food do not pose problems in sample preparation and in such cases sample may have to be only diluted with suitable solvent for final assay.[3] Sometimes hydrolysis is required by acid, alkali, or enzymes. The amino acid components of protein in food are determined only after hydrolysis of the protein. Similarly starch in food can be determined by hydrolyzing the starch to sugars with enzymes and then sugar is determined. Specific reactions may also be done, e.g. separation of glycerides from the sterols by saponification.[4]

16.2.2 Clean-up

Interfering substances in the exacted sample must be removed before determination. The choice of clean-up procedure will depend upon the required level of detection and quantification and also the recovery.

The column chromatography clean-up is preferred. Different materials are silica gel, aluminium oxide, polyamide, florisil and sephadex, etc. The extracted sample in suitable solvent is added to the column after which the column is washed with one or more solvent in which components of interest are insoluble or less soluble than the impurities. Then the solvent composition is changed in such a way that components of interest are eluted selectively from the column. The elute is collected and concentrated without affecting the composition of the substance. Occasionally, the components of interest are present in very low level; then a high concentration of component of interest in the cleaned-up extract is required to make detection possible. In such cases, the residue is dried or almost dried and then dissolved in small and known volume of solvent. Derivatization of the components of interest may be required in certain cases.

Disposable mini-columns/cartridges are commercially available to simplify extraction and clean-up. Liquid-liquid extraction and solid-phase extraction are the techniques used to concentrate solutes from a dilute solution or purified concentrated solvent extracts, etc. Elute the substance with different polarities of solvent into separate fractions for micro-analysis, when the quantity of sample available is very low. Determination of 17 colours in food sample was reported using Sep-Pak NH_2 cartridges as a preparative technique.[5] Similarly for determination of antibiotics in foods and food products, clean-up with Sep-Pak C_{18} cartridges is also mentioned.[71] The same cartridge was used for clean-up extract for determination of aflatoxins in liver sample.[6]

16.3 CARBOHYDRATES

Simultaneous separation and the analysis of carbohydrates in food is of high importance and work carried out in this regard have been reported by different workers.[7–11]

Quantitative analysis of malto-oligosaccharides and monosaccharide in beet for controlling brewing processes and characterization of final product by HPTLC using gradient AMD was reported.[12] Analysis of beet and cane molasses was done by HPTLC - AMD.[13]

Quantitative evaluation of inulin in food products by HPTLC was reported. The sample extracts were spotted on silica gel, developed three times, twice with n-propanol - acetone - water (9:6:5) and the third time with n-propanol - acetone - water (5:4:1). Visualization was made by dipping into DAP reagent (2% diphenylamine and 2% aniline in methanol - 85% phosphoric acid) (4:1) for 10 seconds. The coloured spots were seen after heating at 120° C for 10 min and then evaluated by densitometry in transmission mode. The linear working was found between 0.5 and 4.0 μg per spot.[14]

HPTLC of sugars on silica using AMD was reported by Kroh.[15] Detection was done by spraying with aniline - diphenylamine - phosphoric acid and drying for 5 min at 105° C. Quantification was carried out by densitometry at 385 nm.

Detection of glucose and fructose by HPTLC was reported by Klaus et al.[16] on amino modified silica with acetonitrile - water in the ratio of 7:3. Detection was done under UV 365 nm after heating the chromatogram. Comparison was made.

Rapid quantitative HPTLC of glucose, fructose and other common sugars on silica by single or multiple developments with acetonitrile-water in the ratio of 17:3 was reported by Pukl et al.[17] Detection was done by immersion of dried plates for 10 s in DAP (2% diphenylamine and 2% aniline in methanol - 85% phosphoric acid) (8:2) and heating at 120° C for 10 min. Quantification was done by densitometry at 440 and 515 nm.

TLC behaviour of 30 different sugars on silica with 4 solvent systems was published by Yamazaki et al,[18] on silanized Kieselguhr G with acetone-acetonitrile-water in different ratio. Densitometry was done at 450 nm.

Rotation planar extraction and medium-pressure solid-liquid extraction of onion was reported by Vovk.[19] HPTLC of monosaccharides and disaccharides was performed on silica gel with acetonitrile-water-methanol in the ratio of 68:12:1 in unsaturated chamber containing 0.05 diphenylboric acid 2-aminoethyl ester. HPTLC of oligofructans was reported with n-propanol - acetone-water (10:8:5) and n-propanol-acetone-water (5:4:1). The detection reagent was 2% aniline and 2% diphenylamine in methanol containing 20% concentrated orthophosphoric acid.

Determination of starch by HPTLC was carried out by Aranda et al[20] after hydrolysis of starch using alpha-amylase and amyloglucosidase on silica gel pre-washed with methanol and treated by dipping in di-potassium hydrogen phosphate (0.1M in methanol) and then activated for half an hour at 120° C. Plate was developed thrice in a horizontal development chamber with acetonitrile-water (7:3). Detection was made in aniline-diphenylamine reagent and densitometry at 520 nm. The detection and quantification limits for starch was observed to be 0.26 and 0.51 g/100 gm respectively.

Separation of complex fructo-oligosaccharides and inulin mixtures (sucrose, 1-kestose, nystose, and fructosylnystose) on diol phases were done with HPTLC-AMD. Detection was made by derivatization with 4-aminobenzoic acid reagent and quantitation by scanning at 366 nm.[21]

The commercial portion of the plant "Cichorium intybus var" a variety of chicory used as a food was analyzed for guaianolide and sugar contents. The quantitative and qualitative evaluation was done by densitometric HPTLC.[22] A correlation was established between analytical data and bitter taste.

Detection of oligosaccharides in sugar products using HPTLC was reported by Vaccari et al[23] by HPTLC-AMD.

Simultaneously determination of glucose, fructose and sucrose in molasses by HPTLC has been described by Lee et al.[24] The relative standard deviations noticed are 1.1% for sucrose, 2.2% for fructose and 4.3% for glucose. The determination is of interest to many other fields. Quantitative estimation of malto-oligosaccharides by HPTLC was reported by Rsch et al.[25] Nurok and Zlatkis [26, 27] described improvement in separation of malto-oligosaccharides by using HPTLC.

Quantitative analysis of glucose, fructose, lactose or maltose, saccarose and raffinose was optimized using three different plate materials by Patzsch et al.[28] The limit of detection was lowered to 5-30 ng/spot.

Optimized analysis of glucose, fructose, saccharose, lactose and raffinose on HPTLC-amino plates in association with post-chromatographic derivatization by two different procedures was used for their quantitative determination in real samples.[29] The method was applied to the analysis of honey, chocolate and biotechnological suspensions of *Echerichia coli*.

Sugar in beverages was determined[30] using HPTLC silica gel impregnated by spraying with 0.10 M sodium hydrogen sulfite and pH 4.8 citrate buffer solutions and then developed with acetonitrile-water in the ratio of 85:15, quantification was done by densitometry at 515 nm.

16.4 ORGANIC ACIDS

Separation of higher fatty acid methyl esters by reversed-phase HPTLC was done by Pyka et al.[31] Eight methyl esters of higher fatty acids namely methyl laureate, methyl myristate, methyl palmitate, methyl stearate, methyl 12-hydroxystearate, methyl 9, 10-dihydroxystearate, methyl ricinoleate, methyl a-hydroxypalmitate on RP-18 with mixtures of methanol and ethanol in the volume proportions 19:1, 9:1, 17:3, 4:1 and 3:1 were carried out. Visualization was done by exposure to iodine vapor. Further relationships between R_f and mobile phase composition were evaluated.

TLC of benzoic acid and sorbic acid on silica with benzene-dibutyl ether-acetic acid-formic acid in the ratio of 80:35:3:3 was done. Detection made by UV at 254 nm.[32]

Separation and determination of saturated fatty acids by reversed-phase HPTLC was reported by Gattacecchia et al[33] with acetonitrile - isopropanol (2:1). Detection made by UV 254 nm and quantitation by scanning at 254 nm. The detection limit was reported between 7 – 10 ng.

16.5 FOOD GRAINS

Determination of deoxynivalenol in wheat was published by Eppley et al[34] on silica with chloroform - acetone - isopropanol in the ratio of 8:1:1 in unequilibrated tank. Detection was done by spraying with $AlCl_3$ solution and heating for 7 min at 120° C. Quantitation was done by fluorodensitometry and detection limit was 20 ng.

Determination of deoxynivalenol in processed grain products was reported by Trucksess et al.[35] on silica with 15% $AlCl_3$ solution using chloroform-acetone-isopropanol in the ratio of 8:1:1. Quantitation was done by fluorodensitometry after heating the plate at 120° C for 7 min. Limit of detection was 50 ng/g.

16.6 LIPIDS, FATS AND OIL

Analysis of lipids and detection of fatty acids from fats and presence of a foreign fat are very important. Separation and identification of neutral and complex lipid classes have been carried out by numerous workers.[36–40]

Phospholipid hydroperoxides and their parent phospholipids were analyzed as basic oxidation products in cooked turkey meat extracts by HPTLC followed by densitometric quantification at 654 nm.[41] For separation of all animal and plant lecithins and mixtures or fractions of phospholipids, a method was prescribed.[42]

Quantitative evaluation of sanguinarine as an index of argemone oil adulteration in edible mustard oil by HPTLC was reported by Ghosh et al.[43] The quantitation was done by densitometric scanning of the chromatogram in fluorescence/reflectance mode. Dihydrosanguinarine concentration was determined after its conversion to sanguinarine by UV (366 nm), irradiation for 15 min. LOD of sanguinarine in argemone oil was determined as 3 ng per 6 mm band with a signal-to-noise ratio of 3:1 and LOQ was reported on 3 ng per 6 mm band with a signal-to-noise ratio of 10:1. The total content of sanguinarine in argemone oil samples were in the range of 4.84-5.79 mg/ml of oil.

16.7 PROTEIN AND AMINO ACIDS

TLC is a good technique for determination of amino acid components of a given protein or peptide and is reported for separation with high resolution.[44–46]

Determination of nine biogenic amines in fish and squid were reported by Judite et al.[47] HPTLC of dansyl derivatives of biogenic amines viz. agmatine, putrescine, tryptamine, cadaverine, supermidine, histamine, supermine, tyramine and beta-phenylethylamine were done with chloroform-diethyl ether-triethylamine(6:4:1), followed by chloroform-triethylamine(6:1). Quantitation was performed by fluorescence measurement at 330/>400 nm. For tryptamine, tyramine, histamine and beta-phenyl ethylamine, detection limit was 10 ng, whereas for other amines it was 5 ng. The method was found to be an effective and precise analytical tool for separation and determination of biogenic amines.

Planar chromatography has been very useful for evaluation of storage, stability of food and food products. Biogenic amines which are natural antinutrition factors from hygienic view-point have been identified as a causative agent in number of food poisoners, because they initiated various pharmacological reactions. Biogenic amines are produced as a result of action of enzymes produced by a part of microflora or free amino acids as established by different researchers.[48–55] Determination of biogenic amines was reported by TLC.[56, 57] Storage stability of fatty food was also studied using planar chromatography.[58, 59] Minor components of food such as alkaloids, flavonoids, saponins and steroids have been reported to be separated and determined.[60–63]

Biogenic amines were determined in red wine by TLC and fluorescence densitometry was done.[64] Similarly, biogenic amines were extracted from foods and dansyl derivatives were prepared, separated by multiple developments TLC and quantified by densitometry at 254 nm. The limit of detection was approximately 5-10 ng with recovery 86-93%.[65]

Determination of histamine and other biogenic amines in fish was reported by Speer et al.[66] Development solvent was benzene-chloroform-triethylamine(10:6:7 or 16:6:2) in horizontal developing chamber over 90 mm. Quantitative determination was done at 365 nm/400 nm by fluorescence measurement. Benzene is toxic and usually replaced by toluene.

Multiple determination HPTLC analysis of amino acids on cellulose layers was reported by Jolanta Flieger.[67] The retention behaviour of amino acids on cellulose layers was determined. Chromatographic systems with the best separation selectivity were used for analysis of amino acids in reference solutions.

Twenty common protein PTH amino acid derivatives have been separated on silica gel plates by automated multiple development by Belay et al.[68]

16.8 FOOD CONTAMINANTS/ADDITIVES/COLOURS/ADULTERANTS, ETC

Contaminants in food refer to undesirable materials present in small quantity/traces, which are incorporated in food inadvertently or due to negligent handling, processing, storage, transportation, etc. in the food chain. These contaminants, if toxic may be injurious to health even at a low level and their monitoring is a part of food safety quality assurance system. Due to liberalization of world trade, maximum permissible limits of these contaminants play an important role in marketing of food from one country to another country or even within the country. National and International directives are very stringent in these regards.

Analysis of many pesticide residues using TLC has been a good official method in the past but better methods are now available. Considerable work has been carried out for screening multiresidues at a time using HPTLC. Especially screening of more than 250 pesticides in water sample is a landmark work and road map for exploiting such techniques in screening of pesticides in different food samples at an acceptable level of detection and quantification. Such methods are cheaper, convenient and easy to interpret compared to hyphenated techniques. While making an attempt in such direction, analyst would have to validate the method for different food and different pesticides keeping in view sampling, extractions and recovery. HPTLC here is best used as a screening technique to analyse a large number of samples, quickly and economically.

Application of HPTLC for detection/determination of aflatoxins, and other mycotoxins which are highly carcinogenic contaminants in foods at the lowest level is an inspiration for use of this technique, as an official/validated method.

Similarly, presence of heavy metal contaminants which enter the food through water, air, industrial pollution, agricultural technology, use of utensils during processing of foods, etc. make food toxic and unfit for use. Maximum permissible limits have been prescribed in various foods under several national and international food laws. These metals constitute lead, copper, arsenic, zinc, tin, cadmium, mercury, methyl mercury, nickel and chromium. The most accepted technique presently available is AAS-GF/ICP/ICP-MS. Though, work has been carried out on HPTLC in detection/determination of metals, but a lot more needs to be done specially in food, looking into ultra-low limits. Such methods need to be developed for different foods, validated and established as routine use for determination of metals in food.

In view of foregoing, HPTLC can be established as a single tool for screening, detection and determination of different food contaminants, preservatives, antioxidants, adulterants, establishing the food composition and food labelling facts finding tool in the domain of food analysis. This technique is no doubt cheaper, convenient, easy to handle and affordable in an analytical laboratory especially in developing and under developed countries, where presence of food contaminants in agricultural and food commodities pose major threat in assessing the quality and safety of foods.

Food preservatives are the substances capable of inhibiting, retarding or arresting the process of fermentation, acidification or other decomposition of food. There are number of classes I and class II preservatives permitted under the law. Maximum permissible limits have been prescribed for addition of class II preservatives. Several methods are available, of which TLC is the most simple and convenient technique, e.g. detection of benzoic acid and sorbic acid, p-hydroxy benzoate and its esters, volatile acids such as propionic acid and acetic acids, etc. in food, detection and determination of aspartame, acesulfame, saccharine, cyclamate, natural food colours, permitted coaltar food colours, detection of antioxidants like BHT, BHA, TBHQ in fats, etc. as well as detection of flavouring agents like vanillin, ethyl vanillin and coumarin.

Planar chromatography technique for determination of food additives[69, 70] and food preservatives[71, 72, 185] have been reported by numerous workers.

16.9 PESTICIDE RESIDUES

Burger, working in Bayer, Germany was the pioneer of HPTLC in pesticide residue analysis, so much so that his work led to the acceptance of HPTLC-AMD method becoming the first official German method for pesticide residue analysis(DIN-). Burger's work led to the invention of automated multiple development instrument.

General review of pesticides in plant food using AMD was made.[73] Determination of organophosphorus pesticides was published by Tsunoda[74] on silica and alkyl-bonded silica with various solvent systems. Detection was done by UV spraying with 0.1% $PdCl_2$ in 10% HCl and with 2% 4-nitrobenzylpyridine in acetone spraying with 10% tetraethylenepentamine in acetone. Determination of pesticide residues from foodstuffs[75] on silica with hexane - acetone (2 : 1) was reported. Detection was done by spraying with o-toulidine reagent. Sensitivity was observed as 0.05 mg/kg with recovery of 65-80%.

Determination of pesticide residues from foodstuffs[76] on alumina with benzene - acetone (95:5) was done. Detection was done by spraying with p-dimethylaminobenzaldehyde reagent with sensitivity at 0.05 mg - 0.1 mg/kg and recovery of 75-93%.

Method to monitor pesticides in ground and drinking water on HPTLC was reported by Zietz et al.[77] Separation of 13 pesticides was possible using HPTLC and development by a 30-step gradient (methanol - chloroform) using AMD technique. Detection was done under UV reflectance mode at 6 wavelengths. CV $\pm$ 2.2% at a concentration range between 20 and 500 ng was observed.

HPTLC of several pesticides and their major environmental by-products was published by Judge et al[78] on silica with different solvent systems. Detection was made by immersion into a 1 mm solution of rhodamine 6G, Nile red, Merocyanine 540, 1-pyrene-carboxaldehyde, TNS-chloride, mansyl-chloride, NBD-chloride, or 1, 6-diphenyl-1, 3, 5-hexatriene, placed into dark or exposed to short-wave and/or long-wave UV light for 1 to 24 h. Quantification was made by densitometry in fluorescence/ absorbance modes. Screening of 265 pesticides in water by HPTLC with automated multiple development was reported by Butz et al.[79] In total, 283 pesticides were analyzed with application of gradients. 18 pesticides exhibited an instrumental detection limit of more than 100 ng applied and hence cannot be analyzed from 1 litre drinking water samples with AMD-HPTLC without further treatment. La Vigne et al[80] used HPTLC with AMD for the identification and determination of pesticides in water. Screening of organophosphorus pesticides in water was done by Hamada.[81] TLC of 7 organophosphorus pesticides, (azinophos methyl, chlorpyriphos, diazinon, fenthion, methamidophos, methidathion, omethoate) was carried out on silica gel prewashed with methanol - chloroform (1:1) with n-hexane - acetone (15:6). Detection was done under UV 254 and 366 nm and quantitation by fluorescence quenching at 220 nm with scanning densitometer in reflectance mode. The method was found suitable for rapid screening of pesticide residues.

1-naphthylmethyl carbamate is widely used as an insecticide in fruits, vegetables, etc. A simple method was developed by Daundkar et al[82] which could detect up to 5 μg of the carbaryl using spray reagent 5% sodium hydroxide solution and then 1:1 mixture of 2% diphenylamine solution and 5% formaldehyde solution. There is no interference with other carbamate and organophosphorus insecticides.

Urea pesticides (15 samples) were separated on RP-18WF$_{254}$ plates with methanol – water and acetonitrile – methanol (1:1 v/v) - 0.1% aqueous orthophosphoric acid mobile phases (RP-HPTLC) and by using silica gel 60 F$_{254}$ plates with benzene – methanol and benzene – ethanol as mobile phases (NP-TLC). Fifteen urea pesticides have been separated on TP-18WF$_{254}$ plates with methanol-water and mixed organic (acetonitrile-methanol, 1:1 v/v)-0.1% aqueous orthophosphoric acid (H_3PO_4) mobile phases (RP-HPTLC), and on silica gel 60F$_{254}$ plates with benzene-methanol and benzene-ethanol mobile phases (NP-TLC). The pesticides could be classified into three groups:

- Monolinuron (1), chlorotoluron (2), diuron (3), isoproturon (4), and linuron (5);
- Dimefuron (6), diflubensuron (7), teflubenzuron (8), and lufenuron (9) and
- Thifensulfuron methyl (10), triasulfuron (11), chlorsulfuron (12), rimsulfuron (13), amidosulfuron (14) and tribenuron methyl (15).

Relationships between R_f values and mobile phase composition were established for the pesticides.[83]

Amitrol (3-amino-1H-1, 2, 4-triazole) is a herbicide of triazole group. It is widely used to control weeds and for fruit protection. It is absorbed through roots and leaves because of high water solubility. Its presence in water may pose threat to aquatic life. To keep a control on presence of plant protecting agents in water samples, determination of Amitrol in water by AMD was carried out by selective post-chromatographic detection with Bratton-Marshall reagent. The method is suitable for monitoring and checking potable water.[84]

HPTLC determination of carbamate residues (pirimicarb, methomyl, carbofuran, carbaryl) in vegetables was performed on silica gel prewashed with chloroform:methanol (1:1) followed by drying at 110° C for half an hour with system I (two-fold development firstly with toluene-acetone (4:1) and secondly with dichloromethane – acetone (4:1) and system II (two-fold development with ethyl acetate-petroleum ether (3:2) first and then with chloroform – petroleum ether (9:1). Quantitation was done by densitometric scanning at 254 and 366 nm.[85]

Planar chromatography technique for determination of pesticide residues has been reported by others.[86, 87]

Development and evaluation of thin-layer chromatography-digital image-based analysis for the quantitation of botanic origin pesticide azadirachtin in agricultural matrixes and commercial formaluations were done by Tanuja et al.[88] TLC of azadirachtin was performed on silical gel with dichloromethane – ethanol (20:1). Detection by spraying with acidified vanillin reagent followed by heating at 110° C for 3 min. Quantitation was done by densitometry.

A new procedure for separation of complex mixtures of pesticides by multidimensional planar chromatography was done by Tuzimski.[89] The silica gel plate was developed in the first dimension with ethyl acetate - n-heptane (1:3), and then turned by 90°. The plate was dried between 5 and 15 min before second development. Detection was done under UV light at 254 nm.

Separation of a mixture of pesticides by 2D-TLC on two adsorbent-layer multi-K SC5 plate was reported by Tuzimski et al.[90] HPTLC on 16 pesticides (propaquizafop, quizalofop-P, triadimefon, tridimenol, fenoxycarb, quinoxyfen, cyromazine, oxyfluorfen, fluoroglycofen, acetochlor, metazachlor, piperonyl butoxide, furalaxyl, pyriproxifen, buprofezin, clofentezine) was done on silica gel and RP-18; two-dimensional separation on dual plates with ethyl acetate – di-isopropyl ether (2.5:97.5) and acetonitrile – water (17:3); or methanol – water (4:1). Detection was done under UV 254 nm.

Comparative study of hydrophobicity parameters of novel 5'-carbamates of zidovudine was carried out by Raviolo et al.[91] HPTLC was done of 5'-carbamates of zidovudine (3'-azido-3'-deoxythymidine) and the thymidine on RP-18 with methanol-buffer pH 7.4 mixtures with methanol contents between 30 and 80%; or acetone-buffer mixtures with modifier contents between 20 and 80% in 5 or 10% increments. Detection was made after drying at 40° C developed under UV radiation.

16.10 MYCOTOXINS

Mycotoxins (Fig. 16.1) are secondary metabolites of moulds. The most important mycotoxins are aflatoxins, deoxynivalenol, ochratoxin A, fumonisins, patulin, zearalenone and T-2 toxins. These may contaminate crop plants, fruits, nuts, and grains before or after harvest. The toxic impacts of mytoxins on human and animal health are proven beyond doubt.

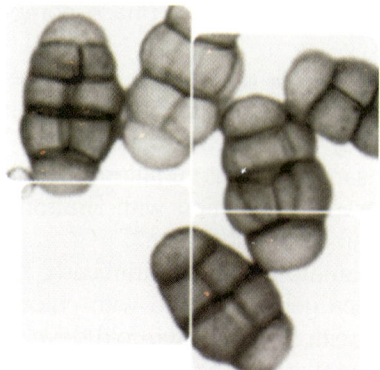

Fig.16.1: Mycotoxins

16.10.1 Aflatoxins

Aflatoxins (Fig. 16.2) are mycotoxins produced by *Aspergillus* species like *A. flavus* or *A. parasiticus*. Primarily aflatoxin B_1, B_2, G_1 and G_2 and the hydroxylated metabolite M_1 are of interest. Aflatoxin B_1 is the most frequently occurring toxin.

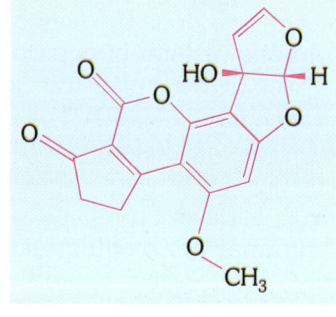
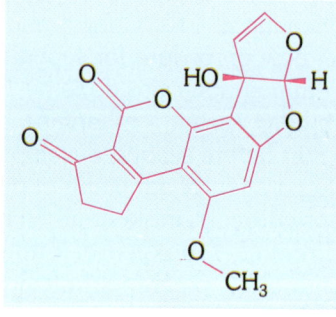

B_1 M_1

Fig 16.2: Aflatoxins

Aflatoxins are found on commodities like cereals, nuts, spices, figs and dried fruit. Aflatoxin M_1, the metabolite of aflatoxin B_1, is found in milk and dairy products. Aflatoxin B_1 is the most potent hepato-carcinogens known and hence levels of aflatoxins in the diet are an important consideration for health.

Many countries have established regulatory limits for either aflatoxin B_1 or for total aflatoxins, i.e. sum of aflatoxin B_1, B_2, G_1, and G_2, and aflatoxin M_1.

16.10.2 Trichothecenes (T-2, DON and others)

Trichothecenes(Fig. 16.3) are a group of sesquiterpenes produced by various *Fusarium* species like *F. graminearum*, *F. sporotrichioides*, *F. poae* or *F. equiseti*. The important structural features causing the biological activities of trichothecenes are the 12, 13-epoxy ring, presence of hydroxyl or acetyl groups at appropriate positions on the trichothecene nucleus and the structure and position of the side-chain. They are found on grains such as wheat, oat or maize.

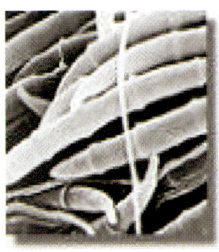

T-2 toxin

Deoxynivalenol

Fig. 16.3: Trichothecenes

This group of structurally related mycotoxins has an impact on the health of human and animals due to their immunosuppressive effects. Type A trichothecenes (e.g. T-2 toxin, HT-2 toxin, diacetoxyscirpenol) are of special interest because they are more toxic than the related type B trichothecenes (e.g. deoxynivalenol, nivalenol, 3- and 15-acetyldeoxynivalenol).

Few countries have recommended levels for these mycotoxins in food and animal feed, but it is necessary to prevent them from entering the food chain. Hence these should be looked for.

16.10.3 Fumonisins

Fumonisins (Fig. 16.4) are a group of mycotoxins produced by *Fusarium* species like *F. moniliforme* and *F. proliferatum*. Corn is the most common affected commodity. Conditions favouring fumonisins are drought stress followed by warm, wet weather. Fumonisins have ill effects on humans and animals, causing pulmonary edema in swine, and are suspected to promote the formation of esophageal cancer in humans.

16.10.4 Ochratoxins

Ochratoxins (Fig. 16.5) A, B, and C are produced by some *Aspergillus* species and *Penicillium* species, like *A. ochraceus* or *P. viridicatum*. Ochratoxin A is the most prevalent and relevant fungal toxin of this group. Ochratoxin A is known to occur in commodities such as cereals, coffee, dried fruit and red wine. It is a human carcinogen. It is of special interest as it accumulates in the meat of animals. Thus meat and meat products can be contaminated with this toxin. Regulatory limits have been established in many countries including European Union.

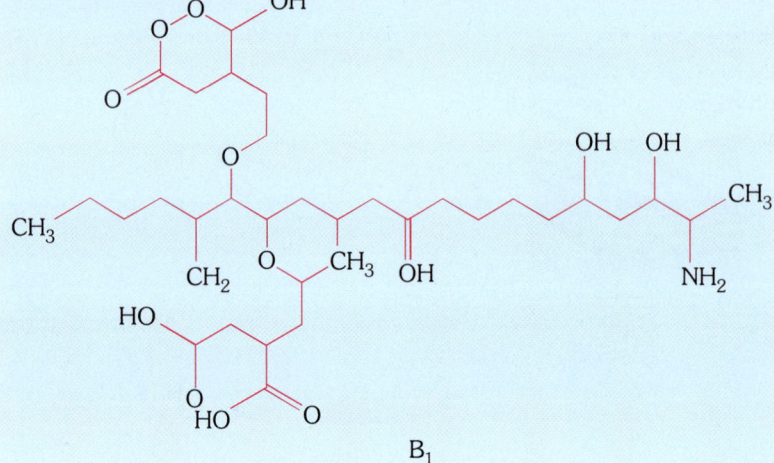

Fig. 16.4: Fumonisins

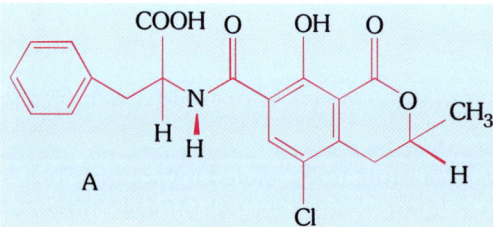

Fig. 16.5: Ochratoxins

16.10.5 Zearalenone

Zearalenone, (Fig. 16.6) a mycotoxin is produced by *Fusarium* species like *F. graminearum*. It occurs largely in grains and cereal products. Zearalenone is not acutely toxic but it has estrogenic effects on mammals. The negative effects on the reproductive system are a matter of concern in animal husbandry.

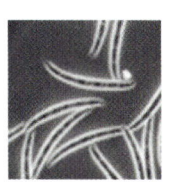

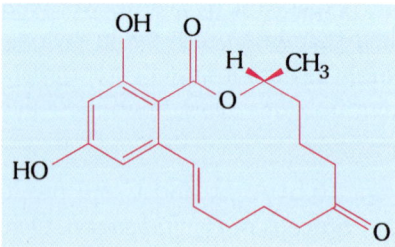

Fig.16.6: Zearalenone

Few countries have recommended levels for this mycotoxin in animal feed but it is tested for to prevent losses in animal husbandry.

16.10.6 Patulin

Patulin(Fig. 16.7) is produced by *Penicillium* species and *Aspergillus* species. The contamination mainly occurs on damaged/rotting fruits like apples, pears, peaches and grapes. Patulin is believed to be a carcinogen. As patulin occurs mainly in fruits and fruit juices which are also used as baby or infant food, some countries have established regulatory limits.

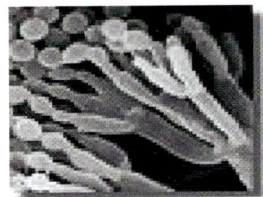 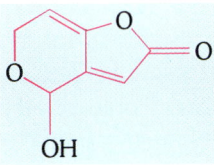

Fig. 16.7: Patulin

16.10.7 Other Mycotoxins

Other mycotoxins (Fig. 16.8) are citrinin, moniliformin, sterigmatocystin and cyclopiazonic acid.

 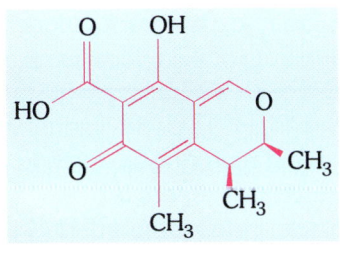 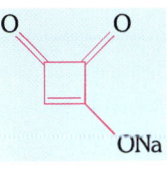

Citrinin   Moniliformin

Fig. 16.8: Other mycotoxins

16.10.8 Analysis of Mycotoxins

Numerous works have been carried out for determination of mycotoxins in food.[92–100] Special mention may be made here of the extensive work done in the lab and field by Dr. R. D. Cocker and others at the Overseas Development of Natural Resources Inst. in UK. The quantitation of ochratoxin in rice was done by densitometry at 365 nm after HPTLC on silica gel developed with the solvent, i.e. diethyl ether-methanol (49:1) in the reverse direction technique and then with toluene-ethyl acetate-formic acid (5:4:1) after cutting off the end of the plate close to the zone and having the cut edge down in the tank with hexane-ethyl acetate-acetic acid (18:3:1) after cutting off again 30 mm from the plate bottom.[101] Other methods for determination of ochratoxin A in other food commodities like cereals, green coffee beans, corn, barley and kidney were reported by different workers.[101–104] The mycotoxin patulin in apples and its detection was reported by Buchalla.[105] Similarly quantification of patulin in apples, apple concentrate, and apple juice with reflectance scanning densitometry at 275 nm or at 412 nm was reported at the detection limit of 30-50 ng/spot.[106]

Fumonisin B in rice[107] up to 16 μg/g, with HPTLC and scanning fluorodensitometry was analyzed. The mushroom nephrotoxin orellanine was evaluated with detection limit of 15 ng by TLC on cellulose plates developed with 1-butanol-acetic acid-water by fluorescence densitometry.[108]

Various techniques are available to determine different mycotoxins but determination of mycotoxin by HPTLC is now a highly recognized technique and this has been a method of choice for aflatoxin analysis. Nesbitt et al[109] and Hartley et al[110] have established four main components of aflatoxin. Two components which gave blue fluorescence under UV light were named as aflatoxin B_1 and B_2 and two other components which gave turquoise fluorescence under UV light were designated as G_1 and G_2. Literature from 1961 to 1984 has been covered in a review on mycotoxins[111] by TLC. Nesheim and Trucksess[112] mentioned the procedures of mycotoxin recognition and methods for best known mycotoxins. Lee et al[113] described determination of multi-mycotoxin using HPTLC whereas Stahr and Domoto[114] described determination technique of 10 mycotoxins. Plenty of literature are available for determination of mycotoxins in various food and feed. Nowotny et al[115] detected citrinin, ochratoxin A and sterigmatocystin in cheese and further work done on screening method of 22 mycotoxins[116] Lee et al[117] determined 13 mycotoxins by HPTLC. By use of seven continuous multiple developments with two solvent systems of different polarity, separation of sterigmatocystin, zearalenone, citrinin, ochratoxin A, patulin, penicillic acid, luteoskyrin and aflatoxins (B_1, B_2, G_1, G_2, M_1 and M_2) was obtained.

One and two-dimensional TLC methods were used to identify fungal[118] Filtenborg et al[119] developed a screening method for brevianamide A, c treoviridin, cyclopiazonic acid, luteoskyrin, penitrem A, roquefortine C, sterigmatocystin, verruculogen, viomellein and xanthomegnin. Quantitative determination of aflatoxin in spiked corn samples for determining its recovery was done[120]

Blanc et al[121] carried out reverse phase HPTLC with fluorimetric detection for binding of aflatoxin M_1 for milk products. Another rapid and reproducible method for extraction and determination of aflatoxin M_1 in milk and dairy products was reported[122]

Determination of cyclopiazonic acid in foods by HPTLC was undertaken[123] Monomeric and dimeric naphtha-y-pyrones, extracted from the mycelium of *Aspergillus niger* were revaluated on HPTLC plate[124] The components were identified by their colour, fluorescence under long wave UV light and the colour after spraying with Gibbs reagent.

Assessment of aflatoxin in different commodities like amla whole (*Emblica officinalis*), amla powder, castor seed (*Ricinus communis*), shikakai (*Acacia Concinna*), mahua (*Madhuca indica Linn*) and tamarind (*Tamarindus indica*) was reported by Jaiswal et al.[125] Assessment of aflatoxin in groundnut (*Arachis hypogia*)[126], anar dana (*Punica granatum Linn*), ajwain (*Trachyspermum ammi*), clove (*Syzygium aromaticum*) and coconut (*cocos nucifera Linn*)[127] was carried out by Jaiswal et al. Assessment of aflatoxin in mustard oil (*Brasica campestiris*) and ghee was done by Jaiswal et al.[128]

A simple and rapid method for determination of aflatoxin M_1 in milk powder was reported by Vasilikiotis et al[129] with ether – methanol – water (96:3:1). Quantitation was done by densitometry at 439 nm. The detection limit was 50 ppt. Determination of aflatoxin M_1 in powdered milk was undertaken by Serralheiro et al[130] on silica with ether – methanol – water in the ratio of 95:4:1. Visualization was made directly under 366 nm or after spraying with sulfuric acid – water in the ratio of 1:3. Detection limit was 0.5 μg/kg. Determination of aflatoxin M_1 in milk and milk powder was reported by Bijl et al[131] with two-dimensional TLC on silica with ether – methanol – water (95:4:1) in the 1st direction and chloroform – acetone (7:3) in the 2nd direction. Quantitation was done by fluorescence densitometry and detection limit was 0.005 ng/g for milk and 0.05 ng/g for milk powder.

Rapid TLC procedure for mycotoxin determination on silica with different solvent combinations was reported by Dutton et al.[132] Detection was done under long wave UV light and spraying with different reagents.

Rapid clean-up method for analysis of aflatoxins and comparison with various methods was made by Kaminura et al.[133] Aflatoxins B_1, B_2, G_1, G_2 and M_1 were determined on silica with chloroform – acetone (9:1) or ether – methanol – water (94:4.5:1) for aflatoxins B_1, B_2, G_1, G_2 and chloroform – isopropanol – acetone (85:5:10) for aflatoxin M_1. Quantitation was done by densitometry. Comparison was also made and this method was found to be more effective for quick analysis.

Determination of aflatoxins in vegetable oil was reported by Miller et al[134] on silica with chloroform - acetone (9:1) in the first dimension and ether – methanol – water (96:3:1) in the second dimension. The visualization was done under 366 nm UV light.

Rapid screening method for aflatoxin in corn for quantitative use was reported by Spilmann[135] on silica with chloroform – acetone (9:1) followed by fluorescence scanning.

Determination of aflatoxin in peanut products was reported by Tosch et al[136] on silica with chloroform – acetone (88:12). Quantitation was done by fluorodensitometry.

Determination of aflatoxin B_1 and ochratoxin A in black olives was reported by Tutour et al[137] on silica with benzene – hexane (3:1) and toluene – ethyl acetate – formic acid (60:30:15). Detection was done under UV at 375 nm. Quantitation was done by spectrophotometry. Detection limit for aflatoxin B_1 was 5-7 $\mu g/kg$ and for ochratoxin A as 20 $\mu g/kg$.

Rapid clean-up method for analysis of aflatoxins was done by Kamimura et al[133] on silica with chloroform – acetone (9:1) and ether – methanol – water (94:4.5:1.5) or chloroform – isopropanol – acetone (85:5:10). Detection was done by UV.

Analysis of aflatoxins in different foods and fodder was published by Varadi[138] on silica with ethyl acetate – chloroform – THF (15:10:1). Densitometry was done by fluorescence at 366/ >440 nm.

A comparative study of phenyl bonded phase, CB and Romer clean-up procedures for determining aflatoxin levels in maize was reported by Bradburn et al.[139] Separation of aflatoxins was done in aqueous acetone extracts of maize on silica. Evaluation was made and range of aflatoxin concentrations was reported between 3.4 and 901 $\mu g/kg$, CV in the range of 1.7-10.8% with mean recoveries of 92-99%.

Development and evaluation of analytical methodology for the determination of aflatoxins in palm kernels was reported by Nawaz et al[140] on silica with ether for 17 min and with chloroform – xylene – acetone (6:3:1) for 20 min. Quantification was done by fluorodensitometry. Detection limits were 3.7, 2.5, 3.0, 1.3 $\mu g/kg$ respectively.

Determination of aflatoxins B_1, B_2, G_1, G_2 in corn and peanut products was published by Park et al[141] on silica with chloroform – acetone (9:1). After evaporation of solvent for 5 min in a fume hood and 2 min in 50° C forced draft oven, examination was done under long-wave UV. Quantification was made by fluorodensitometry 365/>430 nm.

Method for determination of aflatoxin in food has been standardized by Horwitz[142] and FAO.[143] A simple method is reported for rapid determination of aflatoxin in corn, buckwheat, peanuts and cheese by Kamimura et al[144] and comparison with various methods was also made.

A method is described for simultaneous determination of 13 mycotoxins using HPTLC by Lee et al.[145] In seven consecutive multiple developments with two solvent systems of different polarity, a base-line separation of sterigmatocystin, zearalenone, citrinin, ochratoxin A, patulin, penicillic acid, leuleoskyrin, and the aflatoxins B_1, B_2, G_1, G_2, M_1, and M_2, was observed. It was possible to determine 13 mycotoxins from one spot by insitu scanning of the plate and nanogram detection limits were observed by UV-visible absorbance (reflectance mode) and in the low picogram range by fluorescence with a relative standard deviation of 0.7 to 2.2% in the nanogram range.

Reversed-phase thin-layer-chromatography of 18 common mycotoxins viz. aflatoxins B_1, B_2, G_1, G_2, sterigmatocystin, ochratoxin A, citrinin, penicillic acid, patulin, zearalenone, and the trichothecenes diacetoxyscirpenol, HT-2 toxin, rivalenol, neosolaniol, fusarenone-X, T-2 toxin, deoxynivalenol and 3-acetyl-deoxynivalenol was done on C_{18} and diphenyl bonded phase.[146] The mobile phases were neutral with acidified binary mixtures of water and methanol, acetonitrile or tetrahydrofuran.

Aflatoxin contents of cereals and nuts were reported by Saleemullah et al.[147] Storage of cereals and nuts for a prolonged period in warm humid condition should be avoided to eliminate risk of aflatoxin contamination.

A collaborative study for determination of aflatoxin B_1, B_2, G_1 and G_2 in corn and peanut products was reported by Park et al.[148] HPTLC determination of deoxynivalenol, fusarenon-X and nivalenol in barley, corn and wheat was reported by Trucksess et al.[149]

Characterization and biological activity of gangliosides(gs) in buffalo milk was elucidated by Colarow et al.[150] TLC and scanning densitometry showed distinct profiles of hydrophilic and lipophilic GS fractions. This was established that toxin binding properties of buffalo milk GS and anti-inflammatory activity of the lipophilized GS fraction will be important for developing innovative food applications which would be useful for future research.

Determination of cyclopiazonic acid in peanuts and corn was published by Lansden.[151] The developing solvent was ethyl acetate - 2-propanol – NH_3 (50:15:10). Quantitation was done by densitometry.

Screening and quantification of Ochratoxin A in corn, peanuts, beans, rice and cassava was reported by Soars et al.[152]

16.11 ANTIBIOTICS AND DRUGS RESIDUES IN FOOD

Thin-layer chromatography for a long time has been accepted as a methodology for separation of different classes of antibiotics residues in food. Considerable efforts have been made to improve the quantitative aspect of different methods.

A method to determine sulphonamides residues in milk, meat and eggs was developed by Diserens et al.[153]

Practical screening procedure for chloramphenicol in milk at low parts per billion levels was published by Schwartz et al[154] on silica with ethyl acetate. Detection was done by 2 minutes exposure in jar containing solid calcium hypochlorite, 45 sec in jar containing 37% formaldehyde, spray with aqueous solution containing 1% starch +1% KI. Detection limit was ≥60 ng.

Quantitative determination of residues of three nitroimidazoles (ronidazole, dimetridazole, and their major metabolite hydroxydimetridazole) were determined at 2-5 $\mu g/kg$ in pork and poultry muscle tissue by HPTLC multiple development technique with methanol and ethyl acetate and then the detection by spraying with pyridine and observation under 312 nm UV light.[155]

Residues of the quinolonic, antibiotics oxolinic acid and flumequine were quantified in fish tissue by densitometry at 320 nm using HPTLC.[156]

Naidong et al[157] have described a procedure using silica gel plate with fluorescence indicator and reverse phase on silanized silica gel for identification of chlortetracycline, demeclocycline, doxycycline, metacycline, minocycline, oxytetracycline, and tetracycline. The reverse phase plates were examined under UV light at 365 nm. The analysis by reverse phase has shown inferior results as compared to normal phase.

Kang and Ebel[158] have described the procedure for the separation of tetracycline, chlortetracycline, oxytetracycline and demeclocycline with CN-HPTLC plate. The authors used doxycycline as an example for quantitation and the same is determined through densitometry at 254 nm.

Corti et al[159] determined rifaximin and its oxidation products for quality control of raw material, i.e. cream and for the residue in the milk after the cream has been used to treat mastitis in cows.

Detection of antibiotic residues (tetracycline) was reported by several workers.[160–165] The brief procedure of extraction from food was that 5 gm food sample was blended in a high speed blender with 20, 20, and 10 ml of 0.1 M disodium ethylenediaminotetra-acetate (Na_2EDTA) + 614.5 ml of 0.1 M citric acid + 385.5 ml of 0.2 M sodium hydrogen phosphate, pH 4.0. This was centrifuged at 4000 rpm for 5 min. approximately. The supernatant was recentrifuged at 4000 rpm for 15 min. and then filtered. In case of honey, 25 g sample is dissolved in 50 ml of 0.1 M Na_2EDTA-McIlvaine buffer (pH 4.0) and filtered. Clean-up procedure was done through a C18-cartridge pretreated with saturated aqueous Na_2EDTA and the cartridge is washed with 20 ml of water; air dried for 5 min and elutes tetracycline with 10 ml of ethyl acetate followed by 20 ml of methanol-ethyl acetate (5:95). Elute is evaporated to dryness under reduced pressure at 30° C and then dissolve the residue in 0.1 ml of methanol.

HPTLC or TLC silica gel plates were used for separation. The plates were predeveloped in saturated aqueous Na_2EDTA solution and dried in room temperature for 1 hr and then activated at 130° C for 2 hours approximately. 5 μl of sample is applied and different volumes of solution standard (1 mg ml^{-1} methanol) are developed with chloroform-methanol-5% Na_2EDTA solution (65:20:5). C_8 plate is developed up to 9 cm with methanol-acetonitrile-0.5 M oxalic acid solution, pH 2.0 (1:1:4). Then densitometry determination is done at maximum absorbance for tetracycline, oxytetracycline, chlortetracycline, doxycycline, and 4-epitetracycline at nearly 360 nm, anhydrotetracycline and 4-epianhydrotetracycline at nearly 425 nm.

Ampicillin in milk and muscle of food producing animals was reported by Abjean et al.[166] Determination at level of 4 and 50 $\mu g/kg$ of milk and muscle was done by densitometry at 365 nm.

HPTLC analysis of quinolonic antibiotics in fish and fish seed was reported by Vega et al.[167] Due to risk of residues in fish tissues, it is important to develop reliable method for qualitative and quantitative analysis. The method is suitable for identification and quantification of quinolonic antibiotics such as oxolinic acid and flumequine as residues in fish meat, as well as for the quality assurance analysis of spiked fish sample.

A paper was presented by Irena Maria Choma[168] on screening of enrogloxacin and ciprofloxacin residues in milk by HPLC and by TLC with direct bioautography. The TLC method can be used for semi-quantitative determination and method is suitable for testing of many samples in one run and is advantageous over HPLC methods.

Rapid quantitative thin-layer chromatographic screening procedure for sulfathioazole residues in honey was published by Neidert et al[169] on silica with chloroform – pentanol (9:1). Detection was possible after drying of the plate by spraying with fluorescamine solution and quantitation by densitometry. Detection limit was 0.02-0.60 mg/kg. Quantitative TLC method is sensitive.

Residues of sulfadiazine, sulfadimidine, sulfadimetoxine and sulfaquinoxaline were screened in pork and beef muscle at 5-100 µg/kg by extraction with ethyl acetate, following cleaning procedure by silica gel and HPTLC on silica gel with preadsorbent zone using ethyl acetate-hexane (2:1) mobile phase. Nitrofurans were detected by spraying with pyridine and examination under 366 nm UV light. Sulfonamides and chloramphenicol were detected by spraying with stannous chloride, NaOH and fluorescamine solutions.[175]

Determination of pentachlorophenol and cymiazole in water and honey was published by Sherma et al[170] on RP-18 silica with toluene–methanol (9:1) and hexane - acetone–methanol–acetic acid (35:10:5:0.1). Detection was done under UV 254 nm. Quantification was done by densitometry at 215 nm for pentachlorophenol and at 265 nm for cymiazole. Recoveries were made from water at 0.25 - 5 ppm,i.e. 97.7–100 % for pentachlorophenol and 89.5-94.9% for cymiazole, and for honey at 10 and 50 ppm, to the extent of 94.0–96.1% and 91.9–93.7% respectively.

A method to identify and quantify multiple antibiotic residues (chloramphenicol, ampicillin, benzylpenicillin, dicloxacillin and erythromycin) in cow's milk by HPTLC combined with bioautography was developed by Ramirez et al.[171] The test micro-organism used for bioautography was *Bacillus subtilis* ATCC 6633. Antibiotic was extracted by acetonitrile, with elimination of fat by petroleum ether. The residue was isolated with dichloromethane. The method is suitable to detect a limit well below the maximum residue limits (MRL) allowed for milk. Percentage recoveries were reported between 90–100% with C V between 7.2 and 21.3%.

TLC technique for determination of residue of antibiotics[172–174] has been reported by more workers.

Matrix solid-phase dispersion combined with thin-layer chromatography–direct bioautography for determination of enrofloxacin and ciprofloxacin residues in milk was done by Choma et al.[175] TLC of enrofloxacin and ciprofloxacin was carried out on silica gel in sandwich chambers with dichloromethane–methanol–2–propanol–25% ammonia in the ratio of 3:3:5:2. Detection was made by bioautography using nutrient medium and *B. subtilis* spore suspension.

Thin-layer chromatography–direct bioautography of lumequine residues in milk was reported by Choma.[176] TLC of flumequine (9–fluoro–6, 7–dihydro–5–methyl–1–oxo–1H, 5H–benzo[ij] quinolizine–2–carboxylic acid) was performed on silica gel in a sandwich chamber with dichloromethane–methanol–2–propanol–25% aqueous ammonia (3:3:5:2). The plates were developed to the top and then continued for 1 hour. Bioautography with nutrient medium and *Bacillus subtilis* spore suspension. After incubation, the plates were sprayed with MTT-solution.

16.12 PIGMENTS AND DYES

Quantification of capsanthin in paprika red pigments on silica gel and densitometry at 470 nm was reported.[177] Detection and quantification of 12 dyes in food extracts was done using HPTLC on silica gel with 2–propanol:1–propanol:1–butanol:ammonia:water (8:4:4:2:1) and then densitometry at the maximum absorption wavelength of each compound was performed and detection limit was 4–10 ng/zone.[178] Synthetic dyes were identified and quantified in wine at a detection level of <5 mg/l by silica gel and cellulose TLC using densitometry.[179]

Use of modified flatbed scanner for quantification of chromatogram detected by fluorescence quenching was made. HPTLC of 5 dyes namely Crystal Violet, Sudan Black, Sudan II, Sudan III, dimethylaminoazobenzene on alumina using n-octyl alumina with benzene–chloroform (4:1) was done. Detection was made under UV (254 nm) on TLC of 4 platinum complexes with acetonitrile–water (4:1).[180] The n-octyl modified alumina was prepared by reaction of alumina with n-octyltrichlorosilane; plates were prepared from the slurry of the sorbent, containing polyvinylpyrrolidone as binder and an inorganic luminophor (zinc and calcium silicate activated with manganese). The stationary phase has been well characterized.

Identification and determination of some synthetic food colours by HPTLC was reported by Rizova et al[181] on silica with 1–propanol:n-propanol :n-butanol:NH$_3$:water (8:4:4:2:1). Quantification of 12 colours was done by densitometry. Detection limit was 4–10 ng/spot.

Imaging and quantification of Sudan II in wet plates was performed in reflectance and transmission modes. The limit of detection on a dry plate (0.5 ng) was observed to be lower than wet plate (2 ng).[182] Planar chromatography techniques for determination of natural pigments[183, 71] have been reported in books. Identification of synthetic food dyes in alcoholic product was developed by Steele[184] on silica with 1–butanol:2–butanol: acetonitrile:THF:MEK:water:NH$_3$ (10:10:25:15:20:18:2) and 1–propanol:acetonitrile:THF:MEK: ethyl acetate:water:NH$_3$ (20:15:25:10:10:18:2). Densitometric scanning was done by absorbance at 480 nm.

Separation, identification and semi-quantitative analysis of natural pigments were reported by Li et al.[185] The separation of pigment is done reversed phase TLC. The method was applied for separation of various carotenoid standards using various proportion of hexane–ethylacetate–acetone–methanol achieving better separation than normal phase TLC.[186] The pigments in vegetables, flowers, foods and herbs have been separated and identified by TLC. A reversed phase TLC method coupled with scanning densitometry for analysis of gardenic yellow was mentioned using acetonitrile–tetrahydrofuran–0.1 m oxalic acid (7:8:7) as eluent system.[187] Hirokado et al[188] employed C$_{18}$ reversed phase with cellulose TLC for analysis of processed foods. The method is applied to the detection of natural colours in different varieties of processed foods when labels indicated for the presence of natural colours. Separation of rutin was done on silica stationary phase with ethyl acetate:formic acid:water (80:10:10 v/v) as eluent and quantitated by scanner.[189] The colour pigments of capsicum annum were separated by adsorption and reverse phase thin–layer chromatography.[190] Different types of plates were used to separate the colour pigments of paprika. In case of adsorption chromatography, the best separations were achieved on aluminium oxide layers and developed with n-hexane-chloroform mixtures. In case of modified phase chromatography, only acetone–water and tetrahydrofuran-water mixtures were able to transport the colour pigments on cyano plates and on supports made by impregnation of cellulose, silica and diatomaceous earth.

The use of coaltar food colour is being discouraged and now natural colouring is frequently used in preparation of food. Therefore, accurate and rapid analytical methods for natural colours are necessary. Reverse phase TLC with scanning densitometer is very useful for this purpose. This is required for food quality surveillance. Other methods like identification of separated colours by TLC with UV–VIS spectrophotometry is also used, but measurement is laborious and time consuming. By use of reverse phase TLC with scanning densitometer, methods for annatto extract, orange colour, gardenia yellow, paprika colour, tomato colour, marigold colour, β-carotene, turmeric oleoresin (carotenoid colourings), lac colour, cochineal colour (quinoid colourings), carthamus yellow (flavonoid colouring), and red cabbage colour (anthocyanin colouring) have been cited out by different workers[191–197].

Method for determination of carotenoid colour of tomato, orange and marigold was developed by using TLC method.[195] The main component of colour in tomato is lycopene. This yellowish red colour is heat and light resistant. The main component of orange colour from orange fruit is fatty acid ester of β–cryptoxanthin. The main component of marigold colour is fatty acid ester of lutein. These colours were extracted from the samples with ethyl ether. The extract is evaporated, and then the residue is dissolved in methanol. In case of tomato extract, water is added and then it is purified through a C_{18} cartridge before TLC analysis. In case of orange and marigold colours, 2 ml of 5% sodium hydroxide–methanol solution is added. The mixture is stirred and kept for 24 hours at room temperature in dark. pH of the mixture is made 4.5 or less using 1 mol/l of hydrochloric acid. Then the mixture is purified with a C_{18} cartridge before analysis. Reverse phase plate is used. The solvent system is acetonitrile–acetone–n–hexane in the ratio of 11:7:2 and acetone–water in the ratio of 9:1. Absorption spectra of the colours are measured in visible range using scanning densitometry in situ.

β-carotene is an orange dye found in most green leaves and in carrots. Paprika colour is obtained from extract of fruit of red pepper which contains capsanthin and its esters of lauric acid, myristic acid, and palmitic acid in large amounts as its colour components. The method for analysis of β-carotene for paprika colour in food was developed by using TLC and scanning densitometry.[193] β-carotene is extracted with ethyl ether, while paprika colour is extracted with ethyl ether after saponification with sodium hydroxide and methanol. The extracts in both case were cleaned up with a C_{18} cartridge and the separation of colours were done on the reversed phase C_{18} TLC plate using solvent as n-hexane–acetone–acetonitrile in the ratio of 2:7:11 as a mobile solvent. The absorption spectra of the colour are measured with the help of scanning densitometry.

Turmeric oleoresin is a yellow pigment obtained by extraction of whole turmeric with a volatile solvent and then the solvent is removed. The main component "curcumin" is used for colouring of pickle and curry powder, cooked vegetables, etc.

Gardenia yellow is extracted from gardenia fruit and is yellow in colour. The major components are crocin and crocetin. Anatto colour is yellow orange pigment obtained from the seeds of Bixa orellans. The main components are bixin and norbixin which are used for colouring of butter, cheese, etc. Analytical method for turmeric, oleoresin, gardenia yellow and annatto extract (including annatto, water soluble) in foods was developed by using reversed-phase TLC with scanning densitometer.[192] The colours are extracted by water or methanol from food and the extracts are cleaned with C_{18} cartridge after evaporation of extracting solvent. The separation of sample is done on reversed phase C_{18} TLC plate using acetonitrile–tetrahydrofuran–0.1 mol/L oxalic acid (7:8:7) as mobile phase and the visible absorption spectra is measured using scanning densitometer.

Another method for carotenoid colourings, i.e. annatto extract, orange colour, gardenia yellow, paprika colour, tomato colour, marigold colour and β-carotene in food was reported in literature using reversed phase C_{18} TLC plate using acetonitrile:acetone: n-hexane (11:7:2) and acetone:water (9:1). Also normal phase TLC plate using a solvent system of n-hexane–diethylether–acetic acid (4:1:1) and benzene–ethyl acetate–methanol (15:4:1) were used for separation. R_f value of all the colour spots was evaluated under different TLC conditions.

Lac dye, a natural food additive is extracted from stick lac and this is widely used as a colouring in food.[198–200] It is derived from a water–soluble pigment including laccaic acids A, B, C, and E.

Cochineal colouring (red pigment) is obtained from the insect Coccus cacti that live on cactus plants. The main component is carmic acid. This is used for colouring of candy and jelly. Analysis of lac and cochineal colours by reversed phase TLC and scanning densitometer was developed by Itakura et al. [191] The colour is extracted with 0.1 mol/L oxalic acid, 80% methanol and extracts is cleaned up with C_{18} cartridge after evaporation of methanol. Separation is achieved using methanol–

0.5 mol/L oxalic acid (5.5:4.5) on the RP HPTLC plate. Visible spectra were recorded by scanning densitometry.

Carthamus yellow is a yellow dye extracted from flower of Carthamus tinctorius L with water. The dye is derived from flavonoids, saffronin A and saffronin B. The colour is yellow under acidic media and reddish yellow under basic conditions. This is used for colouring of juice, candy, jelly, chewing gum, fruit wine, chocolate, etc. A method for analysis of carthamus yellow was described by Watanable et al[197] using reversed phase TLC method with scanning densitometry. The colour is extracted with water and is cleaned up with C_{18} cartridge. Separation of colour was achieved by using 2–butanone–methanol–5% sodium sulfate–5% acetic acid (3:2:5:5) as mobile phase in a reversed phase C_{18} TLC plate. The colours are measured using a scanner.

Red cabbage colour (red dye) is extracted from red leaf of Brassica oleracea L. var. capitata DC with water under weak acidic conditions. The red colour is derived from derivatives of cyanidin acylglucoside. The colour is red purple under acidic media. This is used for colouring juice, candy, jelly, chewing gum, etc. Latest method was developed by Itakura et al[195] for analysis of red cabbage colour using the reversed–phase TLC and scanning densitometry. The colour is extracted using 0.1% trifluoroacetic acid from foods and it is cleaned up using C_{18} cartridge with 5 ml of methanol– 0.1% trifluoroacetic acid (9:1). The colours are separated on reversed phase C_{18} TLC plate by use of acetonitrile–0.2 mol/L trifluoroacetic acid (1:2).

16.13 ESSENCES

Determination of vanillin and related compounds in natural vanilla extracts and vanilla–flavoured food using automated multiple development was reported by Belay et al.[201]

HPTLC of vanillin, 4–hydroxy–3–methoxycinnamaldehyde and pinoresinol on silica gel with petroleum ether–ether (1:1) or chloroform–acetonitrile (2:1) was reported by Carpinella.[202] Detection was done under UV. After cutting the plate in two pieces, spores of F. verticillioides were sprayed directly, on one half of the plate and the other half was sprayed with vanillin sulfuric acid reagent followed by heating. TLC and preparative TLC of phenolic compounds, e.g. rosmarinic acid, on silica gel with chloroform–methanol–water (13:7:2) was carried out. Detection was done after drying and then spraying with 2.5% iron (III) chloride solution.[203]

Identification of 5–(hydroxymethyl)–2–furfural in vanilla extracts was published by Kiridena et al[204] on silica with chloroform–ethyl acetate–propane–1–ol (94:2:4). Quantification was made by densitometry at 280 nm.

Determination of vanillin and related flavour compounds in natural vanilla extracts and vanilla– flavoured foods by HPTLC and AMD was reported by Belay et al[205] on silica by two different solvent gradients. Identification was made by in situ measurement of the UV spectra. Determination was made by densitometry at 280 nm. Detection of coumarin in vanilla flavourings by quantitative HPTLC was done by Sherma et al[206] on silica with toluene–methanol (97:3). Detect by spraying with 5–10% solution under UV 360 nm. Determination limit was 5 ng for fluorescence and 500 ng for absorbance. A method has been developed for the separation of principal polar aromatic flavour components of natural vanilla extracts and spurious vanilla extracts by Poole et al.[207]

HPTLC of vanilla extracts (5 hydroxymethyl 2–furfural, 4–hydroxybenzoic acid, 4– hydroxybenzaldehyde, vanillic acid, vanillin) on silica with AMD was reported by Poole et al.[208] Quantification was done at 255 and 280 nm.

16.14 WINES

Determination of histamine in wines by quantitative HPTLC was reported by Sherma et al[209] on silica with acetonitrile–acetone–NH_3 (15:5:5). Detection was done by spraying with ninhydrin solution followed by heating for 10 min at 110° C. Quantification was done by densitometry at 520 nm.

Quantitative HPTLC determination of carboxylic acids in wine and juice was reported by Lin et al[210] with ethyl acetate–toluene–water–formic acid (60:20:20:15). Detection was done by staining with xylose–aniline reagent. Densitometry evaluation was made by absorbance at 546 nm. The method is fast, sensitive and of wide dynamic range.

HPTLC analysis of red wine pigments was performed by Lambri et al[211] to give qualitative and quantitative information on red wine pigments. The samples were prepared by solid phase extraction on C_{18} cartridges. Analysis was done on C_{18} silica gel plates with isocratic elution with methanol–water–trifluoroacetic acid (55:45:1)v/v. The method enables clear separation of the three classes of anthocyanins with good repeatability and reproducibility for wines of different vintages.

16.15 FRUITS AND VEGETABLES

Anthocyanins (flavonoids) are present in fruits and vegetables as glycosides. Gradient preparative thin-layer chromatography and RP-HPTLC were used for analysis of anthocyanin compounds in extract and were separated by five step gradient elution on silica gel $60F_{254}$ PLC plates using mobile phases having varied amounts of methyl tert-butyl ether (MTBE) as modifier. Preparative chromatography was followed by re-chromatography using mobile phase gradients comprising of methanol, water and formic acid on RP-HPTLC plates.[212]

HPTLC determination of imidacloprid, fenitrothion and parathion in Chinese cabbage was reported by Cao et al.[213] The silica plate was prewashed with methanol and activated at 110° C for 30 min. The solvent used was hexane–acetone (7:3) in an unsaturated twin–trough chamber and quantitation was done by absorbance measurement at 287 nm.

Chemical and analytical screening of some edible mushrooms was reported by Mallavadhani et al[214] using HPTLC. Correlations were established between occurrence of compounds in mushrooms and consumptions in relation to beneficial health effects. Determination of dhatura and related alkaloids extraction was reported by Mroczek et al[215] in plant extract. HPTLC densitometric assay on silica gel plates was evaluated at 205 nm without derivatization and this included single development (distance 9.5 cm) with acetone–methanol–water–25% ammonia (85:5:5:8) mobile phase for L–hyoscyamine and scopolamine separation. For scopolamine–N–oxide and scopolamine N–methyl bromide, a second development (distance 5.5 cm) with acetonitrile–methanol–85% formic acid (120:5:5 v/v) was used. The method can be tried for detection of dhatura in food. HPTLC determination of caffeine in "power" drinks was reported by Abourashed et al[216] using UV densitometric analysis on precoated HPTLC plates. The solvent system comprised of ethyl acetate-methanol (85:15) and caffeine was detected at 275 nm. The validation of the method was done for specificity, repeatability (CV<5%).

Heterocyclic oxygen compounds in citrus fruit essential oils were determined by Dugo et al.[217] The mobile phase used was n–butyl acetate–hexane (8:2) and chloroform–n–butyl acetate–hexane (9:1:15) and detection was done at 254 and 366 nm.

Determination of capsanthin in paprika red pigments was reported by Ding et al[218] on silica with petrol ether–benzene (3:1) and hexane–benzene–ethyl acetate–ethanol (20:2:5:2). Quantitation was done by densitometry at 470 nm.

16.16 SPICES AND ESSENTIAL OILS

Planar chromatography gives an opportunity for performing online sample preparation and separation of crude spice and flavour extracts without further sample clean-up. This is combined with scanning densitometry. Spectral identification and quantitation is simple.

Quantification of eugenol was done by HPTLC[220] of methanolic extract with toluene: ethylacetate: formic acid (90:10:0.1) by densitometry at 280 nm. This method can be applied for food samples.

L-ascorbic acid was determined after chromatographic separation in aqueous extract of pepper juice on silica gel with glacial acetic acid–acetone–methanol–benzene (3:1:4:14), butanol–formic acid–water (200:10:3), and water glacial acetic acid–ethylmethylketone–ethylacetate (1:2:2:5). Detection was done under UV at 254 nm. Quantitative determination was done by absorbance measurement at 588 nm after oxidation with iodate.[221]

Development and validation of HPTLC–densitometric method for the quantitation of alliin from garlic and its formulations was done by Kanaki et al[222] on silica gel with n-butanol–acetic acid–water (3:1:1). Detection was carried out by dipping in ninhydrin reagent for two seconds followed by heating at 110° C for 5 min. Quantitative determination was done by densitometric evaluation of peak areas at 540 nm.

Quality control of commercial mustard by HPTLC was reported by Marutoiu et al[223] on silica gel with isopropanol–25% ammonia (9:1) containing different volumes of water. After development, the compounds were visualized under UV light at 254 nm or by exposure to iodine vapor.

Determination of curcumin in spices was done by A. Janssen et al[224] on silica with chloroform–acetic acid in the ratio of 8:2. Fluorescence derivatives were made by spraying with boric–oxalic acid reagent and heating at 100° C for 10 min and then scanning by fluorescence. Detection limit was 10 ng.

The quality and botanical origin of the cinnamons of commerce is ascertained by simultaneous detection of cinnamaldehyde, eugenol, coumarin, 2–methoxycinnamaldehyde, cinnamyl alcohol, cinnamyl acetate and cinnamic acid in acetonitrile extracts of powdered cinnamon. True cinnamon can be differentiated from cassia with the help of relative ratio of eugenol and coumarone to cinnamaldehyde.[219]

Capsaicinoid quantitation by in situ densitometry was developed by Monforte–Gonzales et al.[225] TLC of capsaicinoids from chilli peppers was carried out with capsaicin, dihydrocapsaicin, coumaric acid, vanillin, ferulic acid, and cinnamic acid as standards, on silica gel by twofold development with cyclohexane–chloroform–acetic acid (7:2:1); chloroform–methanol–acetic acid (95:1:5); and cyclohexane–acetone (4:5). Visualization was done under UV light at 254 nm and quantitation by densitometry at 254 nm.

16.17 ALKALOIDS

Purine alkaloids (caffeine, theobromine, theophylline) were determined in daily foods and health drinks by Extrelut column extraction and silica gel HPTLC. Quantification was done by densitometry at 275 nm with p–hydroxybenzaldehyde as the international standard.[226]

Estimation of nicotine from Gutka, an Indian chewable tobacco preparation by HPTLC was reported by Mandal et al.[227] Standard solution (1 mg/ml) was basified with ammonia, extracted with diethyl ether and sample was triturated with 0.5 M KOH, extracted with ether. 5-25 μl of sample/standard was chromatographed with chloroform:methanol:ammonia (60:50:10) and scanned at 255 nm for evaluation.

16.18 VITAMINS

Corti et al[228] used densitometry method for detection of vitamin B_1 with HPTLC–NH_2 layers with the mobile phase of methanol–water (1:0.2). Further densitometric detection was done at 254 nm and 366 nm as excitation wavelength for the thiochrome used. Planar chromatography technique for determination of vitamins [229, 230] has been reported by workers.

TLC method for the analysis of vitamins was reported by Claudia et al.[231] Preliminary identification of hydrophilic vitamins (vitamin C and B complex: B_1, B_2, B_3, B_5, B_6, B_9, B_{12}, and vitamin H), and lipophilic vitamins(vitamin A, D, E and K) in food was carried out.

Separation and identification of eight hydrophilic vitamins using a new TLC method and Raman spectroscopy was developed by Cimpoiu et al.[232] HPTLC of eight hydrophilic vitamins (B_1, B_2, B_3, B_5, B_6, B_9, B_{12} and C) was done on silica gel with mixtures of methanol and benzene in a saturated N–chamber. Detection was made under UV 254 nm. Vitamin B_5 was detected by spraying with ninhydrin reagent (2% in ethanol). Raman spectra were recorded.

16.19 GENERAL

Gallic acid in tea extracts (gallic acid, caffeine, T–catechin and tannic acid as standards) on silica gel in an unsaturated chamber with chloroform–ethylacetate–formic acid (5:4:1) was determined by HPTLC with densitometric measurement at 289 nm. Limit of detection was 0.1 μg per spot.[233]

HPTLC of famotidine on silica gel with ethyl acetate–methanol–toluene–conc. aqueous ammonia (40:25:20:2) with saturation and of caffeine in cola with ethyl acetate–methanol (19:1) were performed. Detection limit of famotidine and caffeine were determined along with precision and linearity.[234]

Quantitative determination of isoflavonoids(puerarin, 3'–methoxypuerarin, daidzin and daidzein) in several kudzu samples were performed using HPTLC by absorbance measurement at 254 nm. Though repeatability and reproducibility were noticed better by HPLC but separation of isoflavonoids by HPLC is time consuming and complex, whereas HPTLC offers a simple and rapid method with simple isolation process.[235]

Propyl, octyl and dodecyl gallates from olive oil and butter were detected and quantified by densitometry. Similarly quantification of hydroxamic acids in wheat and rye using densitometry determination by HPTLC gradient elution technique was made by UV scanning at 260–310 nm for determination of vanillin, vanillic acid, p–hydroxybenzaldehyde, and p–hydroxybenzoic acid in vanilla extracts, coumarin in cinnamon and curcuminoids in turmeric.[238, 239]

Determination of 7–ketocholesterol and 7–hydroxycholesterol in meat samples was made using densitometric detection by Janoszka.[240] TLC of two oxycholesterols, 7–ketocholesterol and 7–hydroxycholesterol (cholesterol, 7–keto–,7α–hydroxy–, 7β–hydroxycholesterol and cholesterol–

5α,6α–epoxide as standards were done on silica gel with chloroform:acetone (9:1) in a horizontal chamber. Detection of 7–ketocholesterol was made under UV at 254 nm and quantitation was done by densitometry after TLC separation.

Determination of dipalmitoyl phospharidylcholine in food products was reported by Traitler et al.[241] HPTLC separation of phospholipids from other lipids in dietary products was reported on silica with twofold developments with hexane–ether (6:4) and in chloroform–ethanol–triethylamine–water (30:35:35:8). Visualization of phospholipids was made with 10 g sodium molybdate dehydrate, 1.8 g ascorbic acid in 100 ml 9 M sodium acid–acetic acid (6:1) and after elution, quantification was done on GC–MS.

Determination of lycopene in tomato ketchup on silica with bezene–petrol ether in the ratio of 3:7 was reported by Li et al.[242] Quantification was done on densitometry at 480 nm.

Jaiswal[243] presented the use of HPTLC in food analysis. The paper highlighted that deliberate adulteration in food for economic benefit or contamination in food is a biggest threat to a consumer from health view–point. Therefore, detection/determination of unknown adulterants in food is an issue which has been a subject of anxiety for analytical chemists. In these directions, HPTLC has been proved to be a boon for chemists for solving such problems.

HPTLC was used for rapid screening and detection for measuring cadaverine for rapid assessment of bacterial quality of fresh and stale mutton.[244] Analysis of more than 150 samples for degrees of freshness and postmortem age indicated that in more than 90% of the samples, no cadaverine was possible to detect when meat was deemed acceptable with <106 orgs/gm. But when meat exhibited incipient spoilage, as judged by off odour and dull colour, cadaverine was detected in range of 1–5 ppm with bacterial load of 107 orgs/g or more. The method is recommended for rapid assessment of bacterial quality of mutton.

Separation of sterols by reversed phase and argentation TLC was developed by Jarusiewlcz et al.[245] HPTLC and TLC of sterols (cholesterol, cholestanol, beta–sitosterol, stigmasterol, ergosterol, compesterol, desmosterol, and brassicasterol) on RP–18, RP–18 W, RP–2, RP–8, amino, cyano, diol, and phenyl bonded phase, hydrocarbon impregnated layers, and silica gel impregnated with 10% silver nitrate, with 25 mobile phases. Optimal separation of sterols was achieved on RP–18 with acetonitrile–chloroform (8:7) or petroleum ether–acetonitrile–methanol (1:2:2). Detection was done by spraying with ethanolic phosphomolybdic acid and heating at 115° C for 10 min.

HPTLC method for trace analysis of 5 heterocyclic aromatic amines (PhIP, MeIQx, 4, 8) in meat samples was reported by Jautz et al.[246] HPTLC silica gel layer was pre-conditioned with ammonia vapour and developed with methanol–chloroform (1:9). Quantitative determination was done by absorbance measurement at UV 262 and 316 nm, fluorescence measurement at UV 366/> 400 nm. Repeatability was better than 3.3% (n = 14). Reproducibility of migration was better than 1.3% (n = 6). LODs of the 5 heterocyclic aromatic amines ranged between 0.4 and 5 ng/band.

Bioluminex assay[247] deals the coupling of bioluminescence to HPTLC directly, thus provide rapid and unique way of toxicity in complex mixtures like food, food additives and dietary supplements. In this technique after chromatography, HPTLC plate is coated with bioluminescent bacteria *Vibrio fischeri* in a simple dip. Separated fractions which inhibit the growth of bacteria are identified selectivity as dark zones on a luminescent back ground. The results are noticed immediately and can be documented. The method is suitable to screen food, and beverages, etc. for the presence of non-traditional chemicals and toxic adulterants. This is a good tool for identifying compounds with potential biological activity. For example, *V. fischeri* is able to detect ochratoxin in canned corn, aflatoxin B$_1$ in honey, dioxin in milk, benzopyrene in celery seed, capsaicin in cayenne pepper, strychnine or monofluoroacetic acid in drinks, and patulin in apple juice.

A general detection technique for HPTLC based on changes in fluorescence was indicated by Mateos et al.[248] The method was found suitable for determination of complex hydrocarbons. It was further shown than polar compounds, e.g. antibiotics and amino acids give negative peaks which correspond to fluorescence quenching.[249, 250] The quenching may be due to net specific interaction which increase the rate constant of non-radiative decay process, thus producing a decrease in quantum yield.

Determination of acrylamide in drinking water was reported by Alpmann et al.[251] It was discussed that polyacrylamide is used as a flocculating agent in treatment of drinking water besides its use in different industries. As it is highly soluble in water, monomer acrylamide(AA) can be found in ground and drinking water. As compared to HPLC–MS/MS for routine analysis, this method based on the derivatization of AA with fluorophor was developed which was more practical.

Several types of foods are derived from different portion of plants. Either these are consumed by human beings, as such as a food or after processing them. Many plants are poisonous or their consumption in bulk over a long period of time leads to toxicity. Details regarding analysis of plants poisonous have been discussed by Gautam et al.[252]

16.20 HYPHENATED COUPLING TECHNIQUE OF HPTLC IN FOOD ANALYSIS

An online TLC-MS interface with computer-controlled extraction of substances from selected spots on HPTLC plate was constructed by Prosek et al.[253] The controlled collection of the samples and its programmed injection into the mass spectrometer is the main advantage of this type of interface. The interface was validated with a standard solution of caffeine as a test substance. The results were compared from previously established and routinely used off-line TLC–SPE–APCI–MS procedure.

An improved online coupling of planar chromatography with electrospray mass spectrometry for the extraction of fractions from HPTLC glass plates was used by Alpmann et al.[254] This modified device permits repeatability of extraction and linearity of its signal. The detection capability was examined along with influence of the elution solvent on the intensity of mass spectroscopy. The extraction was successful for various substances in the field of food analysis.[255–257] In the study, two products of synthesis namely xanthyl ethyl carbamate and dansyl ethylamide were taken up.

Quantification of isopropylthioxanthone (ITX) in milk, yoghurt and fat sample was demonstrated by Morlock et al.[255] In case of soyabean oil and margarine, partitioning of ITX into acetonitrile was used. Fluorescence measurement was performed at 254/>400 nm for quantification. The limit of detection (S/N of 3) was established at 64 pg for ITX and DTX of both type of HPTLC plate. The positive results were confirmed by online ESI/MS in the SIM mode and the level of quantification was found as 128 pg by DART/MS involving a minimal employment of MS device which is advantageous of HPTLC.

Simultaneous determination of riboflavin, pyridoxine, nicotinamide, caffeine and taurine in energy drinks by planar chromatography–multiple detection with confirmation by electrospray ionization mass spectrometry was carried out by Aranda et al.[258] Ten samples of energy drinks and six samples of beverages containing caffeine were made by degassing in an ultrasonic bath for 20 minutes. After chromatography, scanning was done by UV-absorbance measurement at 261 nm for nicotinamide and 275 nm for caffeine, fluorescence measurement at 366/>400 and 313/>340 nm for riboflavin and pyridoxine respectively and VIS-absorbance measurement at 525 nm for taurine, after post-chromatographic derivatization with ninhydrin reagent. Mass confirmation was performed by single quadrupole MS in positive electrospray ionization (ESI) scan mode for all substances except taurine (negative mode). The method presents a better alternative for routine analysis as it is simple and reliable.

Planar chromatography coupling to mass spectrometry for employment in trace analysis was reported by Jautz et al.[257] HPTLC/MS by a plunger-based extraction device was found to be a suitable technique for quantitative planar chromatography in trace analysis. Reproducibility extraction from silica gel phase in lower–pg range distinguished this technique from other techniques, repeatability of the MS signal demonstration of RSD of 12.5% and analytical response within a plate and over various plates showed determination coefficients of 0.9915 and 0.9488 respectively. Limit of detection with single quadrupole was better than 40 pg and limit of quantitation by a tandem mass spectrometer was better than 20 pg. LOQ/LOD determination was of same magnitude as reported for HPTLC/MS method.

A method was developed to quantify caffeine, ergotamine and metamizol by HPTLC/UV/FLD with mass confirmation by online HPTLC/ESI–MS. The results established that the method was simple and reliable alternative for routine analysis.[259]

Quantitative thin-layer chromatography/mass spectrometry analysis of caffeine using a surface sampling probe electrospray ionization tandem mass spectrometry system was reported by Ford et al.[260] Thin layer chromatography/electrospray tandem mass spectrometry method employed a deuterium-labelled caffeine internal standard and selected reaction monitoring detection up to nine parallel caffeine bands on a single surface scanning experiment requiring 35 min at a surface scan 44 μm/s.

Analysis of native milk oligosaccharides directly from TLC by matrix-assisted laser desorption/ionization orthogonal–time–of–flight mass spectrometry with a glycerol mixture was reported by Dreisewerd et al.[261] In this method, an adoption of procedure for direct coupling of HPTLC with matrix–assisted laser desorption/ionization MS(MALDI–MS) was reported by Dreisewerd for analysis of a complex ganglioside mixtures for the mixtures of native oligosaccharides from human and elephant milk.

Mohan et al[263] developed a method for rapid detection of residues of cardenolides of *Nerium Oleander* (poisonous plant to human beings) using HPTLC. Extraction was done by using accelerated solvent extraction (ASE). Separation of cardenolides was done on HPTLC plate (silica gel 60 F_{254}) with chloroform:acetone:acetic acid (8.5:1:0.5) as mobile phase and densitometric analysis was carried out at 275 nm. The H–NMR spectra were recorded for the separated components and the component corresponding to oleandrin was identified. The method is simple rapid with high resolution of separation and free from interferences from the plants.

16.21 MISCELLANEOUS

Densitometric TLC analysis of seven azaarenes, i.e. acridine, benzo(h) quinoline, benzo(a) acridine, benzo(c) acridine, dibenzo(a, c) acridine, dibenzo(a, j) acridine, and dibenzo(a, h) acridine in grilled meat on RP-18 in a horizontal chamber with dichloromethane-n-hexane-2-propanol (60:40:1) reported by Beata Janoszka.[264] After drying, visualization was made under UV light at 254 and 366 nm. Quantification was done by densitometric fluorescence measurement at 380 nm. Limits of determination were reported from 0.04 to 0.30 ng/zone.

A new fully automated online interface to couple HPTLC with ESI-MS/MS was reported for the first time by Luftmann et al.[265] Among the major features of this interface was the time required for analysis, precision, suited for normal and reversed-phase layers and all plate sizes and carriers besides no post-chromatographic process is required. It can be coupled with all LC-MS ion sources without any adjustment or mass spectrometer modification. The quantitative analysis is performed without any internal standard with a given detectability at the low-nanogram and at pictogram level. The validation results for caffeine quantification in energy drinks and pharmaceutical samples, without internal standard has proved the reliability of the interface and its usefulness for quantitative analysis and comparable results were obtained.

Presentation of the coupling of planar chromatography with direct analysis in real-time time of flight mass spectrometry (DART-TOF-MS) was made by Morlock et al.[266]

By cutting the plate within a track led to substance zones positioned on the plate edge and the intersted zones were directly introduced into the DART gas stream to obtain the mass signals instantaneously within few seconds, giving the detectability in the very low ng/zone-range on the example of isopropylthioxanthone. The coupling was perfectly suitable for identification and qualitative purposes. Whereas for quantification of results, the analytical response and the repeatability were strongly dependent from proper manual positioning of the HPTLC plate into the excited-state gas stream of the ion source. By use of stable isotope-labelled standards, the drawback can be overcome and demonstrated with the example of caffeine, and the analytical response (R2 of 0.9892) and repeatability (RSD < ± 5.4%, n = 6) were improved to a great extent. The spatial resolution by an in-house-built plate holder system was better than 3 mm; the decay of the signal was also observed. On comparison of the efficacy of this new coupling to a plunger-based extraction device for HPTLC/electrospray ionization-MS, the detectability of latter showed to be down to the pg/zone-range. The limit of quantification for isopropylthioxanthone was 100 pg/zone. The repeatability was comparable (RSD ± 6.7%). However, without the need of internal standard correction, the analytical response was slightly better (R2 pf 0.09983). The spatial resolution was 2 or 4 depending on the plunger head used.

HPTLC of 3 phenolic acids (caffeic acid, p-coumaric acid, isoferulic acid) and 4 flavonids (pinocembrin, pinocembrin-7-methyl ether, chrysin, tectochrysin) was done by Josipa et al[267] on silica gel with chloroform-methanol-formic acid (88:7:5) with chamber saturation. Detection was made by spraying with 1% ethanolic aluminium chloride solution with quantification by scanning densitometry in absorbance mode.

HPTLC of trehalose was carried out by Ranganathan et al[268] on silica gel, impregnateg with phosphotungstic acid of pH 2.5, with n-butanol-pyridine-water (8:4:3) and detection by spraying with a solution of 6.5 mM N-(1-naphthyl)-ethylenediamine dihydrochloride in methanol, containign 3% sulfuric acid. The hR$_f$ values of raffinose, trehalose, maltose, surcrose, glucose and fructose were 30, 41, 46, 53, 55 and 59 respectively.

HPTLC of suralose was undertaken by Morlock et at[269] in silica gel, impregnated with phosphotungstic acid of pH 2.5, with n-butanol - pyridine - water (8:4:3) and detectin by spraying with a solution of 6.5 mM N-(1-naphthyl)-ethylenediamine dihydrochloride in methanol, containing 3% sulfuric acid. The hR$_f$ values of raffinose, trehalose, maltose, sucrose, glucose, fructose was also separatd on amino phases with acetonitrile-water (3:1). Detection was performed by dipping in 2-naphthol sulfuric acid reagent and aniline diphenylamine ortho-phosphoric acid reagent, followed by heating at 120° C. Post-chromatographic dericatization on aluminium-backed amino phases was performed by heating the plate 190° C for 20 min. Evaluation was made under UV light at 366 nm. For fluorescence enhancement, the amino phase was dipped into a (1:2) solution of paraffin in n-hexane. Densitometric evaluation was done by fluorescence measurement at 500 and 405 nm.

HPTLC of unsaponifiable constituents of rice bran oil was done by Afinisha et al[270] on silica gel in two stage separation: First separation was with benzene-chloroform (12:1) for sterols, oryzanols, and tocols. Quantitative determination was done by absorbance measurement at 206 nm for sterols (1), 325 nm for oryzanols (2), and 297 nm for tocols (3). Second separation was carried out with petroleum ether-diethyl ether (50:1) for steryl esters (4), wax (5) and squalene (6). Detection was reported by dipping in 5% methanolic sulphuric acid followed by heating at 110° C for 1 hour and quantitative determination by absorbance measurement at 439 nm. The hRf values were 12 for (1) , 21 for (2), 39 for (3), 36 for (4), 46 to (5), and 74 for (6). Linearity was between 150 and 1200 ng/zone for the first separation and was between 400 and 1200 ng/zone the second the second separation. The limits of detection and quantification were noticed 6 and 20 ng/zone for (1), 1 and 4 ng/zone for (2), 11 and 38 ng/zone for (3), 22 and 73 ng/zone for (4), 19 and 65 ng/zone for (5), and 3 and 10 ng/zone for (6), respectively. Intra-assay precision was between 0.52 and 1.94% and inter-assay precision was between 0.87 and 2.27%. Recoveries ranged from 93.5 to 101.9%.

TLC of b-sitosterol in Pumpkin seed oil was reported by Starek et at[271] on silica gel with toluene-ethyl acetate glacial acetic acid (15:4:1) with chamber saturation for 30 min. Visualization by spraying with anisaldehyde reagent and heating at 90° C for 5 min. Densitometric quantitation was done at 525 nm.

HPTLC of absinthin in absinthe beverage (from the wormwood plant Artemisia absinhium L using sensory evaluation and HPTLC analysis of the bitter principle absinthin on silica gel was reported by Lachenmeier[272] with acetoneacetic acid (98%)-toluene-dichloromethane (1:1:3:5). Detection made by dipping into a solution of acetic anhydride sulphuric acid-ethanol (1:1:10) then followed by heating for 5 min and 104° C. Quantitative determination was made by absorbance measurement at 554 nm. The hR_f value of absinthin was 64 and selectively regarding matrix was given. Linearity was between 0.1 and 10 g/L. The precision was better than 13.5% (intraday) and 15.8% (interday). The limit of detection and quantification for absinthin was 0.05 and 0.11 g/L, respectively.

Chemical and analytical screening of some edible mushrooms was carried out by Mallavadhani et at[273] HPTLC of nicotinic acid (1) and pyrazole-3(5)-carboxylic acid (2) of Volvariella volvacea on silica gel was done with chloroformmethanol (15:3) with one drop of formic acid added. Quantitative determination made by absorbance measurement at 190 nm for (1) and 262 nm for (2). The hR_f values for (1) and (2) were 30 and 40 respectively. Linearity was between 400 and 7000 ng/zone (1) and 200 and 2500 ng/zone for (2). The limits of detection and quantification were 50 and 400 ng/zone for (1) and 20 and 200 ng/zone for (2). Recoveries of both compounds were between 96 and 102%.

Monitoring the origin of wine was repoeted by Cimpoiu et at[274] by reversed-phase thin-layer chromatography. TLC of the colour pigments from different sorts of red wine (Cabernet Sauvignon, Merlot, and Burgundy) was done on RP-18 with acetonitrile-water-formic acid (20:29:1) in a saturated chamber. Evaluation made in visible light and under UV light at 366 nm, and by spraying with a methanolic solution of 0.5 mg/mL DPPH (2, 2. diphenyl-1-picrylhydrazyl). RP-TLC is a tool for monitoring wine, for identification of the origin, and for detection of adulteration.

Aflatoxin in Myrobalan, Karanj seed and Powad seed[275] and in Groundnut oil and soyabean oil[276] by HPTLC was reported.

References

1. Aurand LW, Woods AE, Wells MR. *Food Composition and Analysis*, Van Nostrand Reinhold, New York, 1987
2. Pomeranz T, Meloan CE. *Food Analysis: Theory and Practice*, AVI Publishing, Westport, CT, 1971
3. Boyles S, Method for the analysis of inorganic and organic acid anions in all phases of beet production using gradient ion chromatography, *J Am Soc Brew Chem* **50**; 61, 1992
4. FAO, Food and Nutrition paper, Manuals of Food Quality Control, 8, Food Analysis: Quality, Adulteration and Tests of Identity, Publication Division, United Nations Food and Agriculture Organization, Rome, 1986
5. Ishikawa F, Saito K, Nakazato M, Fujinuma K, Moriyasu T, Nishima T. Determination of color additives by ion–pair high performance liquid chromatography, Annual Report of Tokyo Metropolitan Research Laboratory of Public Health, **41**; 101, 1990
6. Naguib KM, Shalaby AR, Badawy A. Rapid micro–method for the extraction and determination of aflatoxins in animal or human liver tissue, Bull. Nutr. Inst., Cairo, Egypt, **13**; 45, 1993
7. Prosek M, Bukl M, Jamnik K, Carbohydrates in Handbook of Thin Layer Chromatography, Sherma, J., Fried. B. eds.. Marcel Dekker, New York, 439, 1991
8. Pruden BB, Pineault G. A thin layer chromatographic method for the quantitative determination of D–mannose, D–glucose and D–galactose in aqueous solution, *J Chromatogr* **115**; 477, 1975
9. Hansen SA. Thin layer chromatographic method for the identification of mono–, di–and trisaccharides, *J Chromatogr* **107**; 224, 1975
10. Mansfield CT, Use of thin layer chromatography in determination of carbohydrates, Quantitative Thin Layer Chromatography, Touchstone, J. C. ed., John Wiley and Sons, New York, 79, 1973
11. Menzies IS, Seakins J, Sugars WT, Chromatographic and Electrophoresis Techniques, Vol. I, Smith, I. and Seakins, J. W. T. eds. William Heinemann Medical Books Ltd., London, 183, 1976
12. Gaugain M, Abjean JP. *J Chromatogr A* **737**; 343, 1996
13. Bruun–Jensen L, Colarow L, Skibsted LH. *J Planar Chromatogr–Mod TLC* **8**; 475, 1995
14. Prosek M. *J Planar Chromatogr* **16**; 58–62, 2003
15. Kroh LW. Chromatographic In COM Sonderband, 139–147, 1996
16. Klaus R, Fiuscher W, Hauck HE. *Chromatographia* **28**; 364–366, 1989
17. Pukl M, Prosek M. *J Planar Chromatogr* **3**; 173–176, 1990
18. Yamazaki M, Miyazaki H, Sato S, Bunseki Kagaku. **37**; T121–T127, 1988
19. Vovk I. *J Planar Chromatogr* **16**; 66–70, 2003
20. Aranda MB, Vega MH, Villegas R F. *J Planar Chromatogr* **18**; 285–289, 2005
21. Bernardi T, Elena Tambjurini, Vaccari G. *J Planar Chromatogr* **18**; 23–27, 2005
22. Poli F, [a] Sacchetti G, [b] Tosi B, [b] Fogagnolo M,[c] Chillemi G, [d] Lazzarin [d] Bruni A[b],
   [a] Department of Evolutionary and Experimental Biology, University of Bologna, Via Irnerio 42, I–40126 Bologna, Italy
   [b] Department of Biology–Pharmaceutical Biology Lab., University of Ferrara, C.so Porta Mare 2, I–44100 Italy.
   [c] Department of Chemistry, University of Ferrara, Via Luigi Borsari 46, I–44100 Ferrara, Italy
   [d] Veneto Agricoltura, Centro Sperimentale Ortofloricolo "Po di Tramontana:, Via Moceniga, 7, I–45010 Rosolina, Rovigo, Italy.
   Downloaded from website.
23. Vaccari G, Lodi G, Tamburini E, Tosi S. Chemistry Department, University of Ferrara, Via L. Borsari 46, 44100 Ferrara, Italy; Downloaded from website
24. Lee KY, Nurok D, Zlatkis A. *J Chromatography* **174**; 187–193, 1979
25. Rsch W, Roulet *J Chromatography* **244**; 177–182, 1982
26. Nurok D, Zlatkis A. *Carbohydr Res* **65**; 265, 1979
27. Nurok D, Zlatkis A. *Carbohydr Res* **81**; 167, 1980
28. Patzsch K, Netz S, Funk W. *J Planar Chromatography* **1**; 39, 1988
29. Patzsch K, Netz S, Funk W. *J Planar Chromatography* **1**; 177–178, 1988
30. Sherma J, Zulick DL. *Acta Chromatogr* **6**; 7, 1996

31. Pyka A, Niestroj A, Sliwiok J. *J Planar Chromatogr* **16**; 227–229, 2003
32. Anonymous, MSZ (Hungarian Norm) 15, 485–7–82, p.4
33. Gattacecchia E, Tonelli D, Bertocchi G. *J Chromatogr* **260**; 517–521, 1983
34. Eppley RM, Trucksess MW, Nesheim S, Thorpe CW, Pohland AE. *J A O A C* **69**; 37–40, 1986
35. Trucksess MW, Flood MT, Page SW. *J A O A C.* **69**; 35–36, 1986
36. Fried B. Lipids in Handbook of Thin Layer Chromatography, Sherma J, Fried B, eds. Marcel Dekker, New York, 593, 1991
37. Christie WW, ed. Lipid Analysis, Pergamon Press, Oxford, 1976
38. Dallas MSJ, Morris LJ, Nichols BW. Chromatography of lipids in Chromatography: A laboratory handbook of Chromatographic and Electrophoretic Methods, Heftmann E. ed. Van Nostrand Reinhold, New York, 527, 1975
39. Privett OS, Dougherty KA, Erdahl WL. Quantitative analysis of lipid classes by thin layer chromatography via charring and densitometry, in quantitative Thin Layer Chromatography. Touchstone JC, ed. John Wiley and Sons, New York, 57, 1973
40. Lake BD, Goodwin HJ. Lipids in Chromatographic and Electrophoretic Techniques, Vol. 1, Smith I, Seakins JWT, eds., William Heinemann Medical Books Ltd., London, 345, 1976
41. Stan HJ, Wippo U. GIT Fachz. Lab., 40, 855, 1996
42. Ding L, Chen L. *J Chromatogr* (Sepu) **13**; 295, 1995
43. Ghosh P, Krishna Reddy MM, Sashidhar RB. *Food Chemistry* **91**; 757–764, 2005
44. Bhushan R, Amino acids and their derivatives, Handbook of Thin Layer Chromatography, Sherma, J. and Fried, B., Eds, Marcel Dekker, New York, 353, 1991
45. Berry HK, Detection of amino acid abnormalities by quantitative thin layer chromatography, in Quantitative Think Layer Chromatography, Touchstone, J. C. Ed., John Wiley and Sons, New York, 1973, 113
46. Ersser RS, Smith I. Amino acids and related compounds, in chromatographic and Electrophoretic Techniques, Vol. I, 4th ed., Smith, I. and Seakins, J. W. T. Eds. William Heinemannedical Books Ltd., London, 75, 1976
47. Judite LG, Pickova J. *J Chromatogr* A 1045(1–2), 223–232, 2004
48. Stratton JE, Hutkins RW, Traylor SL. Biogenic amines in cheese and other fermented foods–a review, J. Food Protection, **54**; 460, 1991
49. Shalaby AR. Survey on biogenic amines in Egyptian foods, J Sci Food Agric **62**; 291, 1993
50. Shalaby AR. Separation, Identification and estimation of biogenic amines in foods by thin layer chromatography, *Food Chem* **49**; 305, 1994
51. Shalaby AR. Multidetection, semiquantitative method for determining biogenic amines in food, *Food Chem* **52**; 367, 1995
52. Tawfik NF, Shalaby AR, Effat BA. Biogenic amine contents of Raw cheese and incidence of their bacterial producers, Egyptian *J Dairy Sci* **20**; 219, 1992
53. Mietz JL, Karmas E. Polyamines and histamine content of rockfish, salmon, lobster and shrimp as an indicator of decomposition, *J Assoc off anal Chem* **61**; 139, 1978
54. Staruszkiewicz WF, Bond JE. Gas chromatographic determination of cadaverine, putrescine and histamine in foods, *J Assoc off Anal Chem* **64**; 584, 1981
55. Shalaby A R. Correlation between freshness indices and degree of fish decomposition, Ph. D. Thesis, Fac. Agric., Ain Shams Univ., Cairo, Egypt, 1990
56. Hwang KT, Regenstein JM. Characteristics of mackerel mince lipid hydrolysis. *J Food Sci* **58**; 79, 1993
57. Anglo AJ, James C. Jr. Analysis of lipids from cooked beef by thin layer chromatography with flame ionization detection, *JAOCS*, **70**; 1245, 1993
58. Tyihak E, Vagujfalvi D. Thin layer chromatography of alkaloids, in Progress in Thin Layer Chromatography and Related Methods, Vol. III, Niederwieser. A and Patqaki, G., Eds. Ann Arbor Science Publishers, Ann Arbor, MI, 71, 1972
59. Abu–Raiia SH, Shalaby AR. Influence of fenugreek seed extracts on aflatoxins production by Aspergillus parasiticus, New Egyptian *J Medicine* **6**; 635, 1992
60. Ng KG, Price KR, Fenwick GR. A TLC method for the analysis of quinoa (chenopodium quinoa) saponins, *Food Chem* **49**; 311, 1994

61. Szepesi G, Gazdag M. Steroids in Handbook of Thin Layer Chromatography, Sherma, J. Fried, B., Eds. Marcel Dekker, New York, 907, 1991
62. Koch A. Deutsche Apotheker Zeitunci, 137, 4155, 1997
63. Shalaby AR. *Food Chem* **65**; 117, 1999
64. De Leenheer AP, Lambert WE, Nelis NJ. Lipophilic vitamins, in Handbook of Thin Layer Chromatography, Sherma, J., Fried, B., Eds. Marcel Dekker, New York, 993, 1991
65. Ruggeri BA, Watkins TR, Gray RJH, Tomlins RI. Comparative analysis of tocopherol by thin layer chromatography and high performance liquid chromatography, *J Chromatogr* **291**; 377, 1984
66. Speer K, Kretzschmar S, Sibylle N, Huebner DCBS. **95**; 2–4, 2005
67. Jolanta Flieger *J Planar Chromatography*–Modern TLC, **19**; 108, 2006
68. Belay MT, Poole CF. 43–50
    (1) Department of Chemistry, Wayne State University, 48202 Detroit, Ml, USA
69. Tyman JHP. Phenols, aromatic carboxylic acids, and indoles, Handbook of Thin Layer Chromatography, Sherma, J. Fried, B., Eds. Marcel Dekker, New York, 757, 1991
70. Kreuzig F. Antibiotics, Handbook of Thin Layer Chromatography, Sherma, J. Fried B., Eds., Marcel Dekker, New York, 407, 1991
71. FAO, Food and Nutrition Paper, Manual of Food Quality Control, 7, Food Analysis: General Techniques, Additives, Contaminants and Composition, Publication Division, United Nations Food and Agriculture Organization, Rome, 1986
72. Hendrickx S, Roets E, Hoogmartens J, Vanderhaeghe H. Identification of penicillin's by thin layer chromatography. *J Chromatogr* **291**; 211, 1984
73. Rizova V, Stafilov T. Anal Lett **28**; 1305, 1995
74. Tsunoda N, Jap Hygienic Chem **32**; 447–454, 1986
75. Anonymous, MSZ (Hungarian Norm), 14475, 40–86, 1–4
76. Anonymous, MSZ (Hungarian Norm), 14475, 39–86, 1–4
77. Zietz E, Ricker, I. *J Planar Chromatogr* **2**; 262–267, 1989
78. Judge DN, Mullins DE, Young RW. *J Planar Chromatogr* **6**; 300–306, 1993
79. Butz S, Stan HJ. *Analytical Chemistry* **67(3)**; 620–629, 1995
80. Ulf De La Vigne; Janchen, DE, Camag. Sonnenmattstgrasse 11, 4132 Muttenz Switzerland, Zweckverband Landeswasserserversorgung, 7907 Langenau Germany
81. Hamada M. *J Planar Chromatogr* **16**; 4–10, 2003
82. Daundkar BB, Mavle RR, Malve MK, Krishnamurthy R. *J Planar Chromatography*– Modern TLC, **19**; 112, 2006
83. Miszczyk M, Pyka A. *J Planar Chromatography*–Modern TLC, **19**; 107, 2006
84. Ernst plaB, Determination of amitrol in water by AMD, Camag, **96**; 2, 2006
85. Tang F, Ge S, Yue Y, Hua R, Zhang R. *J Planar Chromatogr* **18**; 28–33, 2005
86. Horwitz W, Official Methods of Analysis of the Association of Official Analytical Chemists, 15th ed., AOAC, Arlington, VA, 1184, 1990
87. FAO, Food and Nutrition paper, manuals of Food Quality Control, 10, Training in Mycotoxins Analysis, 1990
88. Tanuja P, Venugopal N, Sashidar RB. *J Assoc off Anal Chem* **90**; 857–863, 2007
89. Tuzimski T. *J sep Sci* **30**; 964–970, 2007
90. Tuzimski T, Wojtowicz J. *J Liq Chromatogr Relat Technol* **28**; 277–87, 2005
91. Raviolo MA, Margarita CB. *J Liq Chromarogr Relat Technol* **28**; 2195–2209, 2005
92. Shepard GS. *J Chromatogr* A **815**; 31, 1998
93. Valenta H, *J Chromatogr* A **815**; 75, 1998
94. Lin L, Zhang J, Wang P, Wang Y, Chen J. *J Chromatogr* A **815**; 3, 1998
95. Nawaz S, Coker RD, Haswell SJ *J Planar Chromatogr.*,– Mod. TLC **8**, 4, 1995
96. Cespedes AE, Diaz GJ. *J AOAC* Int. **80**; 1215, 1997
97. Domagala JJ. Pol. *J. Food Nutr Sci.*, **7**; 117, 1998
98. Takaskashi, T., Aflatoxin contamination of nutmeg: analysis of interfering TLC spots, *J. Food Sci.*, **58**, 197, 1993
99. Kunugi A, Tabei K. *J High Resolut Chromatogr* **20**; 456, 1997
100. Vega M, Rios G, Saelzer R, Herlitz EE, *J Planar Chromatogr.–Mod.* TLC **8**; 378, 1995

101. Dawlantana M, Coker RD, Nagler MJ, Blunden G. *Chromatographia* **42**; 25, 1996
102. Biancardi A, Riberzani A, Lip J. *Chromatogr Rel Technol* **19**; 2395, 1996
103. Milanez TV, Sabino M, Lamardo LCA. *Rev Microbiol* **19**; 79, 1995
104. Heenan CN, Shaw KJ, Pitt JI. *J Food Mycol* **1**; 67, 1988
105. Buchalla B. Elelmezesi Ipar **50**; 303, 1996
106. Vero S, Vasquez A, Cerdeiras MP, Soubes M. *J Planar Chromatogr–Mod* TLC **12**; 172, 1999
107. Dawlatana M, Coker RD, Nagler MJ, Blunden G. *Chromatographia* **41**; 187, 1995
108. Oubrahim H, Richard JM, Cantin–Esnault D, Seigle–Murandi F, Trecourt F. *J Chromatogr* A **758**; 145, 1997
109. Nesbitt BF, O' Kelly J, Sargeant K, Sheridan A. Nature (London) **195**; 1062, 1962
110. Hartley RD, Nesbitt BF, O'Kelly J, Nature (London) **198**; 1056, 1963
111. Betina V. *J Chromatogr* **334**; 211, 1985
112. Nesheim S, Trucksess MW. Modern Methods in the Analysis and Structural Elucidation of Mycotoxins, Academic Press, New York, 239, 1986
113. Lee KY, Poole CF, Zlatkis A. *Anal Chem* **52**; 837, 1980
114. Stahr HM, Domoto M, Adv. Thin Layer Chromatogr. (Proc. Bienn. Symp), 2nd, **405**; 1980, 1982
115. Nowotny P, Baltes W, Kroenert W, Weber R. Lebensmittelchem Gerichtl Chem **37**; 71, 1983
116. Nowotny P, Baltes W, Kroenert W, Weber R. Chem Mikrobiol Technol Lebensm **8**; 24, 1983
117. Lee KY, Poole CF, Zlatkis A. *Anal Chem* **52**; 837, 1980
118. Paterson RRM. *J Chromatogr* **368**; 249, 1986
119. Filtenborg O, Frisvad JC, Svendsen JA. *Appl Environ Microbiol* **45**; 581, 1983
120. Zemnie T M. *J Liq Chromatogr* **7**; 1383, 1984
121. Blanc B, Lasuber E, Sieber R. Microbial Aliment Nutri **1**; 163, 1983
122. Girilli G, *Microbiol Aliment Nutr* **1**; 199, 1983
123. Siz Sehweiz Lab.–Z **44**; 473, 1987: C. A., 108, 73902d., 1988
124. Ehrlich KC, DeLucca AJ, Ciegler A. *Appl Environ Microbiol* **48**; 1, 1984
125. Jaiswal PK, Mankar SN, Banerjee D, Vaidya PS, Husain MI. *J Inst Chemists(India)* **77**; 6, 69–170, 2005
126. Jaiswal PK, vaidya PS, Mankar SN, Husain MI. J Inst . *Chemists(India)* **77**; part–6, 184–185, 2005
127. Jaiswal PK, Vaidya PS, Mankar SN, Husain MI. *J Institution of Chemists(India)* **76**; 4, 126–127, 2004
128. Jaiswal PK, Vsaidya PS. *J Institution of Chemists(India)* under publication
129. Vasilikiotis GS, Papadoyannis IN. *J Microchem* **31**; 170–172, 1985
130. Serralheiro MK, Quinta ML. *J A O A C* **68**; 952–954, 1985
131. Bijl J, Van Peteghem C. *Anal Chem Acta* **170**; 149–152, 1985
132. Dutton MF, Westlake K. *J A O A C* **68**; 839–842, 1985
133. Kaminura H, Nishijima M, Yasuda K, Ushiyama H, Tabata S, Matsumoto S, Nishima T. *J O A C* **68**; 458–461, 1985
134. Miller N, Pretorius HE, Trinder DW, *J A O A C* **68**; 136–137, 1985
135. Spilmann JR. *J A O A C* **68**; 453–456, 1985
136. Tosch D, Waltking AE, Schlesier JF. *J A O A C* **67**; 337–339, 1984
137. Tutour BL, Elaraki AT, Aboussalim A. *J A O A C* **67**; 611–612, 1984
138. Varadi M. Hungarian Scientific Instruments **61**; 23–28, 1986
139. Bradburn N, Coker RD, Jewers K. *Chromatographia* **29**; 177–181, 1990
140. Nawaz S, Coker RD, Haswell SJ. Analyst **117**; 67–74, 1992
141. Park DL, Trucksess MW, Nesheim S, Stack M, Newell, RF. *J Assoc off Anal Chem* **77**; 637–646, 1994
142. Horwitz W. Official Methods of Analysis of the Association of Official Analytical Chemists, 15th ed. AOAC, Arlington, VA, 1990, 1184.
143. FAO, Food and Nutrition paper, manuals of Food Quality Control, 10, Training in Mycotoxins Analysis, 1990
144. Kamimura H, Nishijima M, Yasuda K, Ushiyama H, Tabata S, Matsumoto S, Nishima T. *J Asso off Anal. Chem* 68(3), 1985
145. Lee KY, Colin F, Poole and Zlatkis, A., Department of Chemistry, University of Houston, Houston, Texas 77004
146. Aramson D, Thorsteinson T, Forest D. Agriculture Canada Research Station, 195 Dafoe Road, R3T 2M9 Winnipeg, Manitoba, Canada

147. Saleemullah Amjad I, Iqtidar AK, Hamidullah S. *Food Chemistry* **98**; 699–703, 2006
148. Park DL, Trucksess MW, Neshem S, Stack M, Newell RF.J. AOAC International, **77**(3), 1994
149. Trucksess MW, Flood MT, Mossoba MM, Page SWJ. Agricultural & Food Chemistry, 445–448, 1987
150. Colarow, L.,[a], Turini, M.,[a], Teneberg, S. [b], Berger, A., [c]
    [a]Nestle Research Center, Vers–chez–les–Blanc, CH–1000, Lausanne 26, Switzerland
    [b]Institute of Medical Biochemistry, Goteborg University, SE 405 30, Goteborg, Sweden
    [c]Cytochroma, Inc., 330 Cochrane Drive, Markham, Ontario, L3R 8E4, Canada
    downloaded from website
151. Lansden J A, J A O A C **69**; 964–966, 1986
152. Soares LMV, Rodriguez–Amaya DB J Assoc off Anal Chem **68**; 6, 1985
153. Diserens JM, Renaud–Bezot C, Savoy Perroud MC, Quality Assurance Department, Nestec Ltd. Av. Nestle 55, 1800 Vevey, Switzerland
154. Schwartz D, Mcdonough F. J A O A C **67**; 563–565, 1984
155. Lange R, Fiebig HJ. Fatt/Lipid **101**; 77, 1999
156. Mantovani G, Vaccari E, Dosi G, Lodi G. *Carbohydr polym* **37**; 263, 1998
157. Naidong W, Cachet T, Roets E, Hoogmartens J. Identification of tetracyclines by TLC, *J Planar Chromatogr* **2**; 424, 1989
158. Kang JS, Ebel S. Identification and quantitation of tetracycline antibiotics by cyanophase HPTLC, *J Planar Chromatogr* **2**; 434, 1989
159. Corti P, Corbini G, Dreassi E, Politi N, Montecchi L. Thin Layer Chromatography in the quantitative analysis of drugs. Determination of rifaximine and its oxidation products, Analysis **19**; 257, 1991
160. Oka H, Ikai Y, Hayakawa J, Masuda K, Harada KI, Suzuki M, Martz V, MacNeil D. Improvement of chemical analysis of antibiotics. 18:Identification of residual tetracyclines in ovine tissues by TLC/FABMS with a sample condensation technique. *J Agric Food Chem* **41**; 410, 1993
161. Oka H, Ikai Y, Hayakawa J, Masuda K, Harada KI, Suzuki M. Improvement of chemical analysis of antibiotics. 19: Determination of tetracycline in milk by liquid chromatography and thin–layer chromatography/fast atom bombardment mass spectrometry, *J AOAC Int* **77**; 891, 1994
162. Oka H, Uno K, Harada KI, Hayashi M, Suzuki M. Improvement of chemical analysis of antibiotics. VI. Detection reagents for tetracycline in thin layer chromatography, *J Chromatogr* **295**; 129, 1984
163. Oka H, Ikai Y, Kawamura N, Uno K, Yamada M, Harada KI, Uchiyama M, Asukabe H, Suzuki M. Improvement of chemical analysis of antibiotics. X: Determination of eight tetracycline using thin–layer and high–performance liquid chromatography, *J Chromatogr* **393**; 285, 1987
164. Oka H, Ikai Y, Kawamura N, Uno K, Yamada M, Harada KI, Uchiyama M, Asukabe H, Suzuki M, Improvement of chemical analysis of antibiotics. XII. Simultaneous analysis of seven tetracycline in honey, *J Chromatogr* **400**; 253, 1987
165. Oka H, Ikai Y, Kawamura N, Yamada M, Harada KI, Suzuki M. Improvement of chemical analysis of antibiotics. XIII. Systematic simultaneous analysis of tetracycline in animal tissues using thin–layer and high–performance liquid chromatography, *J Chromatogr* **411**; 313, 1987
166. Abjean JP, Lahogue V. *J AOAC Int* **80**; 1171, 1997
167. Vega M, Rios G, Sawelzer R, Herlitz E. Depto.de Bromatologfa, Nutricion y Dietetica, Facultad de Farmacia, Universidad de Concepcion, Concepcion, Chile
168. Choma IMJ. Planar Chromatography–Modern TLC **19**; 108, 2006
169. Neidert E, Baraniak Z, Sauve A, J A O A C **69**; 641–643, 1986
170. Sherma J, SH C McGINNIS. *J Liquid Chromatogr* **18**; 755–761, 1995
171. Ramirez, A. [a], Gutierrez, R. [a], Diaz, G. [a], Gonzalez, C. [a], Perez, N. [a], Vega, S. [a], Noa, M. [b]
    [a]Departamento de Produccion Agricola y Animal, Universidad Autonoma Metropolitana, Unidad Xochimilco, Calzada del Hueso No. 1100, Col. Villa Quietud, 04960, Coyoacan, Mexico, D. F.
    [b] Departamento de Prevencion, Facultad de Medicina Veterinaria, Universidad Agraria de La Habana, Autopista Nacional Km 23 y Carretera de Tapaste, San Jose de las Lajas, La Habana, Cuba, C. P. 32700
    Downloaded from website.
172. Okuyama D, Okabe M, Fukagawa Y, Ishikura T. Thin layer chromatographic analysis of the OA–6129 group of carbapenem antibiotics in fermentation broths, *J Chromatogr* **291**; 464, 1984

173. Fodor–Csorba K. Pesticides, in Handbook of Thin Layer Chromatography, Sherma, J. Fried, B., eds. Marcel Dekker, New York, 663, 1991

174. Horwitz W, Official Methods of Analysis of the Association of Official Analytical Chemists, 15th ed., AOAC, Arlington, VA, 274, 1990

175. Choma I, Komaniecka I. Liq Chromatogr Relat Technol **28**; 2467–2478, 2005

176. Choma IM. *Liq Chromatogr Relat Technol* **29**; 2083–2093, 2006

177. Damberger V, Wurzinger A, Bandion F. *Mitt Kiosterneuburci* **47**; 173, 1997

178. Abjean J.–P. *J AOAC Int* **80**; 737, 1997

179. Shepard GS. *J Chromatogr* A **815**; 31, 1998

180. Marutoiu C, Filip M, Tigae C, Coman V, Grecu R, Marcu G. *J Planar Chromatogr* **16**; 183–185, 2003

181. Rizova V, Stafilov T. *Anal Letters* **28**; 1305–1306, 1995

182. Lancaster M, Goodall DM, Bergstroem ET, Mccrossen S, Myers P. *J Chromatogr A* **1090 (1–2)**; 165–171, 2005

183. Horwitz W. Official Methods of Analysis of the Association of Official Analytical Chemists, 15th ed., AOAC, Arlington, VA, 1142, 1990

184. Steele, *J J A O A C* **67**; 540–541, 1984

185. Li H, Chen F, Zhang T, Yang F, Xu G. *J Chromatogr A* **905**; 151, 2001

186. Wang H, Ren D, Liu I, Shi S. Zhang *J Shuisheng Shengwu Xuebao* **28**; 380, 2004

187. Krizsan K, Szokan G, Toth ZA, Hoollosy F, Laszlo M, Khlafulla A. *J Liq Chromatogr* Relat Technol **19**; 2295, 1996

188. Hirokado M, Kimura K, Suzuki K, Sadamasu Y, Katsuki Y, Yasuda K, Nishijima M. *F Food Hyg Soc Jpn* **40**; 488, 1999

189. Cobzas S, Cimpan G, Olah N, Gocan S. JPC–*J Planar Chromatogr* **12**; 26, 1999

190. Cserhati T, Forgacs E, Hollo J. JPC–*J Planar Chromatogr* **6**; 472, 1993

191. Itakura Y, Ueno E, Ito Y, Oka H, Ozeki N, Hayashi T, Yamada S, Kagami T, Miyazaki Y, Ohtsuji T, Hatano R, Yamada E, Suzuki RJ. Food Hyg Soc *Jpn* **40**; 183–188, 1999

192. Hayashi T, Ueno E, Ito Y, Oka H, Ozeki N, Itakura Y, Yamada S, Kagami T, Miyazawa T. *J Food Hyg Soc Jpn.*, **40**; 356–362, 1999

193. Ozeki N, Ueno E, Ito Y, Oka H, Hayashi T, Itakura Y, Yamada S, Matsumoto H, Ito T, Maruyama T, Tsuruta M, Miyazawa T. *J Food Hyg Soc Jpn* **41**; 347–352, 2000

194. Hayashi T, Oka H, Ito Y, Goto T, Ozeki N, Itakura Y, Matsumoto H, Otuji Y, Akatsuka HT, Miyazawa T, Nagase II. *J Liq Chromatogr and Rel Technol* **25**; 3151 3165, 2002

195. Itakura Y, Ozeki N, Oka H, Ito Y, Ueno E, Goto T, Hayashi T, Ohno H, Sasaki Y, Mukoyama M, Matsumoto H, Nagase H. *J Liq Chromatogr and Rel Technol* **25**; 1283–1294, 2002

196. Hayashi T, Oka H, Ito Y, Goto T, Ozeki N, Itakura Y, Matsumoto H, Otsuji Y, Akatsuka H, Miyazawa T, Nagase H. *J Liq Chromatogr and Rel Technol* **26**; 819–832, 2003

197. Watanable M, Aoyama T, Takasu Y, Inoue K, Terao M, Ito Y, Oka H, Goto T, Matsumoto H. *J Liq Chromatogr and Rel Technol* **28**; 325–334, 2005

198. Yamada S, Noda N, Mikami E, Hayakawa J, Yamada M. *J Assoc off Anal Chem* **72**; 48–51, 1989

199. Pandhare E, Rama Rao AV, Srinivasan R, Venkataraman K. Tetradedron **8**; (Suppl.), 229–239, 1966

200. Pandhare E, Rama Rao AV, Shaikh IN. *Indian J Chem* **7**; 997–986, 1969

201. Belay MT, Poole CF. Department of Chemistry, Wayne State University, 48202 Detroit, Ml, USA

202. Carpinella MC. *J Agri Food Chem* **51**; 2506–2511, 2003

203. Jayasinghe C. *J Agri Food Chem* **51**; 4442–4449, 2003

204. Kiridena W, Poole SK, Poole CF. *J Planar Chromatogr* **7**; 273–277, 1994

205. Belay MT, Poole CF. *Chromatographia* **37**; 365–373, 1993

206. Sherma J, Schafer SL, Morris K. *J Liquid Chrom* **10**; 3583–3593, 1987

207. Poole SK, Daly SL, Poole CF. *J Planar Chromatography* **6**; 129–137, 1993

208. Poole SK, Kiridena W, Miller KG, Poole CF. *J Planar Chromatogr* **8**; 257–268, 1995

209. Sherma J, Raible D, Brubaker K. *J Planar Chromatogr* **4**; 253–254, 1991

210. Lin L, Tanner H. *J High Resol Chromatogr* **8**; 126–131, 1985

211. Lambri M, Jourdes M, Glories Y, Saucie C. *J Planar Chromatography* **16**; 88–94, 2003

212. Skalska A, Matysik A, Gerkowicz M, Wojciak–Kosior M. *J Planar Chromatography*–Modern TLC **19**; 112, 2006
213. Cao HQ, YueYD, Hua RM, Tang F, Zhang R, Chen HY. *J Planar Chromatogr* **18**; 151–154, 2005
214. Mallavadhani, U. V. [a], Akella V. S. Suxhakar[a], Satyanarayana, K. V. S. [a], Anita Mahapatra[a], Wenkui Li[b], Richard B. VanBreemen[b]
    [a]Centre for Herbal Drugs, Regional Research Laboratory(CSIR), Bhubaneshwar 751 013, Orissa, India.
    [b]Department of Medicinal Chemistry and Pharmacognosy, University of Illinois College of Pharmacy, 833 South Wood Street, Chicago, IL 60612, USA
    (downloaded from website)
215. Mroczek T, Glowniak K, Kowalska J, Department of Pharmacognosy with Medicinal Plants Laboratory, Medical University, 1 Chodzki St., 20–093 Lublin, Poland downloaded from website
216. Abourashed A, Mossa J. Department of Pharmacognozy, College of Pharmacy, King Saud University, Riyadh 11451, Saudi Arabia, downloaded from website
217. Dugo P, Mondello L, Lamonica G, Dugo G. *J Planar Chromatogr*–Mod TLC **9**; 120, 1996
218. Ding L, Chen L. *J Chromatogr* **13**; 295–296, 1995
219. Poole CF, Poole SK, Department of Chemistry, Imperial College of Science, Technology & Medicine, south Kensington, London SW7 2 AY, United Kingdom, Planar Chromatographic methods to determine the Quality of Spices and flavours
220. Dighe VV, Gursale AA, Sane RT, Menon S, Raje SC. *J Planar Chromatogr* **18**; 305–307, 2005
221. Buhl F, Szpikowska–Sroka B, Galkowska M. *J Planar Chromatogr* **18**; 368–371, 2005
222. Kanaki NS, Rajani M. *J Association off Anal Chem* **88**; 156801570, 2005
223. Marutoiu C, Oprean L, Marutoiu OF, Soran ML. *J Planar Chromatogr* **18**; 282–284, 2005
224. Janssen A, Gole T. *Chromatographia* **18**; 546–549, 1984
225. Gonzales MM, Lara FM, Carbajal GG, Flota FV. *J Liq Chromatogr & Relat Technol* **30**; 1697–1704, 2007
226. Brandolini V, Menziani E, Mazzotta D, Cabras P, Tosi B, Lodi G. *J Food Compos Anal* **8**; 336, 1995
227. Mandal RD, Ravishankara MN, Shah SA. Indian *J Param Sci* **63(3)**; 250–253, 2001
228. Corti P, Caricchia AM, Franchi G, Lencioni E, Murratzu C, Corbini G. Controle densitometrique des produits de degradation de la vitamine B$_1$ et calcul des facteurs d' activation. *Amm Pharm Fr* **47**; 117, 1989
229. Isaksen M. Natural pigments, in Handbook of Thin Layer Chromatography, Sherma, J., Fried B., Eds., Marcel Dekker, New York, 625, 1991
230. Shalberg I, Hynninen PH. Thin Layer chromatography of chlorophylls and their derivatives on sucrose layer, *J Chromatogr* **291**; 331, 1984
231. Claudia C, Hosu A. *Liq. Chromatogr & Relat Technol* **30**; 701–728, 2007
232. Cimpoiu C, Casoni D, Hosu A, Miclaus V, Hodisan T, Damian G. *J Liq Chromatogr Relat Technol* **28**; 2551–2559, 2005
233. Hachula U, Anikiel S, Sajewicz M. *J Planar Chromatogr* **18**; 290–293, 2005
234. Campbell A, et al. Dept. of Chem., Lafayette Col. Easton, PA 18042, USA
235. Fang CB, Wan X, Jiang CJ, Cao HQ, *J Planar Chromatogr* **18**; 73–77, 2005
236. Azosanlou R. *Mitt Geb Lebensmittelunters Hyg* **89**; 355, 1998
237. Garcia C, Garcia S, Heinzen H, Moyna P, Niemeyer HM. *Phytochem Anal* **9**; 278, 1998
238. Lavoine S, Arnaudo JF, Coutiere, D., Riv. Ital. EPPOS (Spec. Num), 580, 1998
239. Lavoine S, Arnaudo JF, Coutiere D. *Ann Falsif Expert Chim Toxicol* **91**; 41, 1998
240. Janoszka BJ. *Planar Chromatogr* **16**; 186–191, 2003
241. Traitler H, Ducret P, Perrenoud MC, Fay L, Richi U. *Planar Chromatogr* (Interlaken 1991), Inst. Chromatogr., Bad Durkheim, FRG, 357–365, 1991
242. LI Z, (Li Zongquang) WUJ, (Wu Jitao), GAOY, (Gao Yingheng), (Dep Chem Yangzhou PR China J Food Ferm Ind **2**; 81–84, 1991
243. JaiswalPK. Director of Laboratories, Central Agmark Laboratory, Nagpur, India, paper on 'Analysis of foods by planar chromatography/ HPTLC in India' presented in the "International Symposium for HPTLC" at Berlin, Germany on 9th October to 11th October, 2006
244. Kumudavally KV, Shobha A, Vasundhara TS, Radhakrishna K, Department of Freeze Drying and Animal Products Technology, Defence Food Research Laboratory, Mysore–570 011, India. Downloaded from website
245. Jarusiewicz J, Sherma J, Fried B. *J Liq Chromatogr Relat Technol* **28**; 2607–2617, 2005.

246. Jautz U, Morlock G. *Anal Bioanal Chem* **387**; 1083–1093, 2007
247. Ikenouye L, Hickey S, Verbitski S, Gourdin G. *Camag bibliography Science* **99**; 11, 2007
248. Mateos E, Cebolla VL, Membrado L, Vela J. *Camag Biblographic services* **99**; 9, 2007
249. Galvez E. et al. *Anal Chem* **78**; 3699, 2006
250. Mateos E. et al. *J Chromatogr A* **1146**; 251, 2007
251. Alpmann A, Morlock G, Schwack W. *Camag Bibliographic Services* **99**; 14, 2007
252. Gautam MP, Gautam S. Analysis of Plants Poisons (ISBN–81–89128–07–8), 2006
253. Prosek M, Milivojevic L, Krizman M, Fir M. *J Planar Chromatography* **17**; 420, 2004
254. Alpmann A, Morlock G. Anal Bioanal Chem, D.O.I,10.1007/s00216–006–0692–y, Institute of Food Chemistry, University of Hohenheim, Garbenstrasse 28, 70599 Stuttgart, Germany
255. Morlock G, Schwack W. *Anal Bioanal Chem* **385**; 586–595, 2006
256. Morlock G, Nedele A, Schwack W, Euro Food Chem XIII–Proceedings, 2:513–516, 2005
257. Jautz U, Morlock G. *J Chromatogr A* **1128**; 244–250, 2006
258. Aranda M, Morlock G. *J Chromatography A* **1131**; 253–260, 2006
259. Aranda M, Morlock G. Institute of Food Chemistry, University of Hohenheim, Garbenstrasse 28, 7–599 Stuttgart, Germany
260. Ford MJ, Deibel MA, Tomkins BA, Van Berkel GJ. *Anal Chem* **77**; 4385–4389, 2005
261. Dreisewerd. K. [a], Kolbl S. [a], Peter–Katalinic J. [a], Berkenkamp S. [b], Pohlentz G. [c],
    [a] Institute of Medical Physics and Biophysics, Westfalizche–Wilhelms Universitat Munster, Munster, Germany
    [b] Sequenom GmbH, Hamburg, Germany
    [c] Institute of Medical Physics and Biophysics, Westfalische–Wilhelms Universitat Munster, Munster, Germany
    Downloaded from website
262. Dreisewerd K, Kolbl S, Peter–Katalinic J, Berkenkamp S, Pohlentz G. *Anal Chem* **77**; 4098, 2005
263. Mohan BM, Sanganalmath PU, Gowtham MD, Yogaraje Gowda CV, Nayak VG. Presented in the International Symposium on HPTLC, Berlin (Germany), 9–11 October, 2006
264. Janoszka B, J Planar Chromatogr **20**; 221-26, 2007
265. Luftmann H, Aranda M, Morlock G, Rapid Commun Mass Spectrom, 21, 3772-3776, 2007
266. Morlock G, Ueda Y, J Chromatogr. A 1043(1-2), 243-251, 2007
267. Cvek J, Medic-Saric M, Jasprica I, Mornar AJ. Planar Chromatogr., 20,429-435,2007
268. Ranganathan T, Kulkarni P, Food. Chem **77**; 263-265, 2002
269. Morlock G, Vega-Herrera MA, J Planar. Chromatogr **20**; 411-417, 2007
270. Afinisha L, Soban D, Sundaresan C, Arumughan C. J.Sep.Sci., 30, 1 2786-2793, 2007
271. Starek M, Krzek J, Michnik SJ. Planar Chromatogr **20**; 327-330, 2007
272. Lachenmeier D. Food Res.Int., 40, 167-175,2007
273. Mallavadhani U, Sudhakar A, Satyanarayana K, Mahapatra A. Li,W., ? Breemen,R., Food Chem., 95,58-64, 2006
274. Cimpoiu C, Hosu A, Briciu R, Miclaus VJ. Planar Chromatogr **20**; \ 407-410, 2007
275. Jaiswal PK, Mankar SN, Banerjee D, Singh G, Vaidya PS. J. I of Inst.Chemists (India), 80,(2), 40-41, 2008
276. Jaiswal PK, Kolekar TT, Banerjee D, Saxena VB, Vaidya PS. J. of Inst.Chemists (India), sent for publication 2008

17

Specific HPTLC Analytical Protocols in Food Analysis

I n this chapter, different HPTLC protocols in food for detection/quantitation of adulterants/components/additives, etc. have been cited out. These methods can be put to use for other foods than those specified of course with validation of methods and setting the established detection/determination limits. It may be pointed out that HPTLC has unlimited applications in food analysis. HPTLC is a good method of fingerprinting to correlate the purity/compositional analysis.

METHOD: 17.1 A

QUANTITATIVE DETERMINATION OF SUDAN I, II, III, IV, SUDAN YELLOW, METHYL YELLOW AND PARA RED IN CHILLI POWDER

Aim of Analysis

Quantitative determination of Sudan I, II, III, IV, Sudan yellow, methyl yellow and para red in chilli powder.

Chemicals	:	Toluene, methanol, acetone, water.
Reagent preparation	:	None

Standard Preparation

Dil A: 1) Sudan I, II, III, IV, methyl yellow, para red: (0.5 µg/µl). Dissolve 5 mg each of Sudan I, II, III, IV, methyl yellow, para red respectively in 10 ml of toluene.

2) Sudan Yellow: (0.5µg/µl). Dissolve 5 mg of Sudan yellow in 0.5 ml of DMF. Add 2 ml of methanol and 7.5 ml of toluene.

Dil B: Add 0.5 ml each of Sudan I, II, III, IV, Sudan yellow, para red, methyl yellow in a 10 ml volumetric flask and make up to 10 ml with toluene (0.025 µg/µl).

Sample Preparation

Chilli extract: Weigh 500 mg of chilli powder in 10 ml volumetric flask. Add 5ml of toluene, shake well and sonicate for 30 min and make up to 10 ml with toluene (50 µg/µl).

Stationary Phase
TLC Al sheets silica gel RP18 60F254 precoated 20 × 10 cm, Cat No. 034.5559.

Mobile Phase
Methanol: Acetone: Water (7: 2: 1) (v/v), (Volume = 30 ml)

Sample/standard Application
Apply with the help of Camag ATS-4 or Linomat 5, 2 μl, 4 μl, 6 μl, 8 μl, 10 μl (0.05 μg, 0.1 μg, 0.15 μg, 0.2 μg, 0.25 μg respectively) of Dilute B, 8 mm from the bottom edge. Band length 6 mm. Distance from the side 15 mm.

Development Chamber
Twin-trough chamber of 20 × 10 cm with s. s. lid.

Tank saturation	:	20 min. with filter paper
Plate equilibrium	:	20 min.
Development	:	1st development: 80 mm. Dry the plate. 2nd development: 90 mm (in a second presaturated chamber)
Visualisation	:	Observe in UV cabinet at 254 and 366 nm.
Post-chromatographic derivatisation	:	Not required
Photodocumentation	:	In 254 nm and visible light.
Measurement mode	:	UV absorbance/reflectance.

Scanning
For Quantification
Using scanner 3 with win CATS software, slit-micro, 6 × 0.45 mm. Scan at 400 nm for Methyl Yellow, 494 and 530 nm for Sudan I, II, III and IV, Para red and Sudan yellow.

For Identification
Record and match spectra between 190 to 700 nm.

Image (254 nm)

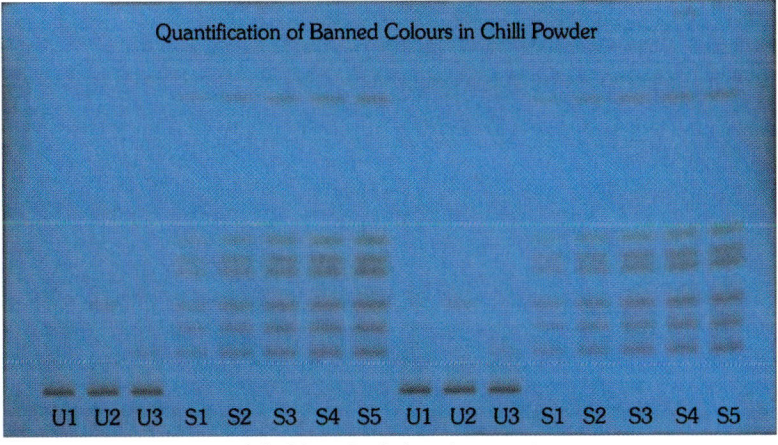

Quantification of Banned Colours in Chilli Powder

U1 U2 U3 S1 S2 S3 S4 S5 U1 U2 U3 S1 S2 S3 S4 S5

(a) After 2nd development

Image (visible)

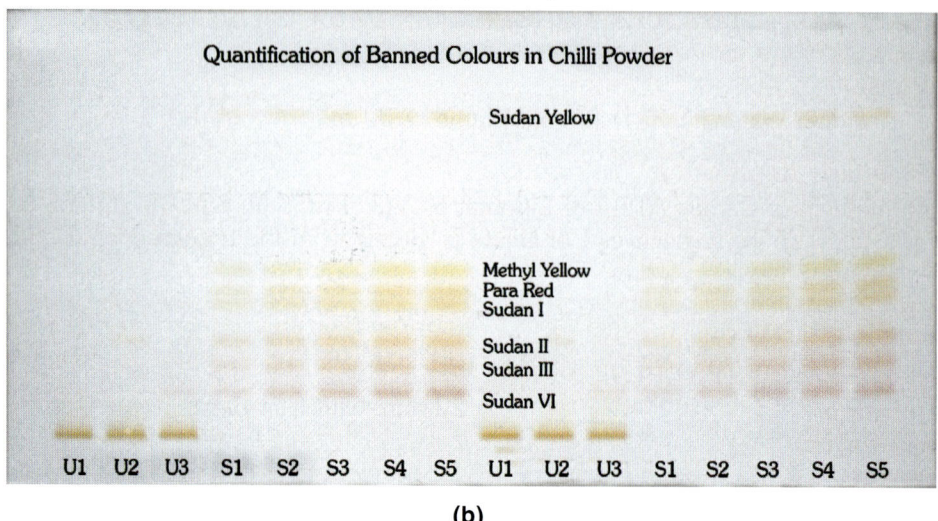

(b)

Fig. 17.1(A) (a, b): Quantitative determination of Sudan I, II, III, IV, Sudan yellow, methyl yellow and para red in chilli powder

U1: Chilli powder 1 (250 µg)
U2: Chilli powder 2 (250 µg)
U3: Chilli powder 3 (250 µg)
S1: Standard banned colours (0.05 µg)
S2: Standard banned colours (0.1 µg)
S3: Standard banned colours (0.15 µg)
S4: Standard banned colours (0.2 µg)
S5: Standard banned colours (0.25 µg)

METHOD: 17.1 B

IDENTIFICATION OF SUDAN I, II, III, IV IN CHILLI POWDER (OTHER METHOD)

Aim of Analysis

Identification of banned Sudan-I, II, III, IV colorants in Chilli powder.

Chemicals and Reagents

n-hexane, n-butyl acetate, toluene

Reagent preparation   :   None.
Standard preparation

Sudan-I

Solution A: Weigh 5 mg of standard Sudan-I powder in a 10 ml Std. volumetric flask, dissolve and make up to 10 ml with toluene (conc. 0.5 mg = 1 ml).

Solution B: Take 1 ml of solution A in a 10 ml Std. volumetric flask and make up to 10 ml with toluene (conc. 0.05 mg = 1 ml).

Sudan-II

Solution C: Weigh 5 mg of standard Sudan-II powder in a 10 ml Std. volumetric flask, dissolve and make up to 10 ml with toluene (conc. 0.5 mg = 1 ml).

Solution D: Take 0.5 ml of solution C in a 10 ml Std. volumetric flask and make up to 10 ml with toluene (conc. 0.025 mg = 1 ml).

Sudan-III

Solution E: Weigh 5 mg of standard Sudan-III powder in a 10 ml Std. volumetric flask, dissolve and make up to 10 ml with toluene (conc. 0.5 mg = 1 ml).

Solution F: Take 0.5 ml of solution E in a 10 ml Std. volumetric flask and make up to 10 ml with toluene (conc. 0.025 mg = 1 ml).

Sudan-IV

Solution G: Weigh 5 mg of standard Sudan-IV powder in a 10 ml Std. volumetric flask, dissolve and make up to 10 ml with toluene (conc. 0.5 mg = 1 ml).

Solution H: Take 0.5 ml of solution G in a 10 ml Std. volumetric flask and make up to 5 ml with toluene (conc. 0.025 mg = 1 ml).

Sample Preparation

Weigh 200 mg of Chilli powder in 10 ml volumetric flask. Add 5 ml of toluene, shake and sonicate for 30 mins. Make up to the mark with toluene. Filter through whatman filter paper 1 and use the filtrate for application (conc. 20 mg = 1 ml).

Stationary Phase

TLC alumina 60 F_{254} plate cut to 20 × 10 cm.

Sample/Standard Application

Apply with the help of CAMAG ATS-4 or Linomat 5, 10 µl (200 µg) of sample, 2 µl (0.1 µg) of solution B, and 2 µl of solution D, F, and H [each 0.05 µg] on precoated layer, 8 mm from bottom edge, band length 8 mm, distance from the side 15 mm.

Development Chamber

ADC 2 device with humidity control

Humidity

At 35.2 % RH

Mobile Phase

n-hexane: n-butyl acetate (9.5 : 0.5) v/v (Volume = 16 ml)

Tank Saturation with Plate Equilibrium

10 min with filter paper. (Use 25 ml of the mobile phase separately for tank saturation).

Development distance   :   80 mm
Visualisation   :   Inspect the plate at 254 nm in UV cabinet 3.
Photodocumentation   :   254 nm and visible

Scanning/Quantification

Using Camag TLC scanner-3 with win CATS software, slit micro-4 × 0.3, scan at 254 nm and at 530 nm.

Image (254 nm)

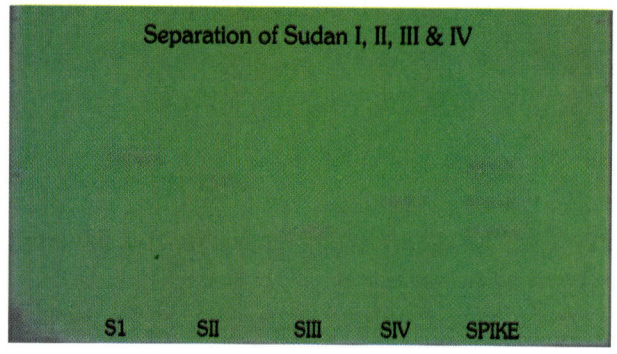

Fig. 17.1 (B): Separation of Sudan on ADC 2 at 35.2% RH

METHOD : 17.1 C

IDENTIFICATION OF FOOD COLOURS IN CARBONATED DRINKS

Aim of Analysis

Identification of Amaranth, Ponceau 4R, Allura red in carbonated drinks.

Chemicals and Reagents

n-butanol, acetone, triethylamine, water, 5% sodium sulphate, 8% sulphuric acid, ammonia.

Reagent Preparation

5% sodium sulphate soln: Weigh 0.75 gm of anhy. sodium sulphate in conical flask and dissolve in 15 ml of water.

8% sulphuric acid(v/v): Add 8 ml of conc. sulphuric acid in 100 ml water in a conical flask.

Dilute ammonia: 1 part of ammonia to 50 parts of water.

Preparation of White Wool for Dyeing (Scouring)

Take 15 mg of non-ionic surfactant in 50 ml beaker. Add 15 ml water. Add 25 cm of wool (62 mg). Keep water bath at 70° C for 10 min and then wash the wool with plenty of water to remove all surfactant (Always prepare fresh).

Standard Preparation

Weigh accurately 5 mg separately of each standard in three different 10 ml volumetric flasks. Dissolve each in 10 ml water (conc.0.5 µg/µl). Pipette out 0.5 ml of each solution separately in three different 10 ml volumetric flasks and make up the volume to 10 ml with (1:50) ammonia solution (conc. 0.025 µg/µl).

Sample Preparation

Take 15 ml of 5% sodium sulphate in 100 ml beaker. Add 1.2 ml of 8% sulphuric acid. Put all scoured wool into this solution. Keep on water bath at 70° C. Stir with glass rod. After 10 min, add 5 ml of carbonated drink sample. Continue on water bath for 45 minutes. Transfer the wool to a 50 ml beaker and strip the colour from the wool by boiling with 10 ml dil. ammonia. This is test solution. Prepare all samples in this manner.

Stationary Phase

TLC Al sheets silica gel 60 F_{254} precoated cut to 20 × 10 cm.

Mobile Phase

n-butanol : acetone : triethylamine : water (6:2:1:1v/v) (Volume = 16 ml)

Development chamber	: Twin-trough chamber of 20 × 10 cm with s. s. lid.
Chamber saturation	: None
Plate equilibrium	: None
Sample/standard application	: Apply with the help of Camag ATS-4 or Linomat 5, 5 µl and 10 µl of each test solution and 5 µl of standard solution on precoated layer 8 mm from the bottom edge. Band length 8 mm. Distance from the side 15 mm.
Development distance	: 80 mm
Visualisation	: Observe under UV cabinet at 254 and 366 nm
Post-chromatographic derivatisation	: Not required.
Photodocumentation	: At 254 nm, visible
Measurement mode	: UV absorbance/reflectance

Scanning
For Quantification

Using Camag scanner 3 with winCATS software, slit-micro, 4 × 0.3 mm, scan at 520 nm.

For Identification

Record spectra between 190-700 nm and ensure match with standard.

Image (254 nm)

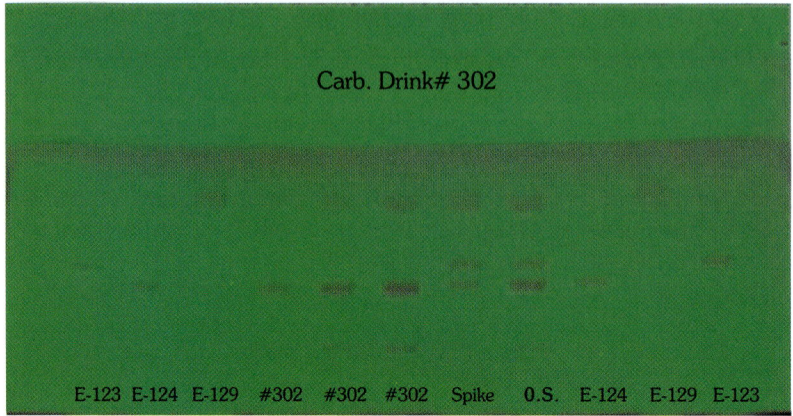

(a)

Image (visible)

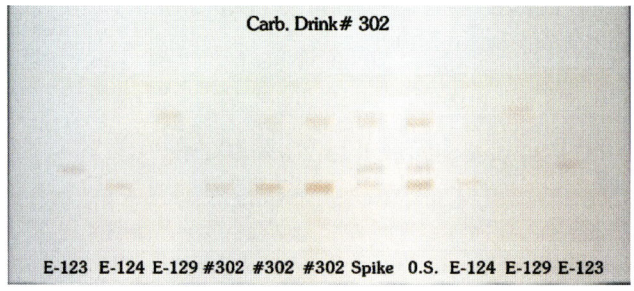

(b)

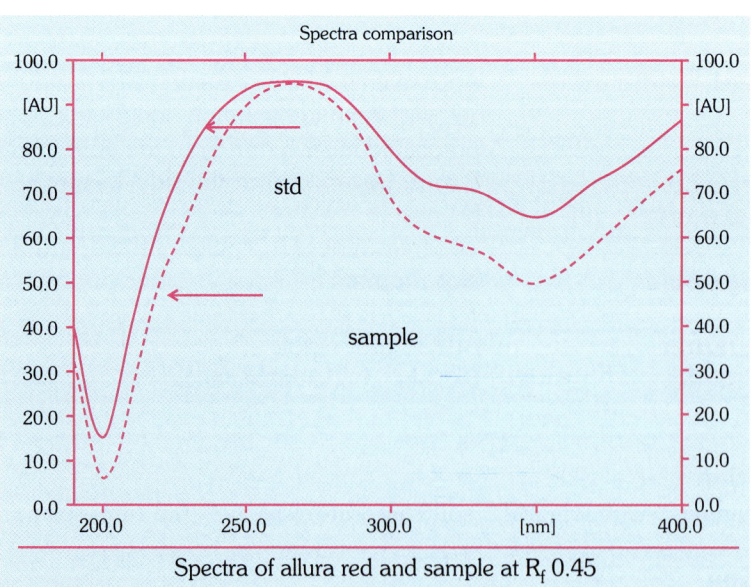

(c)

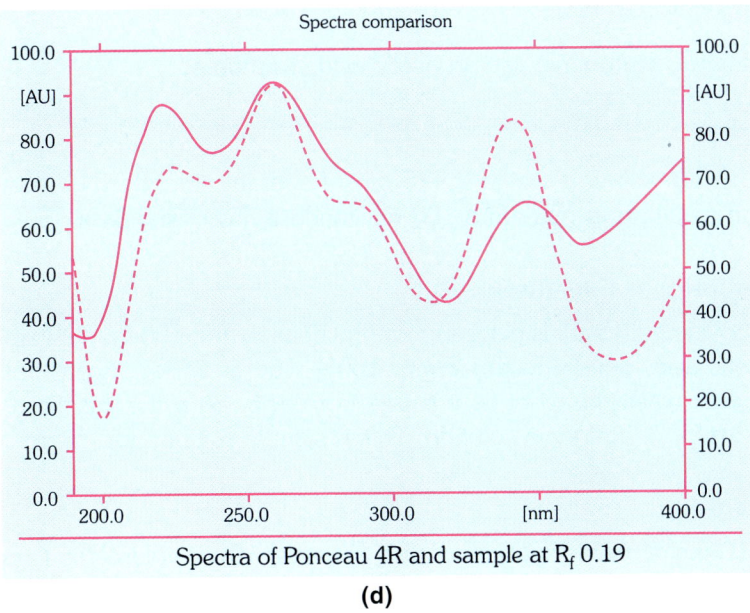

Spectra of Ponceau 4R and sample at R_f 0.19

(d)

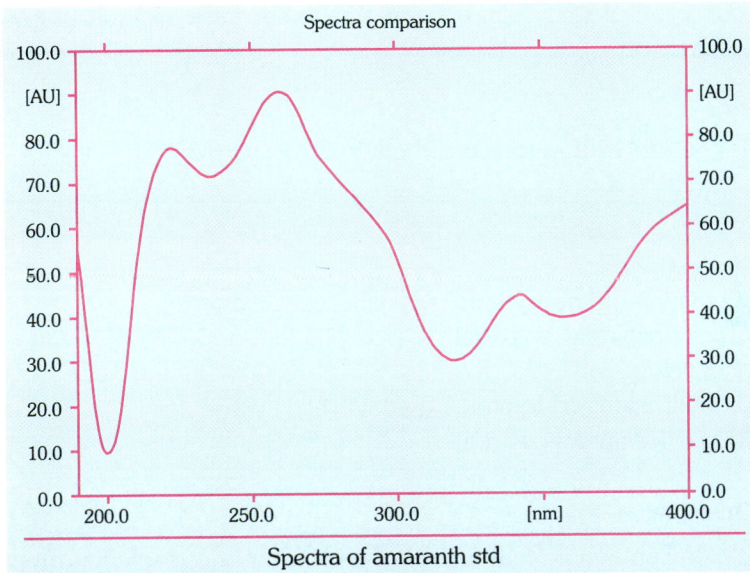

Spectra of amaranth std

(e)

Fig.17.1 (C) (a – e): Identification of colouring in carbonated drink

METHOD: 17.1 D

DETECTION OF FOOD COLOURS

Aim of Analysis

Detection of ponceau 4R, carmoisine, erythrosine, tartrazine, sunset yellow FCF, indigo carmine, brilliant blue FCF, fast green FCF in foods.

Chemicals and Reagents

0.5 M NaCl, methanol, acetonitrile, glacial acetic acid, ammonia.

Reagent Preparation

0.5 M sodium chloride solution–weigh 2.925 gm of sodium chloride (NaCl) and dissolve it in distilled water and make up to mark in a 100 ml standard measuring flask.

Sample Preparation Example–Biscuits

About 20 gm of powdered biscuit is placed in 250 ml beaker, add 150 ml distilled water and 10 ml of 5% glacial acetic acid, boil it. To this 2 gm, 100% wool is added. Again boil for 30 minutes continuously. Cool it, wash the wool till it is free from acid. Strip the adsorbed colour from the wool by adding 2 ml, 5% ammonia solution. This is sample solution for chromatography.

Standard Preparation

Soln. A: Weigh 10 mg of each food colour in a separate 10 ml volumetric flask, add 0.5 ml of ammonia then add 2 ml of distilled water and make up with methanol [Conc.1 µg/µl].

Soln. B: Pipette out 5 ml of soln. A in a 10 ml volumetric flask and make up to the mark with methanol [Conc. 0.5 µg/µl].

Stationary Phase

TLC Al sheets silica gel RP-18 F_{254s} precoated cut to 20 × 10 cm.

Mobile Phase

1. For first development:
   0.5M NaCl: methanol : acetonitrile : glacial acetic acid
   (10: 6:5:0.1) v/v (Volume = 16 ml)
2. For second development:
   0.5M NaCl: methanol : acetonitrile : glacial acetic acid
   (12:5:4:0.2) v/v (Volume = 16 ml)

Development Chamber

Apply with the help of Camag Linomat 5, 1 µl of each food colour solution on precoated layer 8 mm from the bottom edge. Band length 6 mm. Distance from the side 20 mm.

Twin-trough chamber of 20 × 10 cm with s. s. lid.

1st Development

Chamber saturation	:	10 min. without filter paper
Plate equilibrium	:	10 min.
Development distance	:	80 mm from lower edge.
Sample/standard application		

2nd Development

Chamber saturation	:	None
Plate equilibrium	:	None
Development distance	:	90 mm from lower edge of plate
Visualisation	:	Observe under UV cabinet at 254 nm.
Post-chromatographic derivatisation	:	Not required
Photodocumentation	:	254 nm and visible.
Scan mode	:	UV absorbance/reflectance.

Scanning
For Quantification
Using Camag scanner 3 with winCATS software, slit-micro, 4 × 0.3 mm, scan at 254 nm, and visible.

For Identification
Record and match spectra between 190-700 nm.

Image (254 nm)

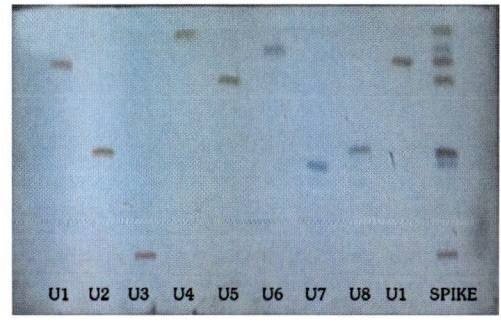

(a) First development

Image (visible)

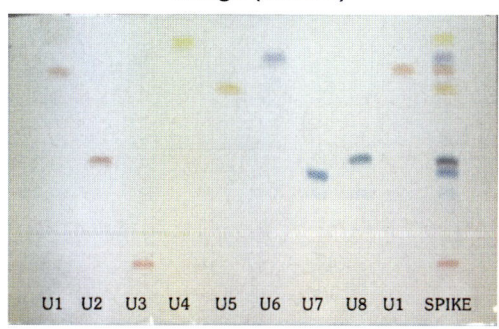

(b)

Image (254 nm)

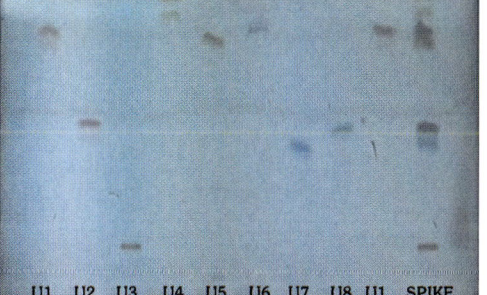

(c) Second development

Image (visible)

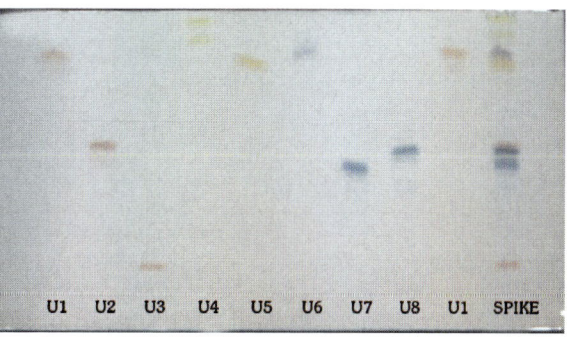

(d)

S. no.	Food colour	R$_f$
U3	Erythrosine	0.15
U7	Brilliant blue FCF	0.58
U8	Fast green FCF	0.60
U2	Carmoisine	0.65
U5	Sunset yellow FCF	0.82
U6	Indigo Carmine	0.88
U1	Ponceau 4R	0.92
U4	Tartrazine	0.95

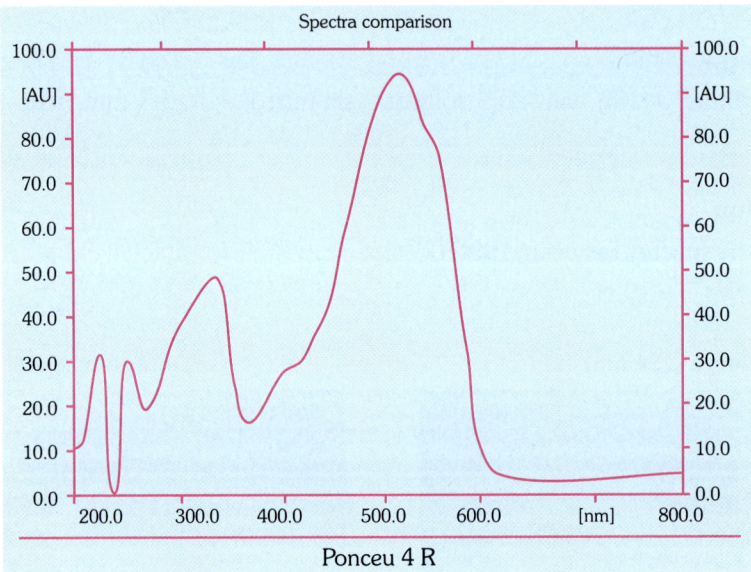

Ponceu 4 R

(e)

Carmoisine

(f)

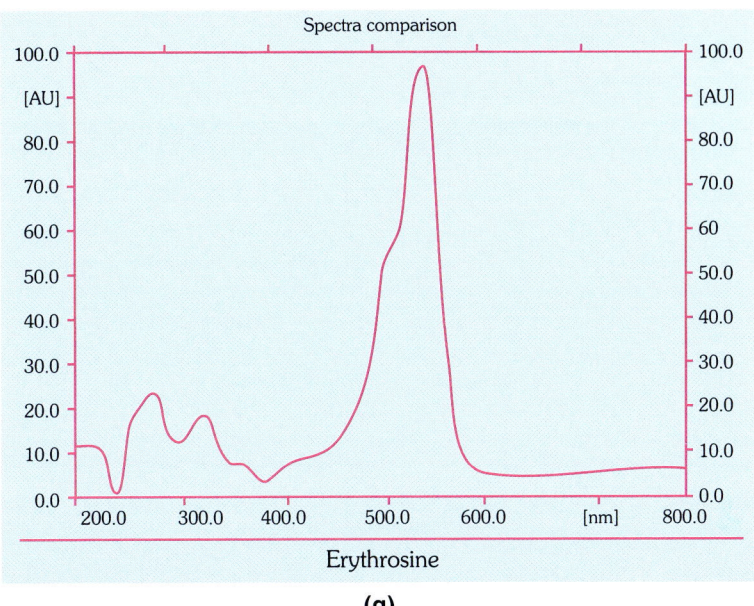

Erythrosine

(g)

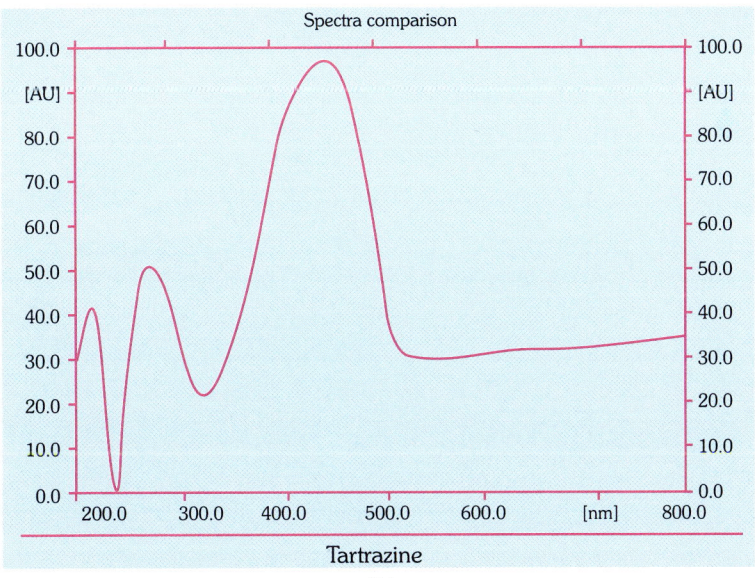

Tartrazine

(h)

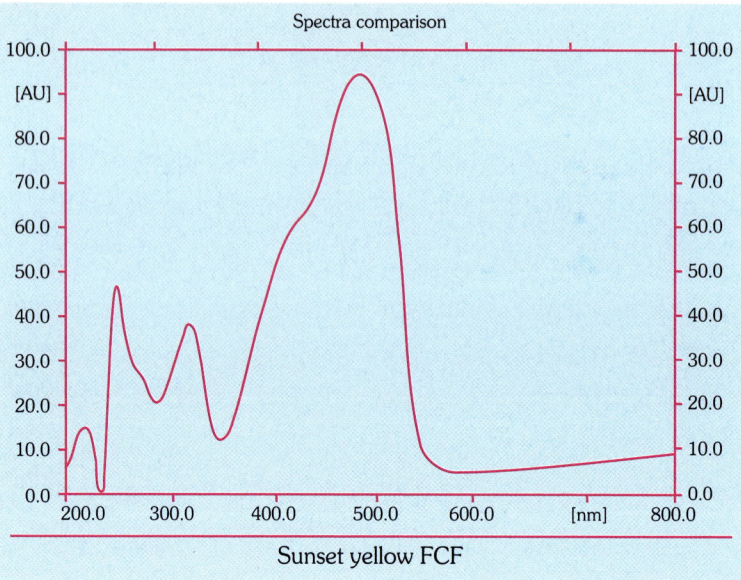

(i)

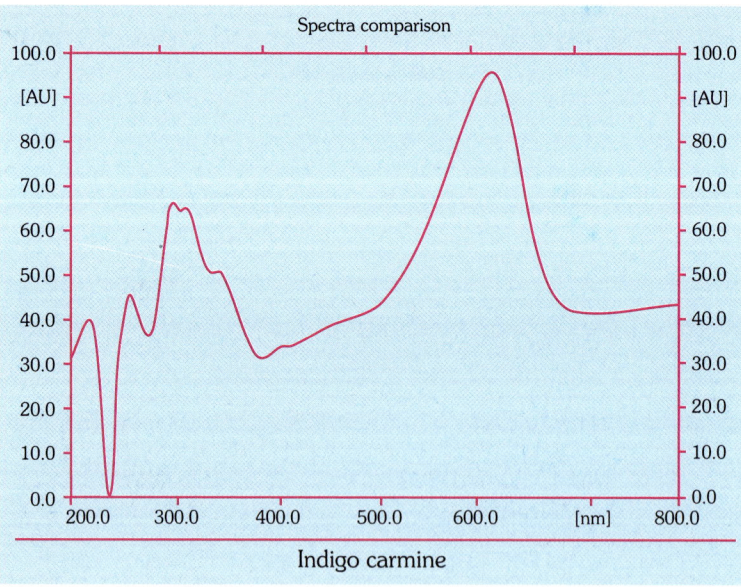

(j)

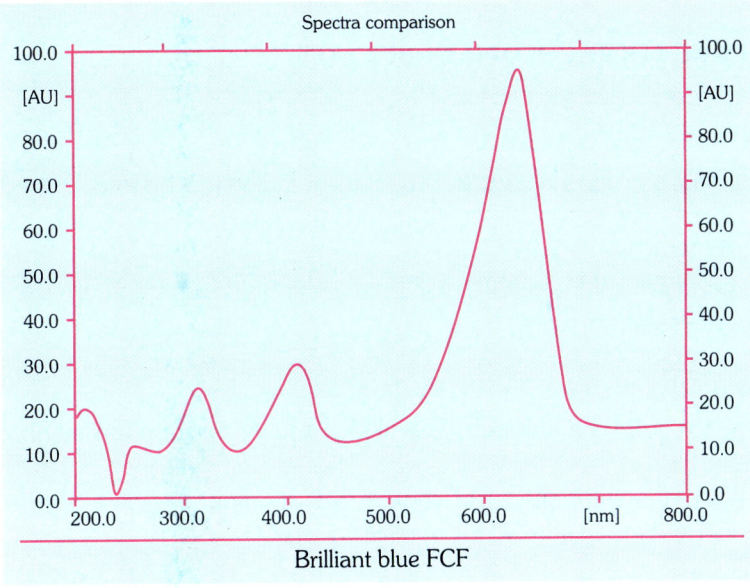

Brilliant blue FCF

(k)

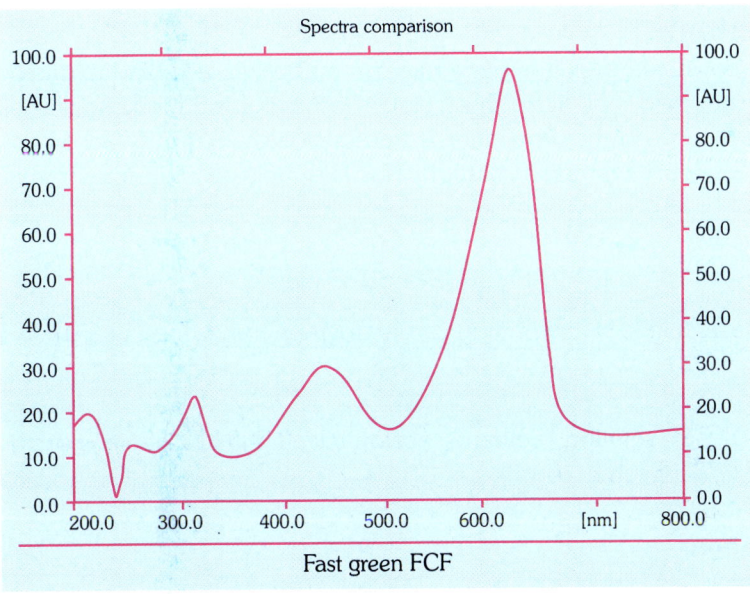

Fast green FCF

(l)

Fig. 17.1 (D) a–l: Separation of food colours

METHOD: 17.1 E

CONFIRMATION OF FOOD COLOUR IN BISCUITS

Aim of Analysis
Confirmation of label claim food colour E-110 (Sunset Yellow) in biscuits.

Chemicals and Reagents
Distilled water, ammonium sulphate (GR), 1-heptane sulphonic acid, sodium salt (GR), methanol, acetonitrile, ammonia solution (25%).

Reagent Preparation
(i) 5% ammonium sulphate solution: Dissolve 5 gm of ammonium sulphate in 100 ml of distilled water.

(ii) 0.005 M 1-heptane sulphonic acid sodium salt solution: Dissolve 50.6 mg of 1-heptane sulphonic acid sodium salt in 50 ml of distilled water.

Standard Preparation
Weigh 2.5 mg of standard Sunset yellow. Add 0.2 ml ammonia solution (25 %) and make up to 100 ml with distilled water.

Sample Preparation
About 20 gm of powdered biscuit is placed in 250 ml beaker, add 150 ml distilled water and 10 ml, 5% glacial acetic acid, boil it. To this 2 gm, 100% wool is added. Again boil for 30 minutes continuously. Cool it, wash the wool till it is free from acid. Strip the adsorbed colour from the wool by adding 2 ml, 5% ammonia solution. This is sample solution for chromatography.

Stationary Phase
TLC Al sheets 20 × 10 cm RP-18 F $_{254}$ (Precoated plate).

Prewashed with acetonitrile by ascending chromatography and dried at 110° C for 20 minutes.

Mobile Phase
5% ammonium sulphate solution: methanol: acetonitrile: 0.005 M 1-heptane sulphonic acid sodium salt solution (3:3:3:7) (volume 16 ml).

Sample Application
Apply with the help of CAMAG ATS-4 or Linomat 5, 100 µl, of biscuit samples.

Standard Application
Apply 5 µl of Sunset yellow (conc. 0.125 µg). 8 mm from bottom edge. Band length 10 mm. Distance from side 15 mm.

Application Pattern

S1 U1 U2 U3 U4 U1 U5 U6 S1 U2 U7 U8 U9 U10 U11 U12 S1

S1 standard Sunset yellow 5 µl
U1 to U12 samples of 100 µl each

Development Chamber

Twin-trough chamber of 20 × 10 cm with s. s. lid.

Tank Saturation

20 minutes with filter paper. Pour all 8 ml of mobile phase in each trough.

Plate equilibrium	:	None
Development distance	:	70 mm.
Visualisation	:	254 nm, visible
Post-chromatographic derivatisation	:	None
Photodocumentation	:	254 nm, visible

Scanning
For Identification

Using Camag TLC scanner 3 with win CATS software, slit-micro, 4 × 0.3 mm, scan at $\lambda_{max} = 486$ nm for Sunset yellow (R_f 0.66).

For Spectral Match

Record the spectra from 400 to 800 nm and compare spectra with standards.

Image (254 nm)

(a)

Image at visible

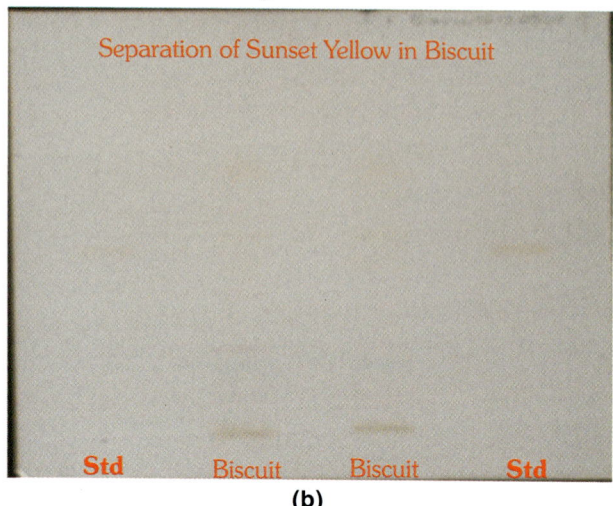

(b)

Track 1, ID: sunset yellow std.(0.125 ug)

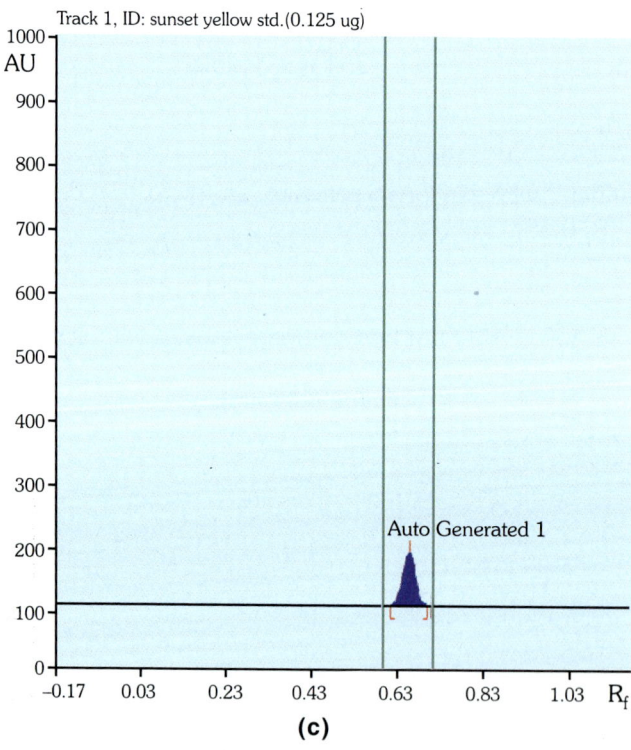

(c)

Track 1, ID: sunset yellow std.(0.125ug)

Peak	Start Position	Start Height	Max Position	Max Height	Max %	End Position	End Height	Area	Area %	Assigned substance
1	0.61 R_f	0.3 AU	0.66 R_f	84.7 AU	100.00 %	0.70 R_f	0.3 AU	1728.2 AU	100.00 %	Sunset Yellow

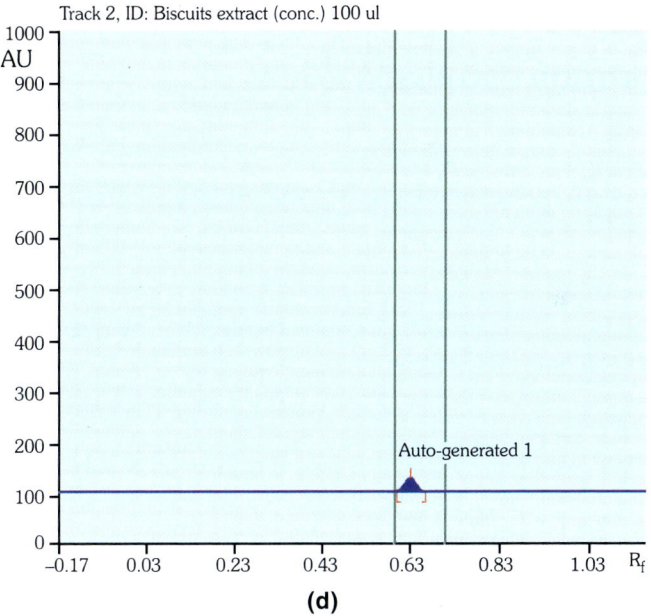

Track 2, ID: Biscuits extract(conc.)100ul

Peak	Start Position	Start Height	Max Position	Max Height	Max %	End Position	End Height	Area	Area %	Assigned substance
1	0.60 R_f	3.4 AU	0.63 R_f	28.7 AU	100.00 %	0.67 R_f	0.2 AU	550.8 AU	100.00 %	Sunset Yellow

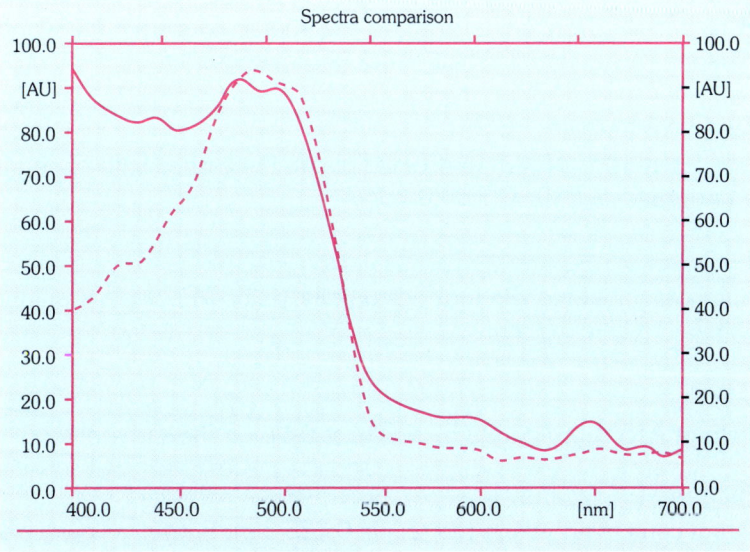

Fig. 17.1(E) a to e: Identification of sunset yellow in biscuit

CONFIRMATION OF FOOD COLOURS IN CANDY

Aim of Analysis

Confirmation of label claim food colours. E-129 (Allura red), E-102 (Tartrazine), E-110 (Sunset yellow), E-132 (Indigo carmine) and E-133 (Brilliant blue) in candy.

Chemicals and Reagents

Distilled water, ammonium sulphate(GR), 1-heptane sulphonic acid sodium salt (GR), methanol, acetonitrile, ammonia solution (25%).

Reagent Preparation

5% ammonium sulphate solution: Dissolve 5 gm of ammonium sulphate in 100 ml of distilled water.

0.005 M 1-Heptane sulphonic acid sodium salt solution: Dissolve 50.6 mg of 1-heptane sulphonic acid sodium salt in 50 ml of distilled water.

Standard Preparation

Weight 2.5 mg of Allura red, 2.5 mg of Sunset yellow, 12.5 mg Indigo carmine, 2.5 mg Brilliant blue and   5 mg of Tartrazine in 100 ml volumetric flask . Add 0.2 ml ammonia solution (25%) and make up to 100 ml with distilled water.

Sample Preparation

About 20 gm of candy is placed in 250 ml beaker, add 150 ml distilled water and 10 ml, 5% glacial acetic acid, boil it. To this 2 gm 100% wool is added. Again boil for 30 minutes continuously. Cool it, wash the wool till it is free from acid. Strip the adsorbed colour from the wool by adding 2 ml, 5% ammonia solution. This is sample solution for chromatography.

Stationary Phase

TLC Al sheets 20 × 10 cm RP-18 $F_{254}S$ (precoated plate). Prewashed with acetonitrile by ascending chromatography and dried at 110° C for 20 minutes.

Mobile Phase

5% ammonium sulphate solution: methanol: acetonitrile: 0.005 M 1-heptane sulphonic acid sodium salt solution (3:3:3:7) (volume 16 ml).

Sample Application

Apply with the help of ATS-4 or linomat 5, 100 μl sample on the same plate.

Standard Application

Apply 5 μl of mixture containing Allura red (conc. 0.125 μg), Sunset yellow (conc. 0.125 μg) and Tartrazine (conc. 0.250 μg), Indigo carmine (conc. 0625 μg), Brilliant blue (conc. 0.125 μg) 8 mm from bottom edge. Band length 10 mm. Distance from side 15 mm.

Application Pattern

S1 U1 U2 U3 U4 U1 U5 U6 S1 U2 U7 U8 U9 U10 U11 U12 S1

S1 standard mixture 5 µl
U1 to U12   sample of 100 µl each

Development Chamber

Twin-trough chamber of 20 × 10 cm with s. s. lid.

Tank Saturation

20 minutes with filter paper. Pour all 8 ml of mobile phase in each trough.

Plate equilibrium	:	None
Development distance	:	70 mm
Visualisation	:	254 nm, visible
Post-chromatographic derivatisation	:	None
Photodocumentation	:	254 nm, Visible

Scanning

For Identification

Using Camag TLC scanner 3 with winCATS software, slit-micro, 4 × 0.3 mm, scan at λ_{max} = 515 nm, for Allura red (R_f 0.52) λ_{max} = 486 nm, for Sunset yellow (R_f 0.65) λ_{max} = 437 nm, for Tartrazine (R_f 0.93) λ_{max} = 610 nm, for Indigo carmine (R_f 0.85) λ_{max} = 637 nm, Brilliant blue (R_f 0.23).

For Spectral Match

Record the spectra from 400 to 800 nm and compare spectra with standards.

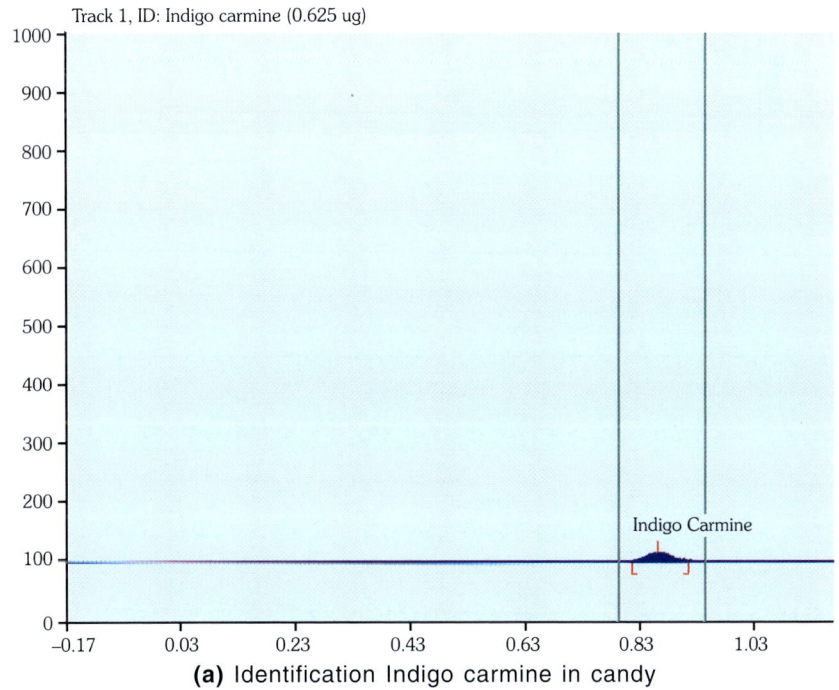

(a) Identification Indigo carmine in candy

Track 2, ID: Indigo Camine(0.625ug)

Peak	Start Position	Start Height	Max Position	Max Height	Max %	End Position	End Height	Area	Area %	Assigned substance
1	0.81 R_f	1.9 AU	0.86 R_f	16.3 AU	100.00 %	0.91 R_f	2.9 AU	561.7 AU	100.00 %	Indigo Carmine

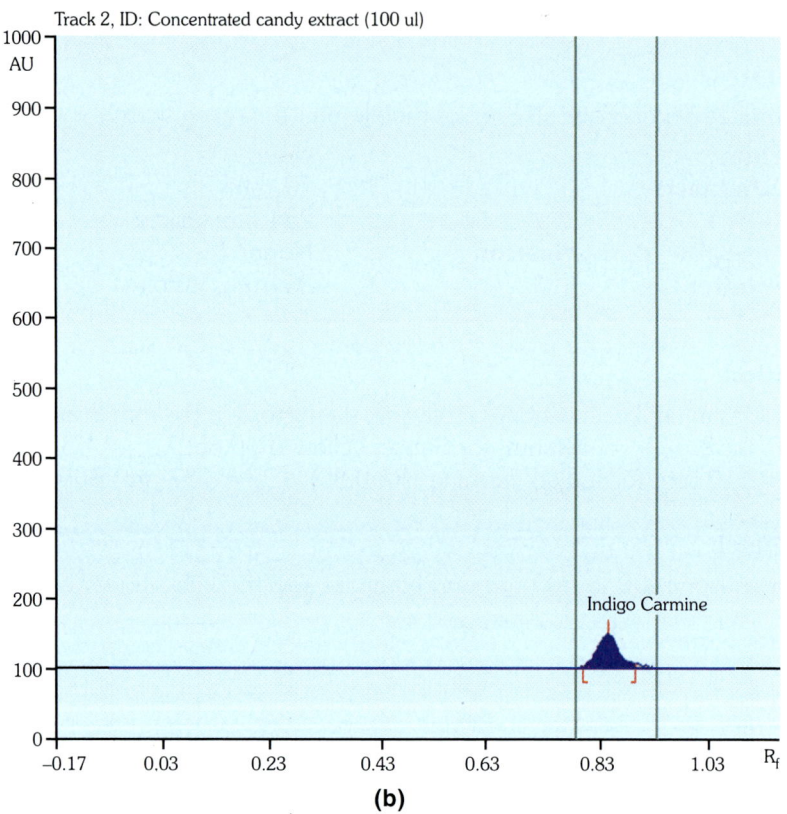

(b)

Track 2, ID: Concentrated Candy Extract (100 ul)

Peak	Start Position	Start Height	Max Position	Max Height	Max %	End Position	End Height	Area	Area %	Assigned substance
1	0.81 R_f	2.0 AU	0.85 R_f	48.1 AU	100.00 %	0.90 R_f	0.5 AU	319.1 AU	100.00 %	Indigo Carmine

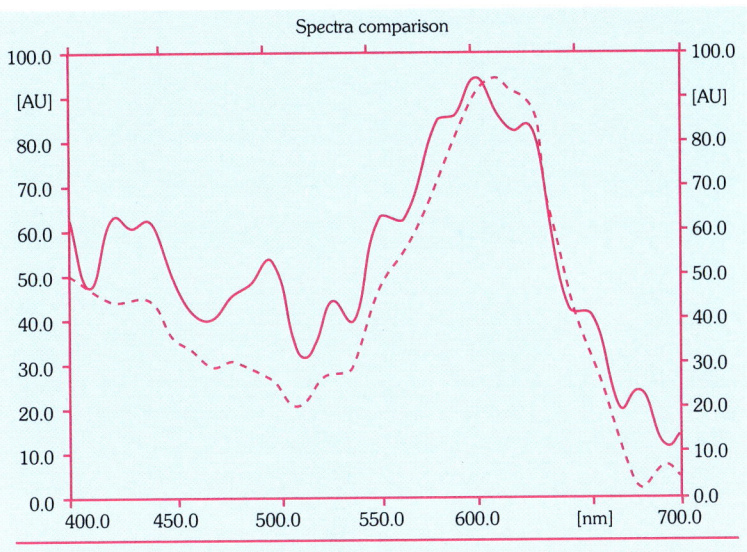

(c) Indigo carmine

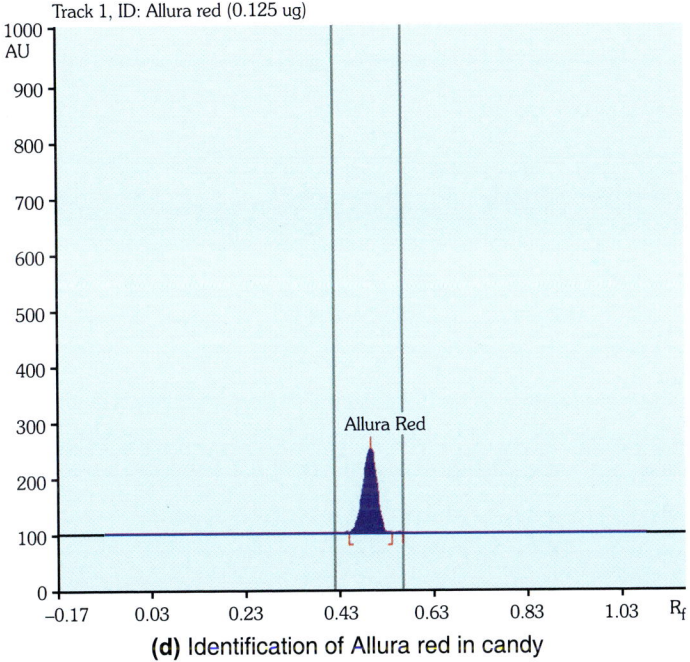

(d) Identification of Allura red in candy

Track 1, ID: Allura red (0.125 ug)

Peak	Start Position	Start Height	Max Position	Max Height	Max %	End Position	End Height	Area	Area %	Assigned substance
1	$0.45\ R_f$	1.0 AU	$0.50\ R_f$	150.1 AU	100.00%	$0.54\ R_f$	1.4 AU	3067.3 AU	100.00%	Allura Red

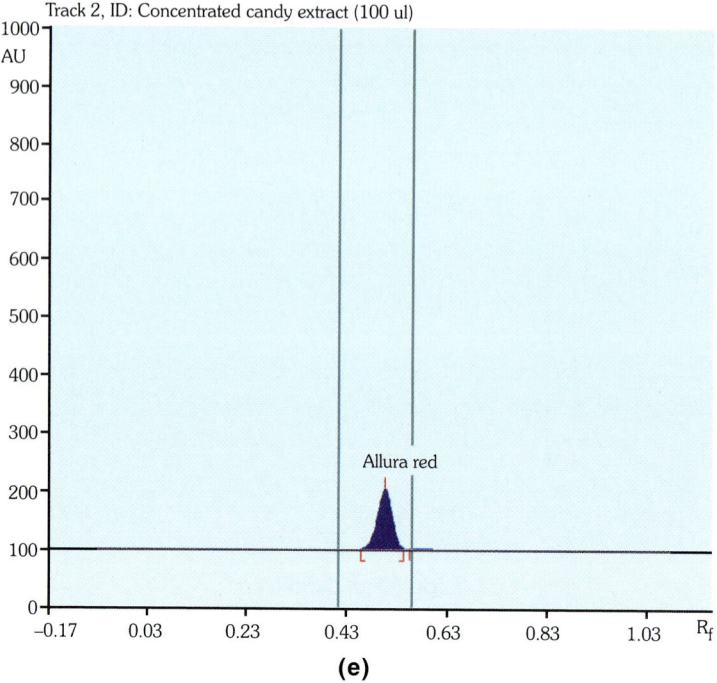

(e)

Track 2, ID: Concentrated candy extract (100 ul)

Peak	Start Position	Start Height	Max Position	Max Height	Max %	End Position	End Height	Area	Area %	Assigned substance
1	0.47 R_f	0.5 AU	0.52 R_f	105.6 AU	100.00 %	0.55 R_f	0.4 AU	2159.2 AU	100.00 %	Allura red

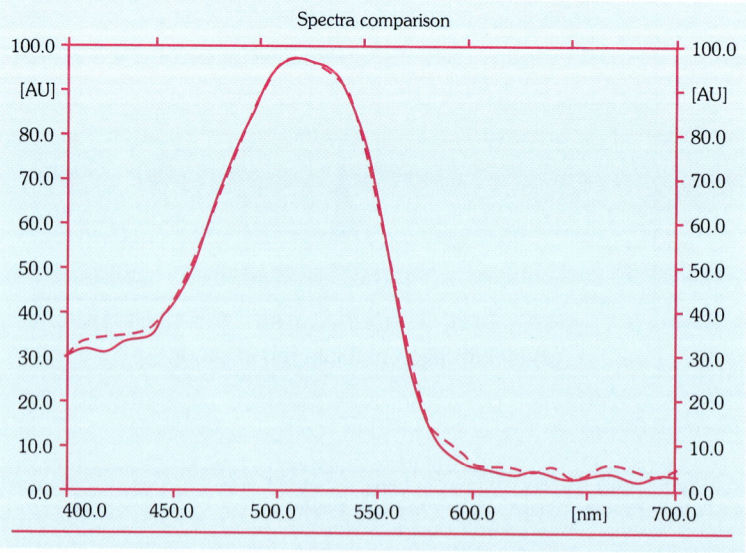

(f) Allura red

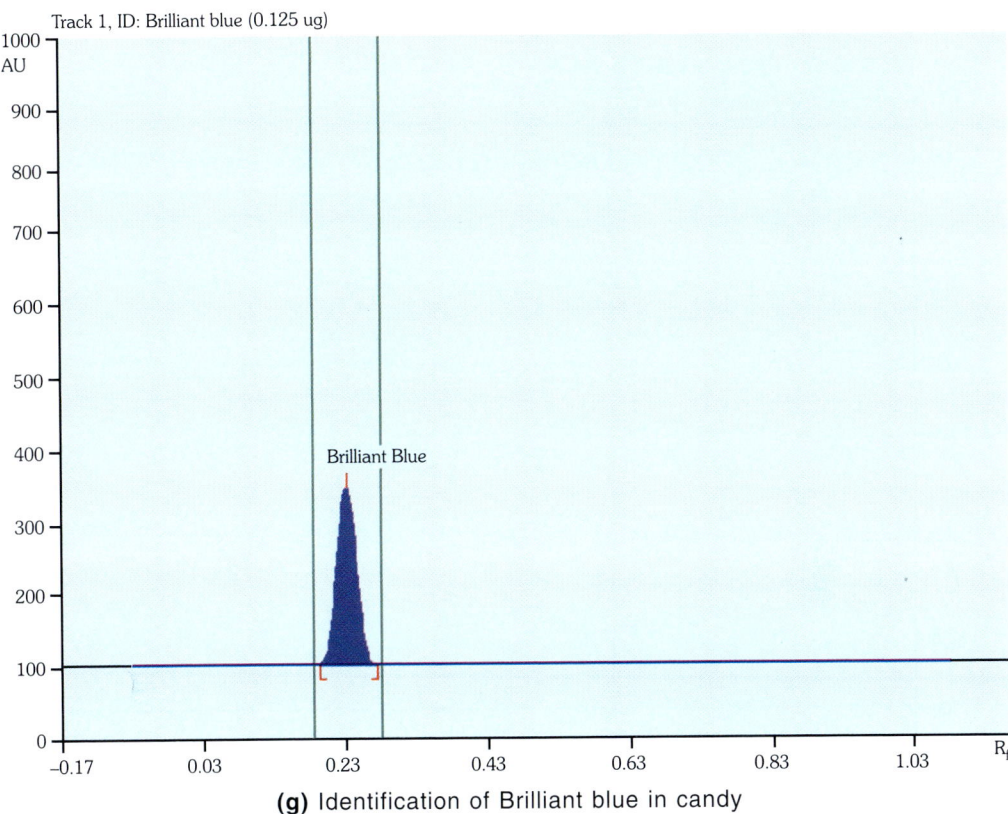

(g) Identification of Brilliant blue in candy

Track 1, DI; Brilliant blue (0.125 ug)

Peak	Start Position	Start Height	Max Position	Max Height	Max %	End Position	End Height	Area	Area %	Assigned substance
1	0.20 R_f	0.3 AU	0.23 R_f	244.1 AU	100.00%	0.28 R_f	0.1 AU	4753.5 AU	100.00%	Brilliant blue

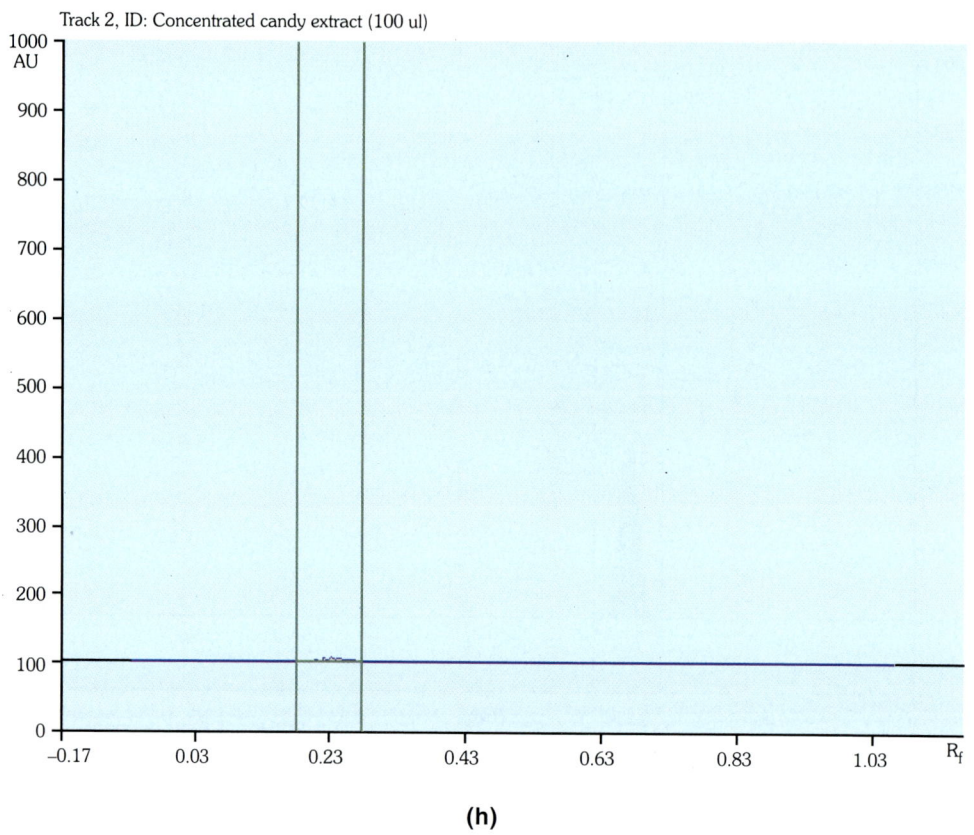

(h)

Track 2, ID: Concentrated candy extract (100 ul)

Peak	Start Position	Start Height	Max Position	Max Height	Max %	End Position	End Height	Area	Area %	Assigned substance

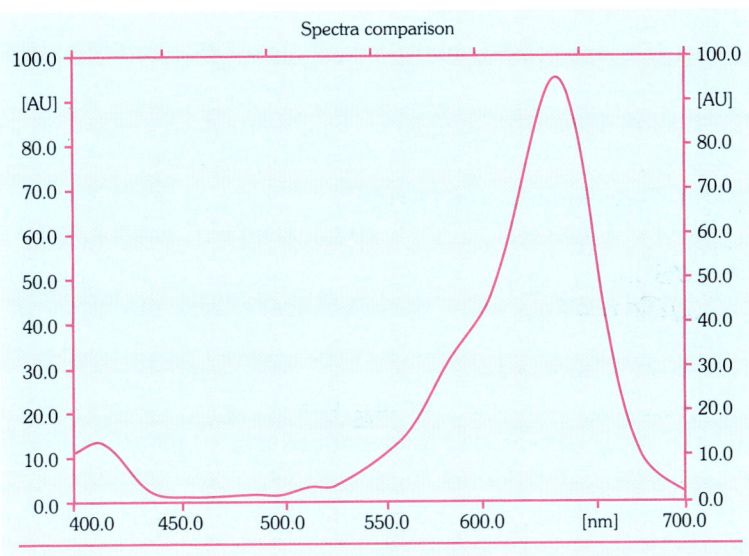

Spectra comparison

(i) Brilliant bule

Fig. 17.1 (F) a to i: Identification of food colours

METHOD: 17.2

DETECTION OF ANTIOXIDANTS IN EDIBLE OIL

Aim of Analysis

Detection of butylated hydroxy anisole (BHA) and butylated hydroxy toluene(BHT), tertiary butylhydroquinone (TBHQ) in ground nut oil.

Chemicals and Reagents

BHA, BHT, TBHQ, ground nut oil, ethanol, phosphomolybdic acid, methanol, petroleum ether (60-80° C).

Reagent Preparation

Phosphomolybdic acid—weigh 250 mg of phosphomolybdic acid and dissolve in 50 ml ethanol.

Standard Preparation

1. Butylated Hydroxyanisole (BHA)

Solution A: Weigh 10 mg of BHA in 10 ml volumetric flask. Dissolve in 10 ml of petroleum ether [Conc: 1µg/µl].

Solution B : 1 ml of solution A in a 10 ml volumetric flask. Add 9 ml of petroleum ether [Conc: 0.1 µg/µl].

2. Butylated Hydroxytoluene (BHT)

Solution C: Weigh 10 mg of BHT in a 10 ml volumetric flask. Add 10 ml of petroleum ether [Conc: 1µg/µl].

Solution D: 1 ml of solution A in a 10 ml volumetric flask. Add 9 ml of petroleum ether [Conc: 0.1 µg/µl].

3. Tertiary Butyl Hydroxyquinone(TBHQ)

Solution E: Weigh 10 mg of TBHQ in a 10 ml volumetric flask. Add 10 ml of petroleum ether [Conc: 1µg/µl].

Solution F: 1 ml of solution A in a 10 ml volumetric flask. Add 9 ml of petroleum ether [Conc: 0.1 µg/µl].

Sample Preparation

1. **Groundnut oil:** Weigh 1000 mg of groundnut oil in a 10 ml volumetric flask. Make up to 10 ml using petroleum ether [Conc: 100 µg/µl]. This is test solution.

2. **BHA in oil sample:** Weigh 1000 mg of groundnut oil and add 1 ml of solution A in a 10 ml volumetric flask and make the final volume to 10 ml with petroleum ether[Conc : 100 µg/µl, 0.1 µg/µl].

3. **BHT in oil sample:** Weigh 1000 mg of groundnut oil and add 1 ml of solution C in 10 ml volumetric flask and make the final volume to 10 ml with petroleum ether [Conc : 100 µg/µl, 0.1 µg/µl].

4. **TBHQ in oil sample:** Weigh 1000 mg of groundnut oil and add 1 ml of solution E in a 10 ml volumetric flask and make the final volume to 10 ml with petroleum ether [Conc: 100 µg/µl, 0.1 µg/µl].

Stationary Phase

TLC aluminium sheets silica gel RP-18 $F_{254}S$(precoated), cut to 20 × 10 cm.

Mobile Phase

Methanol: water (9 : 1) v/v (Volume =16 ml)

Development Chamber

Twin-trough chamber of 20 × 10 cm with s. s. lid.

Chamber saturation	:	None
Plate equilibrium	:	None
Sample/standard application	:	Apply with the help of ATS-4 or linomat 5, 2 µl and 5 µl (qty: 200 µg, 500 µg resp) of each sample and 2 µl and 5 µl (qnty: 0.2 µg and 0.5 µg) of standard soln. B, soln.D, soln. F respectively on precoated layer 8 mm from the bottom edge. Band length 6 mm. Distance from the sides 15 mm.
Development distance	:	80 mm from bottom edge.
Visualisation	:	Observe in UV cabinet at 254 nm.

Post-chromatographic derivatisation	:	The plate is dipped in phosphomolybdic acid reagent for 1 second using Camag immersion device 3. Then heated at 110° C for 10 min.
Photodocumentation	:	254 nm and visible light.
Measurement mode	:	UV absorbance/reflectance.

Scanning

For Quantification

Using TLC scanner 3 with winCATS software, slit-micro, 4 × 0.3 mm, scan at 285 nm.

For Identification

Record and match spectra between 190 and 400 nm.

Image (Visible)

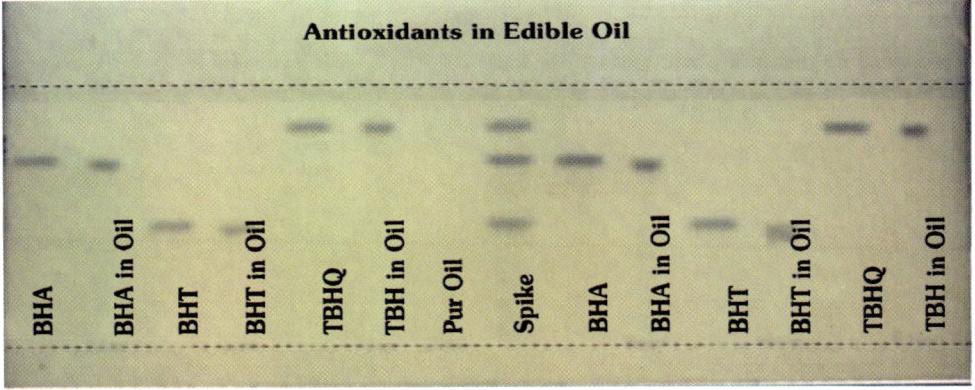

(a)

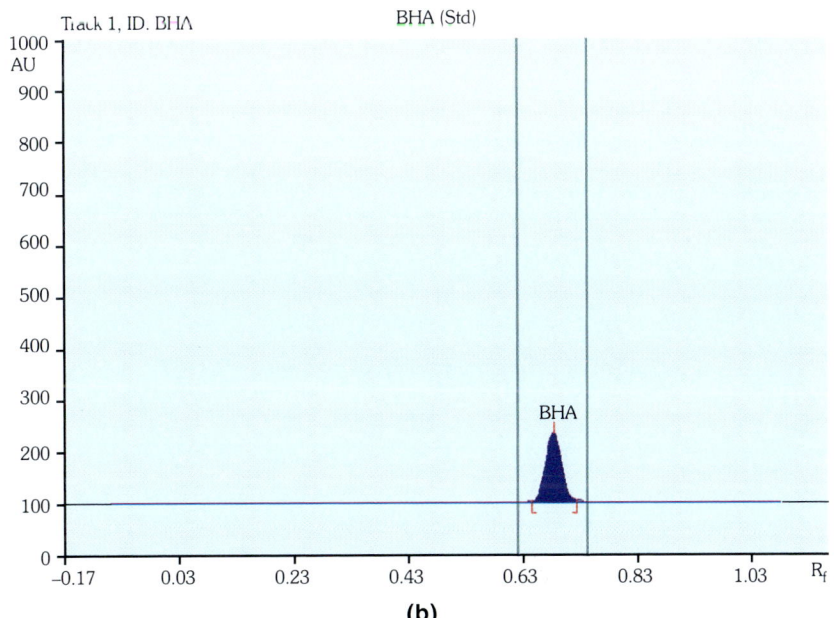

(b)

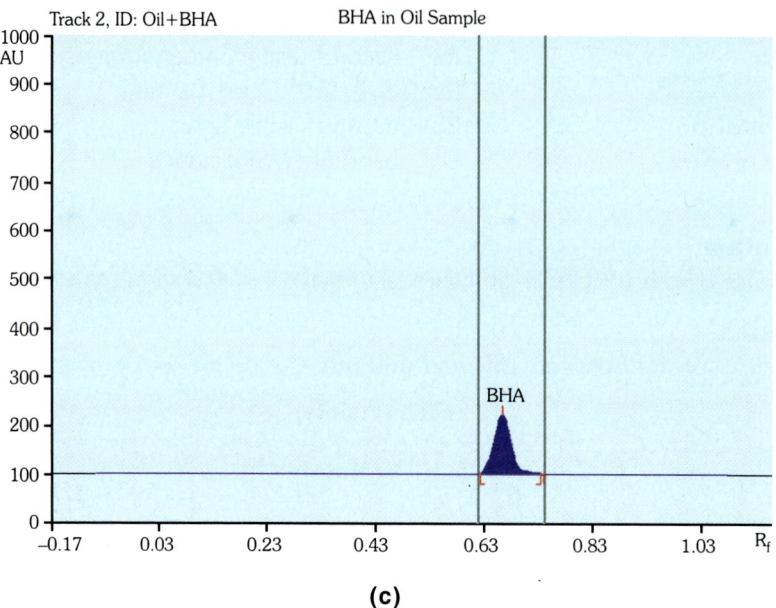

(c)

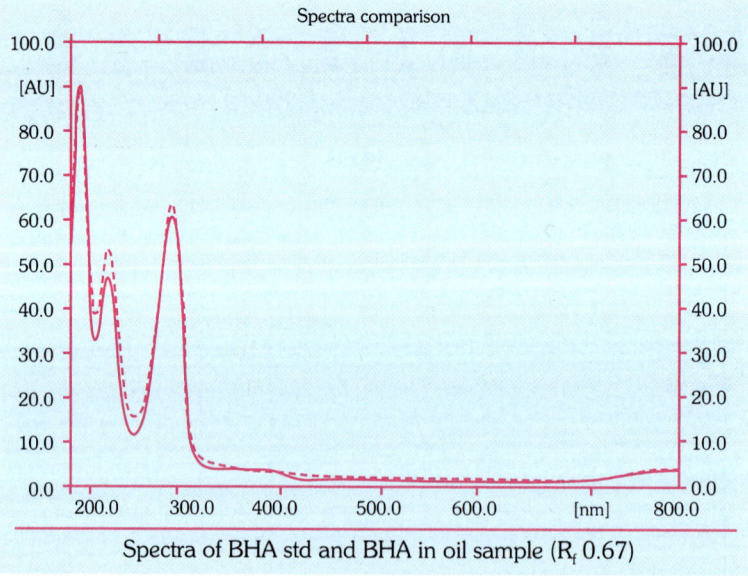

Spectra of BHA std and BHA in oil sample (R_f 0.67)

(d)

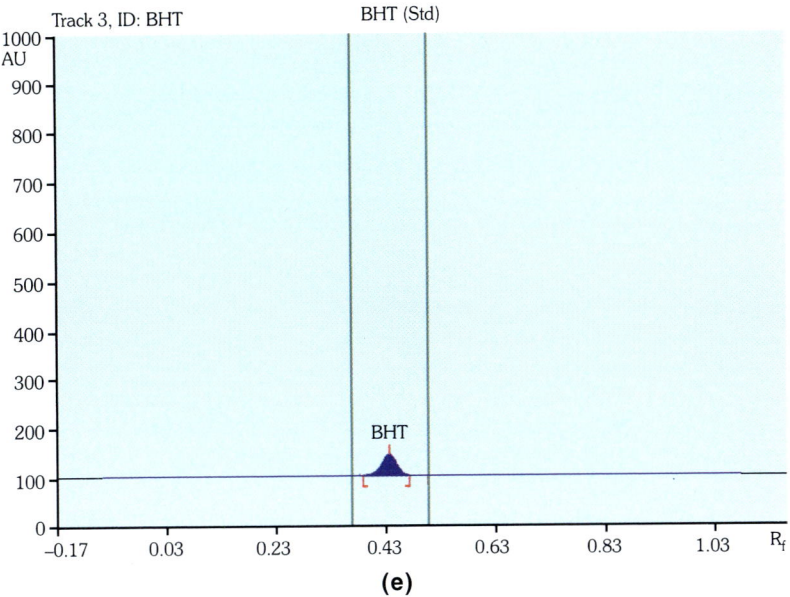

(e)

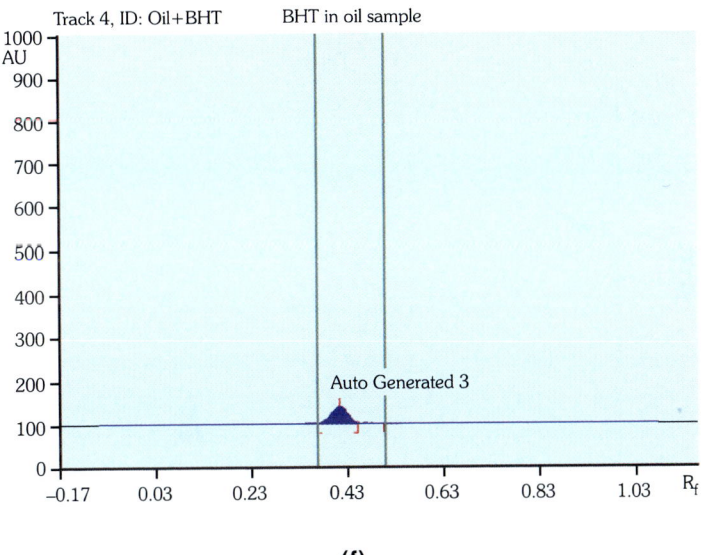

(f)

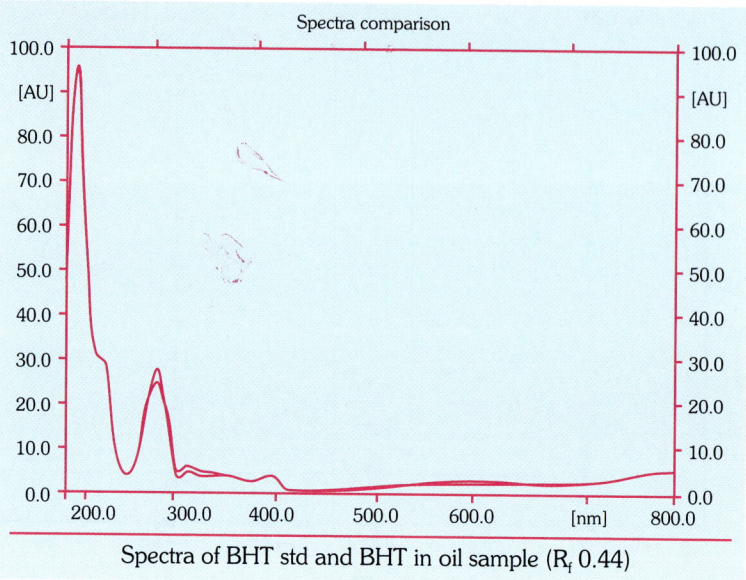

Spectra of BHT std and BHT in oil sample (R_f 0.44)

(g)

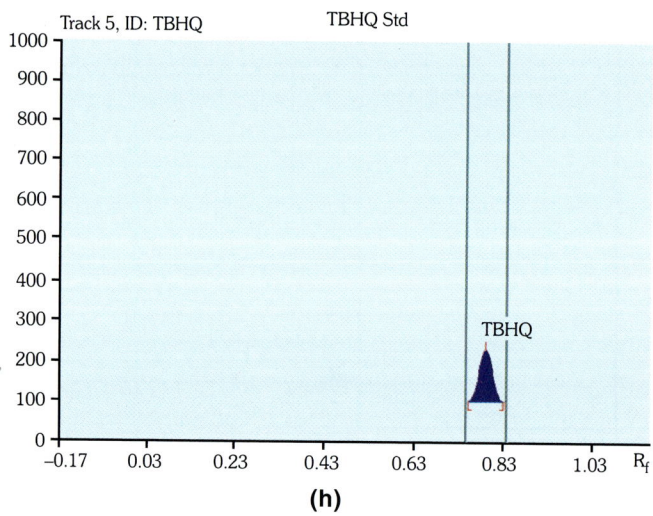

(h)

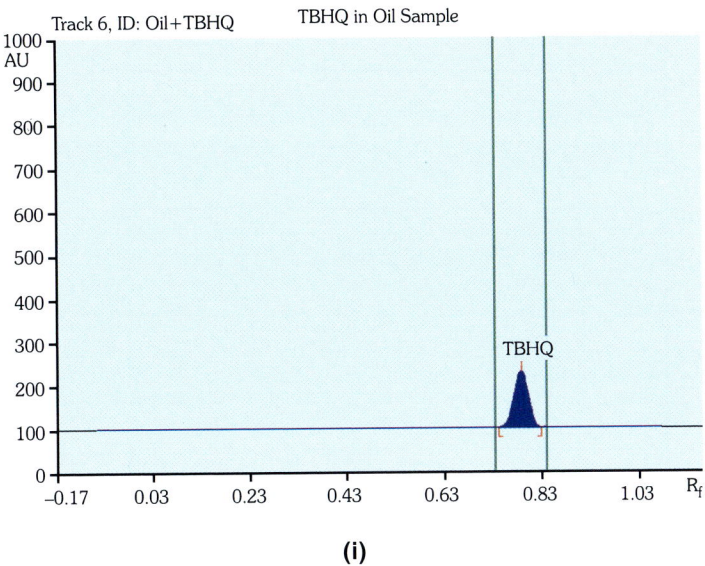

(i)

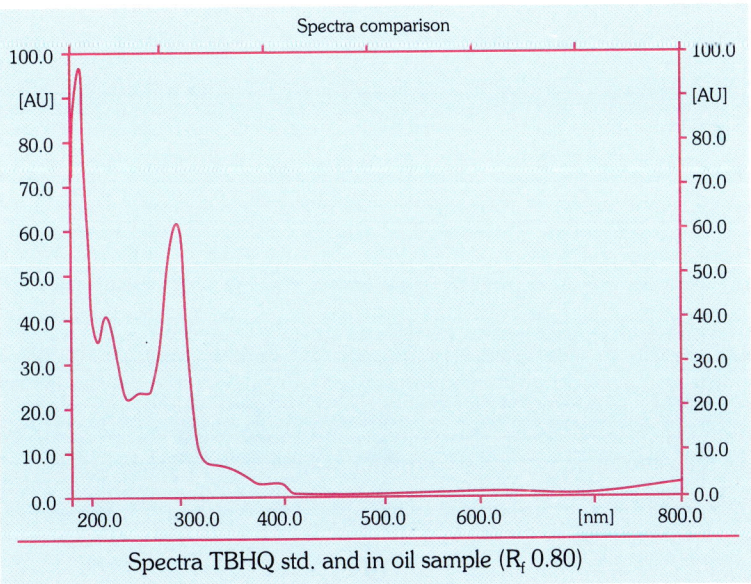

Spectra TBHQ std. and in oil sample (R_f 0.80)

(j)

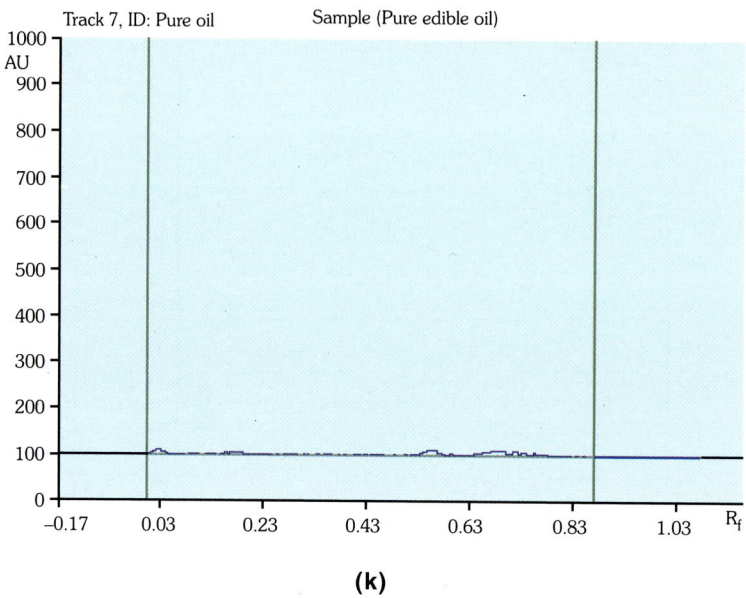

(k)

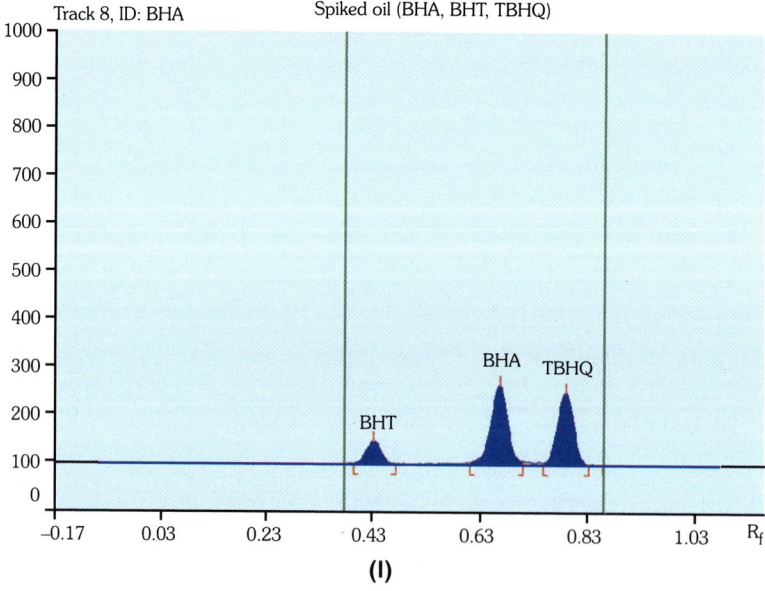

(l)

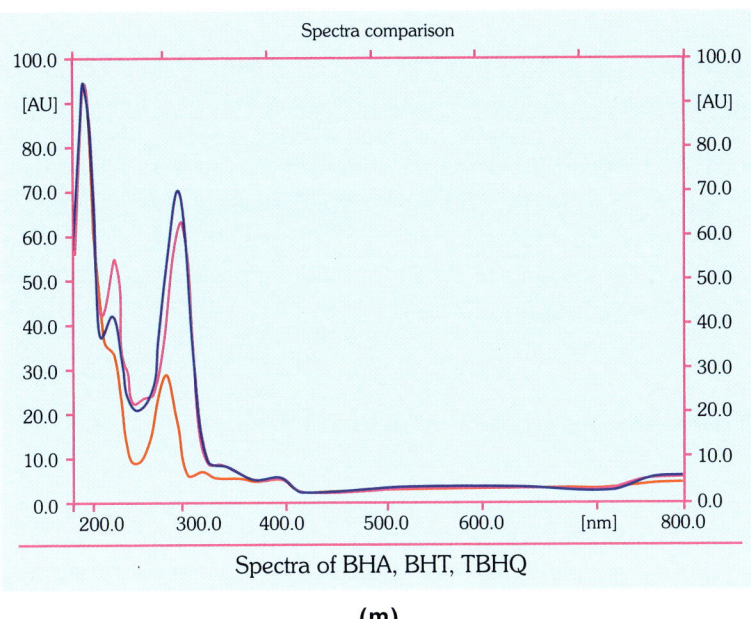

Spectra of BHA, BHT, TBHQ

(m)

Fig. 17.2 a–m: Antioxidants in groundnut oil

METHOD: 17.3

QUANTIFICATION OF CAFFEINE IN TEA

Aim of Analysis

Quantitative estimation of caffeine content in tea.

Chemicals and Reagents

Toluene, acetone, 1.5% potassium permanganate, phosphoric acid, sodium hydroxide, sodium sulphite, ammonium thiocyanate, distilled water, chloroform.

Reagent Preparation

1.5% potassium permanganate solution: Dissolve 1.5 gm $KMnO_4$ in water and dilute to 100 ml.

Dilute phosphoric acid solution: To 15 ml of phosphoric acid, add 70 ml of water.

Reducing solution: Dissolve 5 gm of sodium sulphite (Na_2SO_3) and 5 gm of ammonium thiocyanate (NH_4CNS) in water. Dilute to 100 ml.

Sodium hydroxide solution: Dissolve 2.5 gm of NaOH in 7.5 ml of water.

Standard Solution

Solution A: Weigh accurately 10 mg of caffeine in a 10 ml vol. flask. Dissolve in chloroform (conc. 1 µg/µl).

Solution B: Dilute 1 ml of solution A to 10 ml with chloroform(conc. 0.1 µg/µl).

Sample Preparation

Weigh 75 mg of powdered tea sample in a 250 ml conical flask. Add 3.75 ml of $KMnO_4$ solution. Add 7.5 ml of reducing solution. Add 0.75 ml of phosphoric acid solution. Add 0.75 ml of sodium hydroxide (NaOH) solution. Mix well after addition of each solution. Transfer the solution to 500 ml separating funnel. Add 100 ml chloroform and shake for 5 minutes. Allow the layers to separate and drain the chloroform layer into a 500 ml conical flask after passing it through 3 gm anhyd. sodium sulphate. Re-extract aqueous solution with 50ml of chloroform and drain as before. Repeat this 50 ml extraction. Collect the chloroform layers and concentrate to a final volume of 25 ml in a water bath at 60° C (conc. 3 µg/µl).

Stationary Phase

TLC Al sheets silica gel $60F_{254}$ precoated cut to 20×10 cm.

Mobile Phase

Toluene: acetone (7 : 3) v/v (Volume = 16 ml).

Development Chamber

Twin-trough chamber of 20×10 cm with s. s. lid.

Chamber saturation	: 20 minutes with filter paper.
Plate equilibrium	: None
Sample/standard application	: Apply with the help of Camag ATS-4 or linomat 5, 5 µl of sample (15 µg) and 2, 4, 6, 8, 10 µl of standard caffeine solution 'B' (0.2, 0.4, 0.6, 0.8, 1.0 µg) on precoated layer 8 mm from the bottom edge. Band length 8 mm. Distance from the sides 15 mm.
Development distance	: 70 mm.
Visualisation	: Observe under UV Cabinet at 254 nm.
Post-chromatographic derivatisation	: Not required.
Photodocumentation	: At 254 nm.
Measurement mode	: UV absorbance/reflectance.

Scanning

For Quantification

Using Camag TLC scanner with win CATS software, slit-micro, 4×0.3 mm, scan at 275 nm.

Spectrum

Record and match spectra between 190-400 (R_f = 0.17).

Image (254 nm)

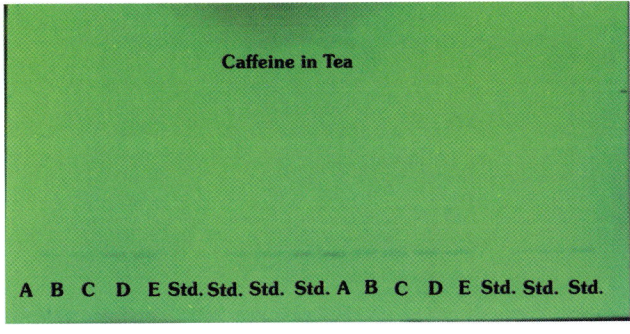

A B C D E Std. Std. Std. Std. A B C D E Std. Std. Std.

(a)

Track 13, ID: Sample C   Sample

Caffeine

(b)

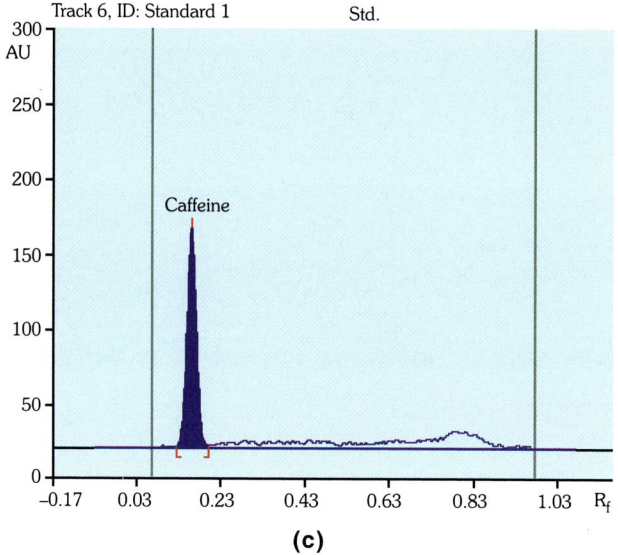

(c)

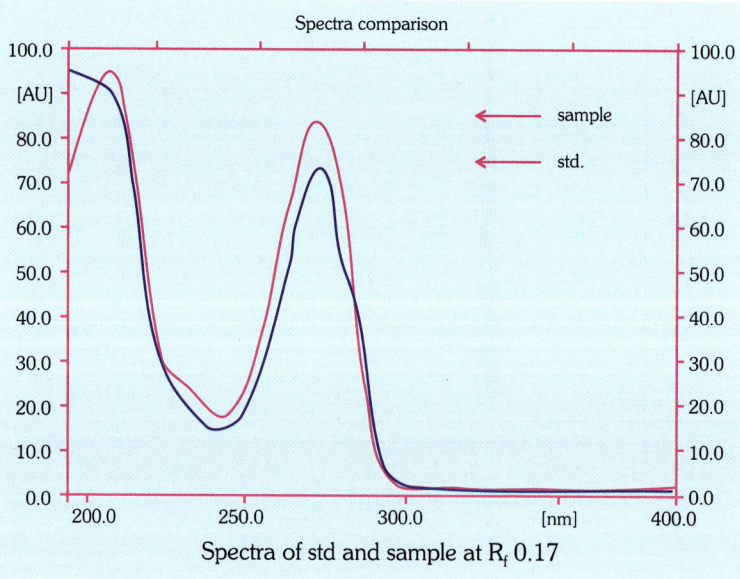

Spectra of std and sample at R_f 0.17

(d)

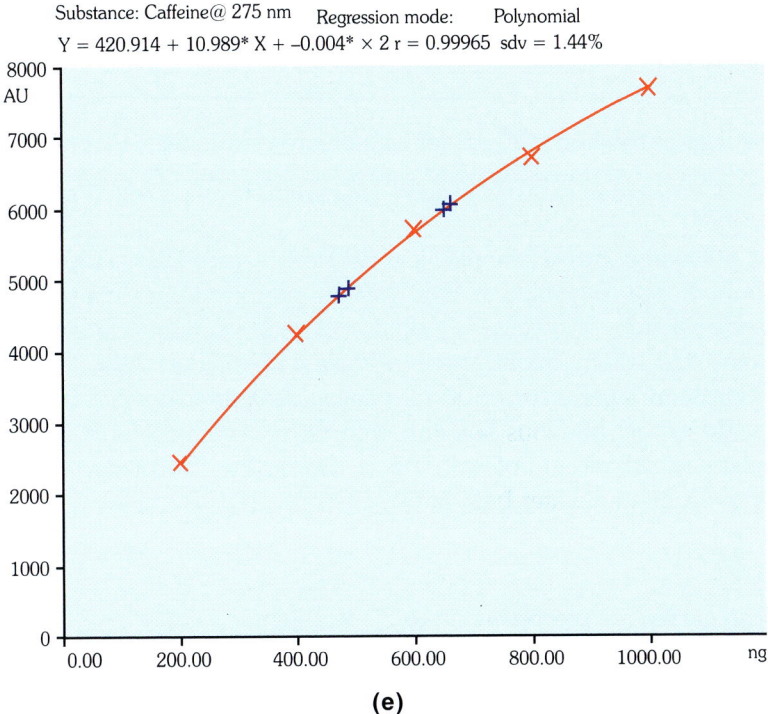

(e)

Fig. 17.3 a–e: Caffeine content in tea powder

METHOD: 17.4

QUANTIFICATION OF CAFFEINE IN COFFEE

Aim of Analysis

Quantitative estimation of caffeine content in coffee.

Chemicals and Reagents

Toluene, acetone, 1.5% potassium permanganate, phosphoric acid, sodium hydroxide, sodium sulphite, ammonium thiocyanate, distilled water, chloroform.

Reagent Preparation

1.5% potassium permanganate solution: Dissolve 1.5 gm $KMnO_4$ in water and dilute to 100 ml.

Dilute phosphoric acid solution: To 15 ml of phosphoric acid, add 70 ml of water.

Reducing solution: Dissolve 5 gm of sodium sulphite(Na_2SO_3) and 5 gm of ammonium thiocyanate(NH_4CNS) in water and dilute to 100 ml.

Sodium hydroxide solution: Dissolve 2.5 gm of NaOH in 7.5 ml of water.

Standard Preparation

Solution A: Weigh accurately 10 mg of caffeine in a 10 ml vol. flask. Dissolve in chloroform (conc 1 µg/µl).

Solution B: Dilute 1ml of solution A to 10 ml with chloroform (conc. 0.1 µg/µl).

Sample Preparation

Weigh 75 mg of powdered coffee sample in a 250 ml conical flask. Add 3.75 ml of $KMnO_4$ solution. Add 7.5 ml of reducing solution. Add 0.75 ml of phosphoric acid solution. Add 0.75 ml of sodium hydroxide solution. Mix well after addition of each solution. Transfer the solution to 500 ml separating funnel. Add 100 ml chloroform and shake for 5 minutes. Allow the layers to separate and drain the chloroform layer into a 500 ml conical flask after passing it through anhydrous sodium sulphate. Re-extract aqueous solution with 50 ml of chloroform and drain as before. Repeat this extraction with 15 ml chloroform, 3 times. Collect the chloroform layers and concentrate to a final volume of 25 ml in a water bath at 60° C [conc. 3µg/µl].

Stationary Phase

TLC Al sheets silica gel $60F_{254}$ precoated cut to 20×10 cm.

Mobile Phase

Toluene: acetone (7 : 3) v/v (Volume = 16 ml).

Development Chamber

Twin-trough chamber of 20×10 cm with s. s. lid.

Chamber saturation	:	10 minutes with filter paper.
Plate equilibrium	:	None
Sample/standard application	:	Apply with the help of Camag ATS-4 or Linomat 5, 5 µl of test solutions (15 µg) and 2, 4,6, 8, 10 *µl* of standard caffeine solution 'B' (0.2, 0.4, 0.6, 0.8, 1.0 µg) on precoated layer 8 mm from the bottom edge. Band length 8 mm. Distance between tracks 10 mm. Distance from the sides 15 mm.
Development distance	:	70 mm from bottom edge.
Visualisation	:	Observe under UV cabinet at 254 nm.
Post-chromatographic derivatisation	:	Not required.
Photodocumentation	:	At 254 nm.
Measurement mode	:	UV absorbance/reflectance.

Scanning

For Quantification

Using Camag scanner with win CATS software, slit-micro, 4×0.3 mm, scan at 275 nm.

Spectrum

Record and match spectra between 190-400 nm (R_f = 0.18).

Image (254 nm)

(a)

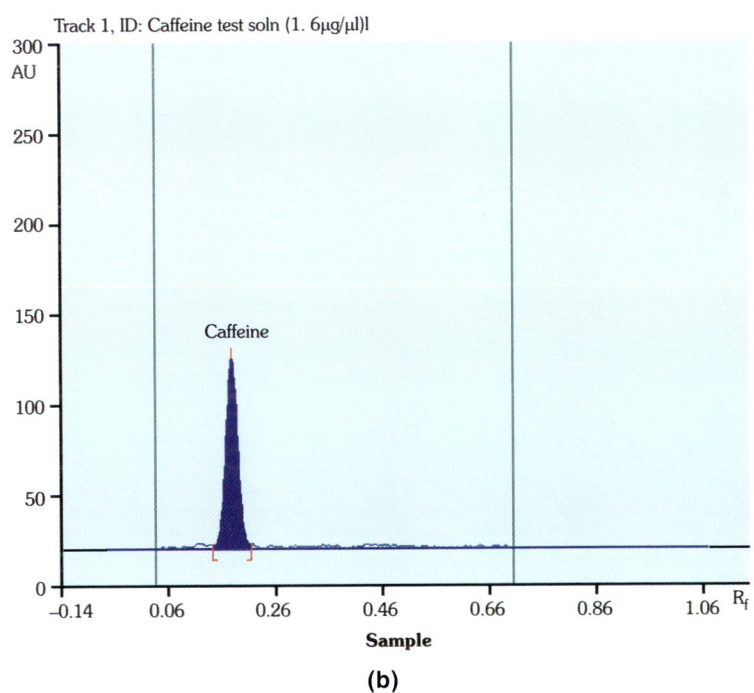

(b)

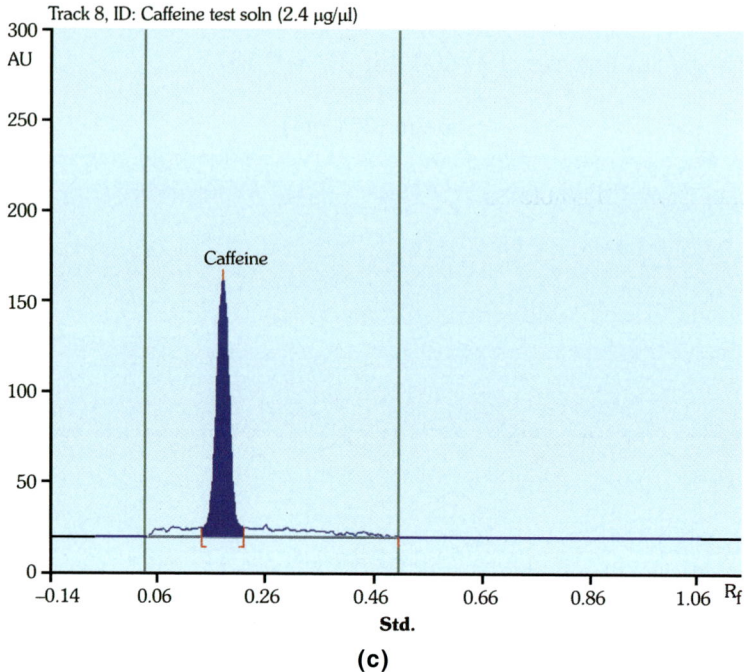

(c)

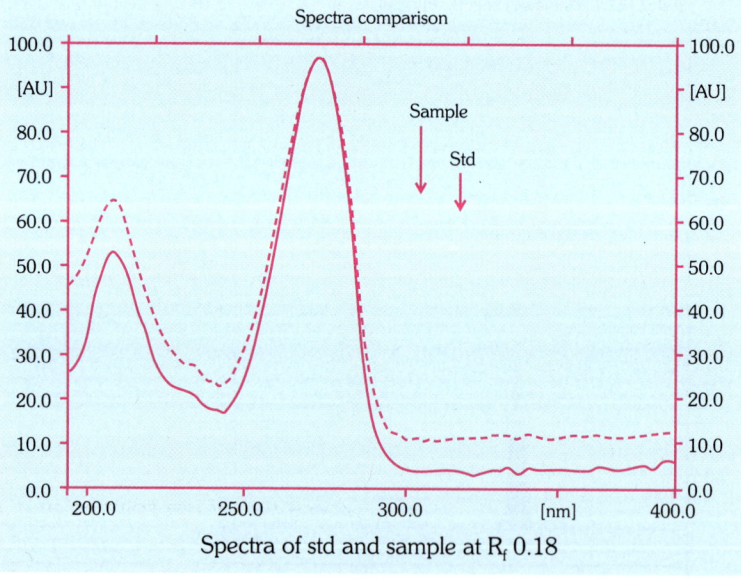

(d)

Fig. 17.4 a–d: Caffeine content in coffee powder

METHOD: 17.5

IDENTIFICATION OF CHOLESTEROL IN EDIBLE OIL

Aim of Analysis

Identification of cholesterol in edible oil.

Chemicals and Reagents

n-hexane, diethyl ether, methanol, toluene, phosphomolybdic acid, cholesterol, ethanol.

Reagent Preparation

Phosphomolybdic acid reagent: Weigh 250 mg of phosphomolybdic acid. Dissolve it in 50 ml of ethanol.

Standard Solution

Weigh 1.1 mg of cholesterol. Dissolve in 1.1 ml of toluene [Conc. 1 µg/µl].

Sample Preparation

Weigh 258.4 mg of ground nut oil and dissolve it in 5 ml of toluene [Conc. 51.68 µg/µl]. This is test solution.

Stationary Phase

TLC Al sheets silica gel $60F_{254}$ cut to 20×10 cm.

Mobile Phase

n-hexane: diethyl ether: methanol (6 :1: 0.1) v/v (Volume = 16 ml).

Development Chamber

Twin-trough chamber of 20×10 cm with s. s. lid.

Chamber saturation	:	None
Plate equilibrium	:	None
Sample/standard application	:	Apply with the help of Camag ATS-4 or linomat 5, 5 µl (258.4 µg) of test solution and 2 µl, 5 µl (Conc. 2 µg, 5 µg resp.) of standard solution on precoated layer 8 mm from the bottom edge. Band length 8 mm. Distance from the side 15 mm.
Development distance	:	80 mm.
Visualisation	:	Observe under UV cabinet at 254 nm and at 366 nm.
Post-chromatographic derivatisation	:	Plate is dipped in phosphomolybdic acid and then heated at 110° C for 10 min.

Photodocumentation   :   At 254 nm, 366 nm, visible.
Measurement mode   :   UV absorbance/reflectance.

Scanning

For Quantification

Using Camag TLC scanner 3 with win CATS software, slit-micro, 4 × 0.3 mm, scan at 200 and 600 nm.

For Identification

Record spectra between 190 to 400 nm and confirm match with standard ($R_f = 0.14$).

Image (254 nm)

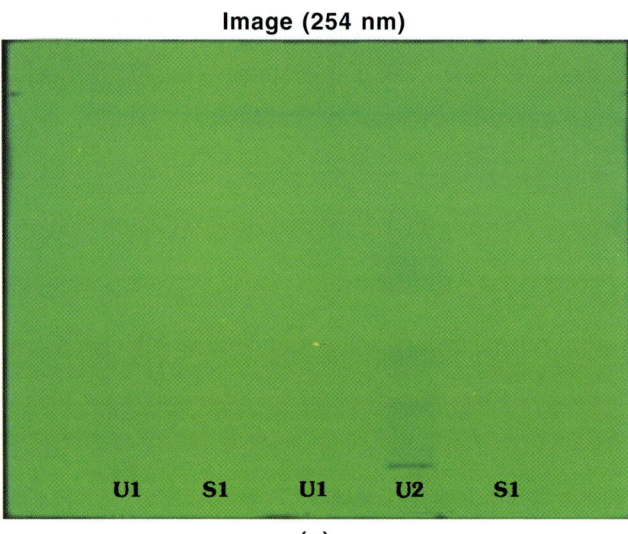

(a)

Image (366 nm)

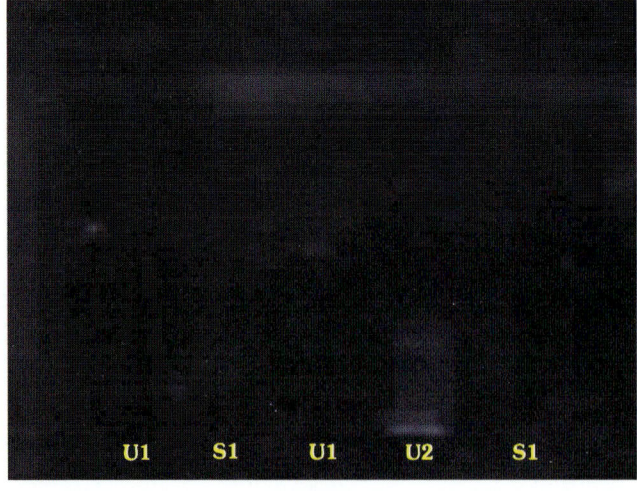

(b)

Image (visible)

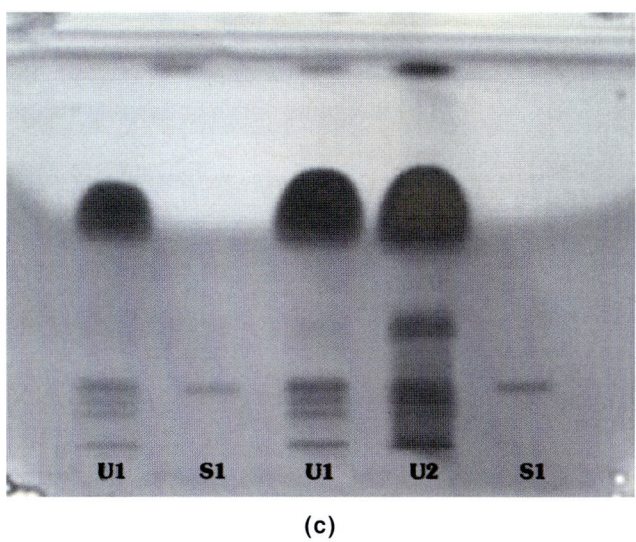

(c)

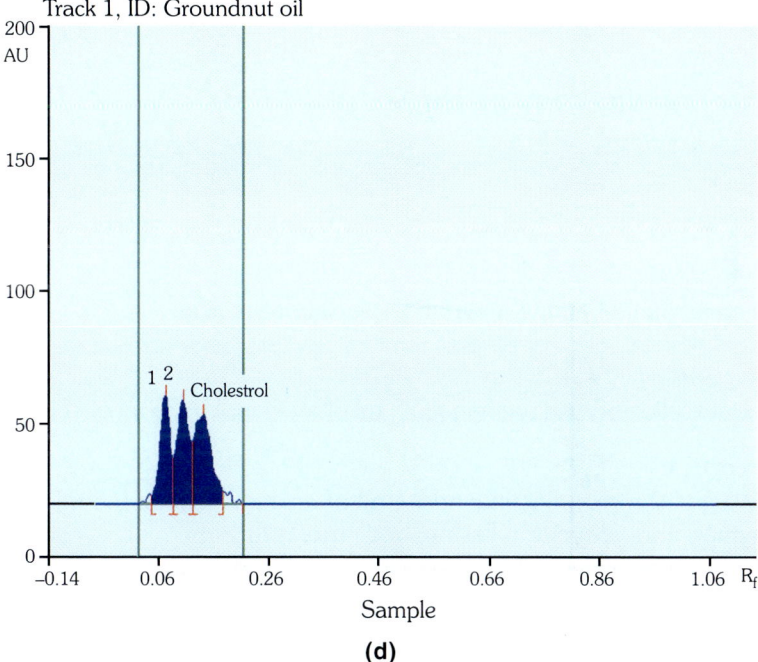

(d)

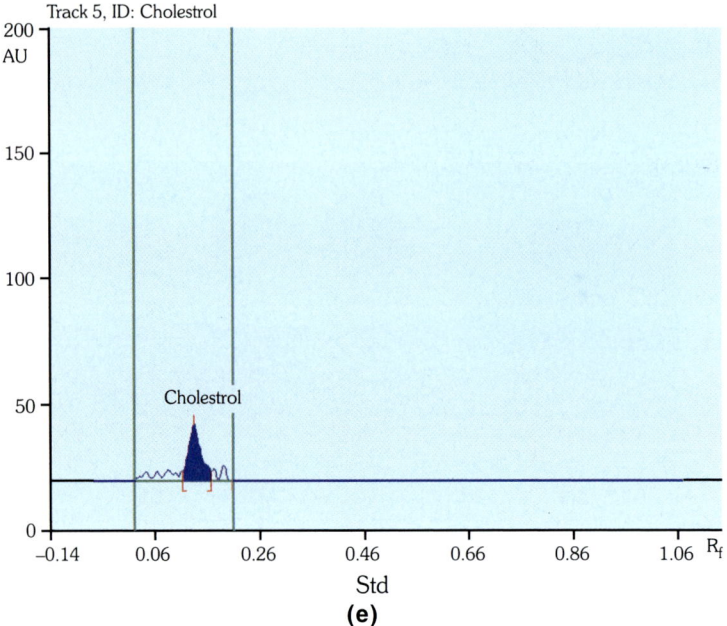

Track 5, ID: Cholestrol

Fig. 17.5 a–e: Identification of cholesterol in edible oil

METHOD: 17.6

QUANTIFICATION OF NICOTINE IN TOBACCO

Aim of Analysis
Quantitative estimation of nicotine content in tobacco.

Chemicals and Reagents
Toluene, ethyl acetate, diethyl amine, methanol, glacial acetic acid.

Reagent Preparation
1:1 mixture of glacial acetic acid and methanol: 25 ml of methanol + 25 ml of acetic acid (Reagent A).

Standard Preparation
Std. soln. A : Weigh 10 mg of nicotine, add 1 ml of reagent A. Add 0.5 ml water and make up to 10 ml with methanol in volumetric flask (Conc. $1\mu g/1\mu l$).
Std. soln. B : Dilute 2 ml of soln. A to 10 ml with methanol (Conc. $1\mu l = 0.2\mu g$).

Sample Preparation
Weigh 40 mg of powdered tobacco in a 10 ml standard flask. Add 1 ml of reagent A. Shake well. Mix the solution for 5 min. Add 0.5 ml of water. Shake well and make the final volume to 10 ml with methanol (conc.4 $\mu g/\mu l$).

Stationary Phase
TLC Al sheets silica gel $60F_{254}$ precoated cut to 20×10 cm.

Mobile Phase

Toluene: ethyl acetate: diethyl amine (7: 2:1) v/v (Vol = 16 ml).

Development Chamber

Twin-trough chamber of 20 × 10 cm with s. s. lid.

Chamber saturation	:	None
Plate equilibrium	:	None
Sample/standard application	:	Apply with the help of Camag Linomat 5 or ATS-4, 10 µl (40µg) of each sample and 2, 3, 4, 5, 6 µl of standard nicotine solution 'B' (0.4, 0.6, 0.8, 1.0, 1.2 µg)on precoated layer 8 mm from the bottom edge. Band length 8 mm. Distance from the sides 15 mm.
Development distance	:	80 mm from bottom edge.
Visualisation	:	Inspect the plate at 254 nm under UV cabinet.
Post-chromatographic derivatisation	:	None
Photodocumentation	:	254 nm.
Measurement mode	:	UV absorbance/reflectance.

Scanning

For Quantification

Using Camag TLC scanner 3 with win CATS software, slit-micro, 4 × 0.3 mm, scan at 262 nm.

Spectrum

Record and match spectra between 190-400 nm (R_f = 0.58).

Image (254 nm)

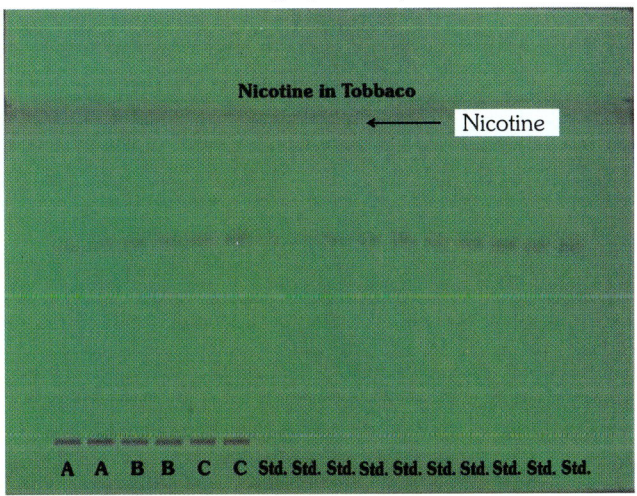

(a)

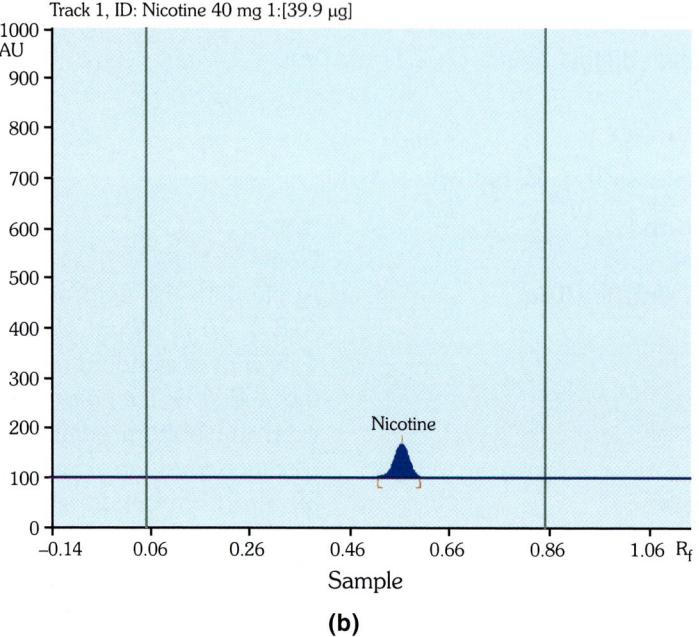

(b)

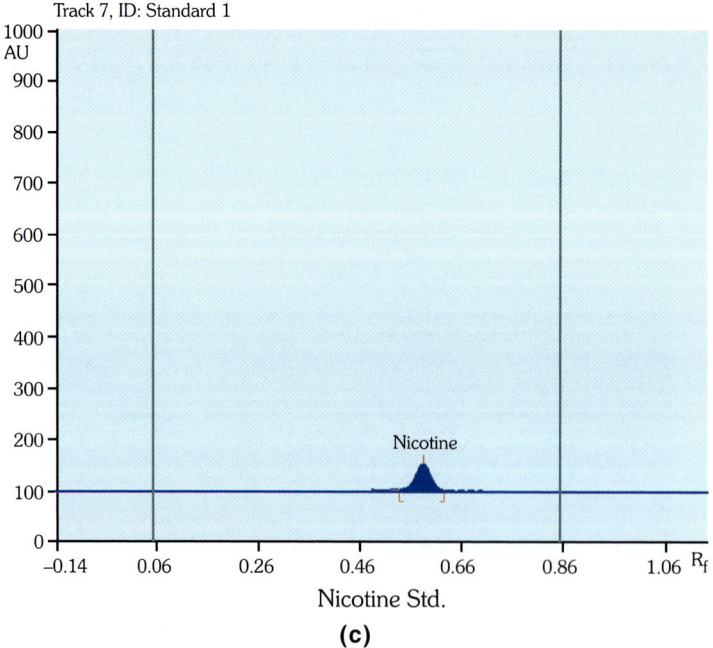

(c)

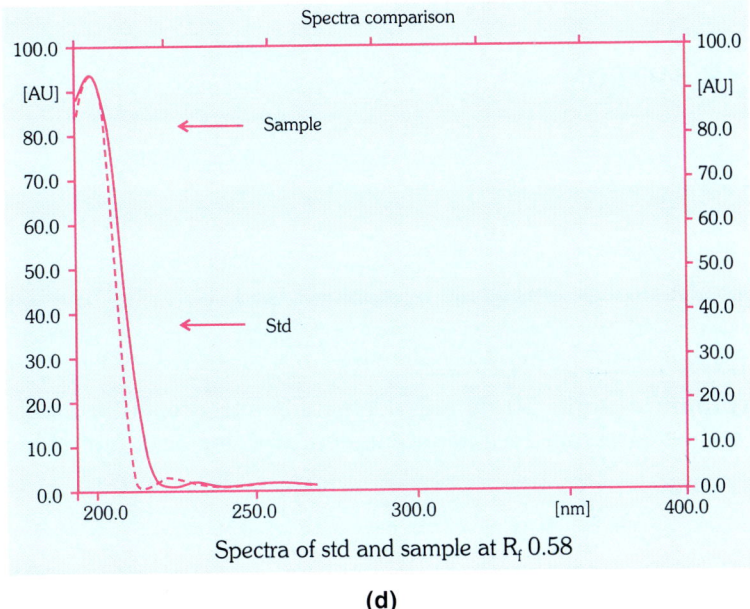

Spectra of std and sample at R$_f$ 0.58

(d)

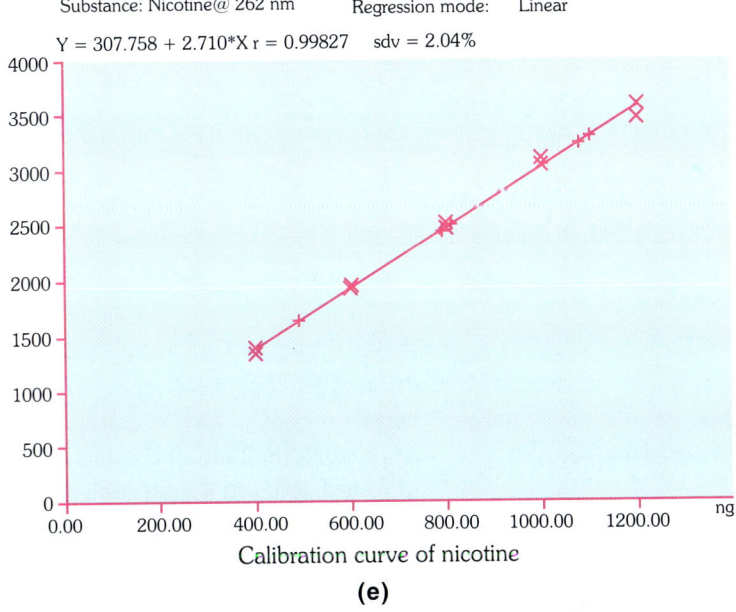

Calibration curve of nicotine

(e)

Fig. 17.6 a–e: Nicotine in tobacco

METHOD: 17.7

FINGERPRINT OF SAFFRON

Aim of Analysis
Comparison of Saffron samples by HPTLC fingerprint.

Chemicals and Reagents
Ethyl acetate, isopropyl alcohol, anisaldehyde, sulphuric acid.

Reagent Preparation
Anisaldehyde sulphuric acid: 10 ml anisaldehyde is mixed with 20 ml glacial acetic acid, followed by 180 ml methanol and 10 ml concentrated sulphuric acid, in that order.

Standard Preparation   :   None

Sample Preparation
10 mg of sample is wetted with 0.2 ml of water, then 1 ml of methanol is added. Filter the solution through Whatman filter paper no.1. The clear filtrate is the sample solution.

Stationary Phase
HPTLC glass plate silica gel $60F_{254}$ precoated 20 × 10 cm.

Prewashing
The plate is prewashed with 2-propanol and then heated at 70° C for 10 minutes.

Mobile Phase
Ethyl acetate : isopropyl alcohol : water (7.5 : 2.5 : 1) v/v (Volume = 16 ml).

Development Chamber
Twin-trough chamber of 20 × 10 cm with s. s. lid.

Chamber saturation	:	20 minutes with filter paper.
Plate equilibrium	:	None.
Sample/standard application	:	Apply with the help of Camag ATS-4 or Linomat 5, 5 μl of test solution on precoated layer 8 mm from the bottom edge. Band length 8 mm. Distance between tracks 14 mm. Distance from the sides 15 mm.
Development distance	:	80 mm from bottom edge.
Visualisation	:	Observe under UV cabinet at 254 nm and white light.
Post-chromatographic derivatisation	:	Plate is dipped in anisaldehyde sulphuric acid reagent for 1 second and then heated at 110° C for 10 min.
Photodocumentation	:	254 nm before derivatisation and 366 nm (visible)after derivatisation.
Measurement mode	:	UV absorbance/reflectance.

Scanning

Using Camag TLC scanner 3 with winCATS software, slit-micro, 4 x 0.3 mm, scan at 254 and 430 nm before derivatisation and at 580 nm after derivatisation.

Spectrum

Record and match spectra between 200 to 700 nm.

After Derivatisation:

Image (254 nm)

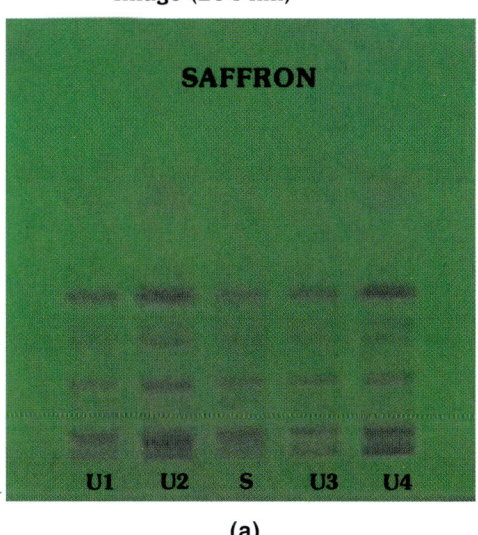

(a)

Image (visible)

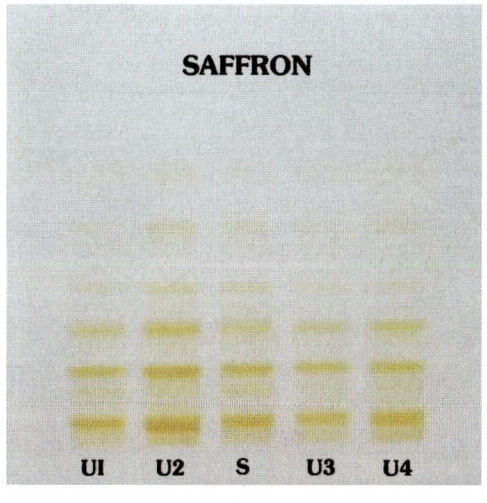

(b)

After Derivatisation:

Image (visible)

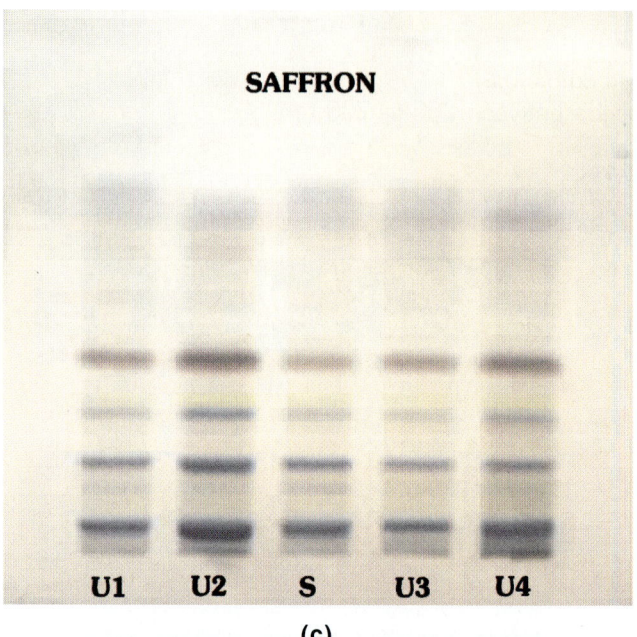

(c)

Std (366 nm)

(d)

Fig. 17.7 a–d: Comparison of saffron samples

METHOD: 17.8

DETECTION OF CASTOR OIL IN EDIBLE OILS

Aim of Analysis

Detection of castor oil in edible oils.

Chemicals and Reagents

Methanol, water, 2% KOH, sulphuric acid.

Reagent Preparation

2 % potassium hydroxide: Weigh 2 gm of potassium hydroxide and dissolve it in 20 ml of water and make it up to the mark in a 100 ml standard measuring flask.

10% methanolic sulphuric acid: Add 10 ml of cold H_2SO_4 to 90 ml of ice cold methanol gradually with stirring.

Standard preparation   :   None

Sample Preparation

1. Saponification of castor oil: Weigh 1000 mg of castor oil and add 2 ml of 2 % potassium hydroxide solution and heat on a boiling water bath for 10 min.
2. Saponification of groundnut oil: Weigh 1000 mg of groundnut oil and add 2 ml of 2 % potassium hydroxide solution and heat on a boiling water bath for 10 min.
3. Saponification of sunflower oil: Weigh 1000 mg of sunflower oil and add 2 ml of 2 % potassium hydroxide solution and heat on a boiling water bath for 10 min.
4. Saponification of mixture of castor oil and groundnut oil : Weigh 1000 mg of castor oil and groundnut oil and add 2 ml of 2% potassium hydroxide solution and heat in a boiling water bath for 10 min.

Sample Application

Apply with the help of Camag Linomat 5, 5 µl of each oil solution on precoated layer 8 mm from the bottom edge. Band length 6 mm. Distance from the side 15 mm.

Stationary Phase

TLC Al sheets silica gel RP-18 $F_{254}S$ precoated cut to 20 × 10 cm.

Mobile Phase

Methanol: water (8 : 2) v/v (Volume = 16 ml)

Development Chamber

Twin-trough chamber of 20 × 10 cm with s. s. lid.

Chamber saturation	: None
Plate equilibrium	: None.
Development distance	: 80 mm from bottom edge.
Visualisation	: Observe under UV cabinet at 366 nm after derivatization.
Post-chromatographic derivatisation	: The plate is dipped for 2 seconds in 10% methanolic sulphuric acid solution then heated at 110° C for 10 min.
Photodocumentation	: Under 366 nm and visible after derivatization.
Measurement mode	: UV absorbance/reflectance Fluorescence/reflectance

Scanning

Using Camag scanner 3 with winCATS software, slit-micro, 4 × 0.3 mm, scan at 366 nm and at 580 nm. Castor oil has a strong band at R_f 0.28 (seen after derivatization). This marker should be absent in the sunflower and groundnut oils.

Image (VIS) after derivatization

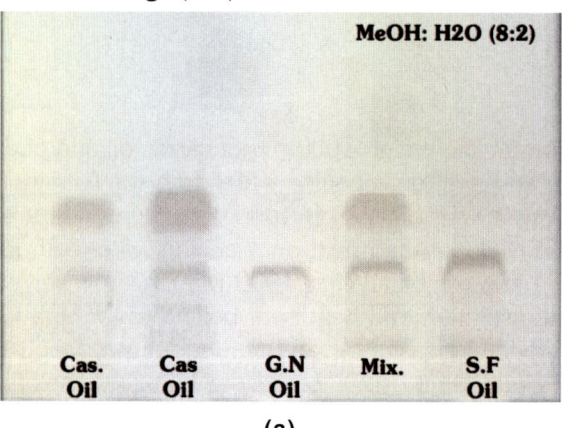

(a)

Image (366 nm)

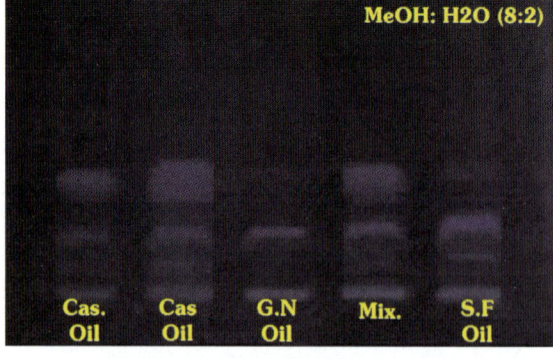

(b)

Fig. 17.8 a, b: Castor oil in Edible oil

METHOD: 17.9

DETERMINATION OF EUGENOL IN CLOVE OIL

Aim of Analysis

Detection of eugenol in clove oil.

Chemicals and Reagents

Toluene, ethyl acetate, formic acid, vanillin sulphuric acid.

Reagent Preparation

Vanillin sulphuric acid: 1% ethanolic vanillin solution (solution I) and 10% ethanolic sulphuric acid (solution II). ·

Standard Solution

Soln. A: Weigh 10 mg of eugenol and dissolve it in 10 ml of methanol [Conc. 1µg/µl].

Soln. B: Take 1 ml of solution A and dilute to 10 ml with methanol [Conc. 0.1µg/µl].

Sample Preparation

Dil A: Take 10 gm of cloves and crush it, then add 50 ml of methanol to it, boil for 5 min and then allow to stand for one hour with occasional shaking. After one hour, filter the solution through Whatman filter paper no.1 containing anhydrous sodium sulphate. Filtrate is evaporated to 5 ml [Conc. 2000 µg/µl] (Soln A^1).

Dil B: Pipette out 1 ml of solution A^1 and dilute to 10 ml with methanol [Conc. 200 µg/µl] (Soln B^1).

Dil C: Pipette out 1 ml of solution B^1 and diluted to 5 ml with methanol [Conc. 40 µg/µl] (Soln C^1).

Stationary Phase

TLC Al sheets silica gel $60F_{254}$ precoated cut to 20 × 10 cm.

Mobile Phase

Toluene : ethyl acetate : formic acid (9.9 : 0.1) v/v (Volume = 16 ml).

Development Chamber

Twin-trough chamber of 20 × 10 cm with s. s. lid.

Chamber saturation	: None
Plate equilibrium	: None
Sample/standard application	: Apply with the help of Camag ATS-4 or Linomat 5, 5 µl of sample and 5 µl of standard solution on precoated layer 8 mm from the bottom edge. Band length 8 mm. Distance from the sides 15 mm.

Development distance	: 80 mm from bottom edge
Visualisation	: Observe under UV cabinet at 254 nm.
Post-chromatographic derivatisation	: The plate is sprayed with 1% ethanolic vanillin solution and then dipped the plate in a 10% ethanolic sulphuric acid solution, then heat the plate at 110° C for 10 min.
Photodocumentation	: At 254 nm, visible (after derivatization).
Measurement mode	: UV absorbance/reflectance.

Scanning

For Quantification

Using Camag TLC scanner 3 with winCATS software, slit-micro, 4 × 0.3 mm, scan at 285 nm before derivatization and 580 nm after derivatization.

For Identification

Record and match spectra between 190 to 400 nm (R_f = 0.45).

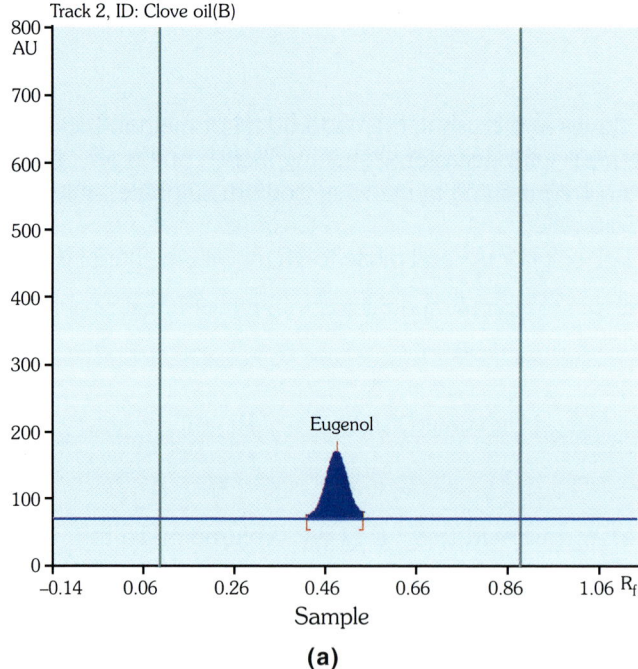

(a)

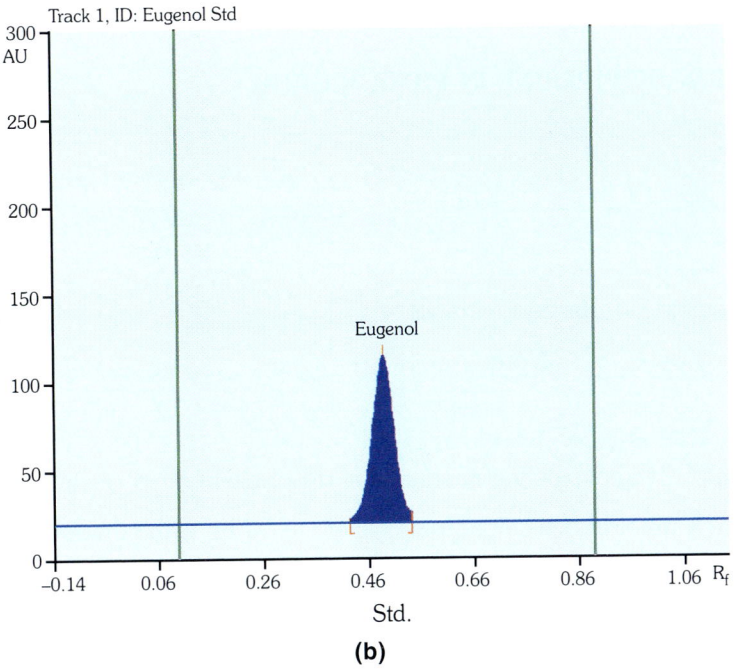

(b)

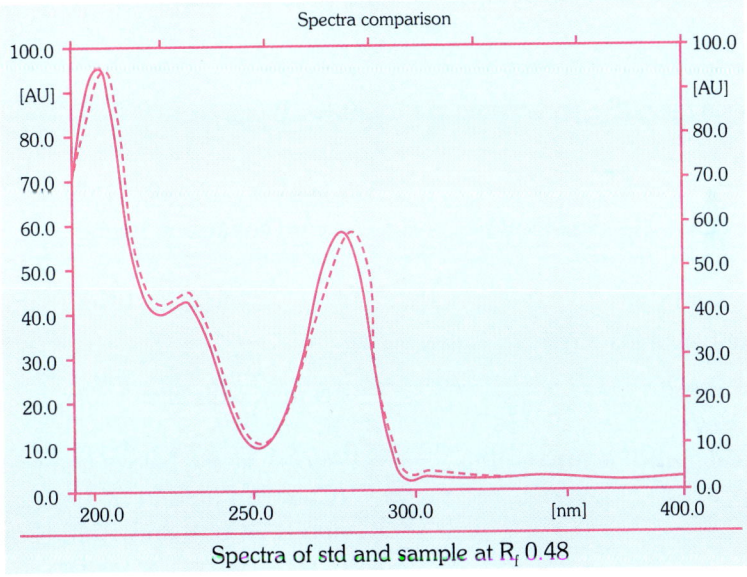

Spectra of std and sample at R_f 0.48

(c)

Fig. 17.9 a–c: Detection of eugenol in clove oil

METHOD: 17.10

QUANTIFICATION OF PIPERINE IN PIPER NIGRUM

Aim of Analysis

Quantification of piperine in piper nigrum.

Chemicals and Reagents

Toluene, diethyl ether, 1-4 dioxane, ethanol

Reagent preparation :   None

Standard Preparation

Solution A: Weigh 5.2 mg of piperine standard and dissolve it in 10 ml of ethanol[Conc. 0.52 µg/µl].

Solution B: Take 1 ml of above solution and dilute to 10 ml with ethanol[Conc. 0.052 µg/µl].

Sample Preparation

Reflux the powdered sample 0.5 g with 50 ml of ethanol for 30 min. Filter the resulting solution and reflux with 50 ml of ethanol. Evaporate the combine filtrate under vacuum to about 25 ml. Cool the resulting solution to room temperature and make up the volume to 100 ml[Conc. 5 µg/µl].

Stationary Phase

TLC Al sheets silica gel $60F_{254}$ precoated cut to 20×10 cm.

Mobile Phase

Toluene: diethyl ether: 1-4 dioxane (14.1:4.8 :3.6) v/v (Volume = 16 ml).

Development Chamber

Twin-trough chamber of 20×10 cm with s. s. lid.

Chamber saturation	:	20 min. with filter paper.
Plate equilibrium	:	10 min.
Sample/standard application	:	Apply with the help of Linomat 5, 5 µl and 10 µl (Conc: 25 µg and 50 µg resp.) of test solution and 2, 4, 6, 8 and 10 µl (Conc: 0.104, 0.208, 0.312, 0.416, 0.52 µg resp. of standard solution on precoated layer 8 mm from the bottom edge. Band length of 8 mm. Distance from the sides 15 mm.
Development distance	:	80 mm.
Visualisation	:	Observe under UV cabinet at 254 and 366 nm.

Post-chromatographic derivatisation	:	Not required.
Photodocumentation	:	At 254 and 366 nm.
Measurement mode	:	UV absorbance/reflectance.

Scanning

For Quantification

Using Camag TLC scanner 3 with winCATS software, slit-micro, 4 × 0.3 mm, scan at 334 nm.

For Identification

Record and match spectra between 190 to 400 nm ($R_f = 0.38$).

Image (254 nm)

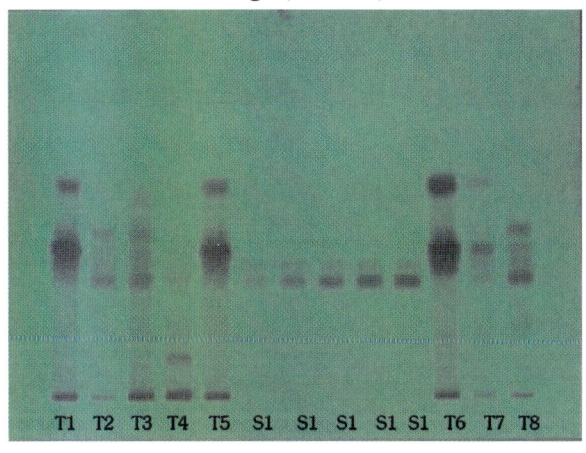

(a)

Image (366 nm)

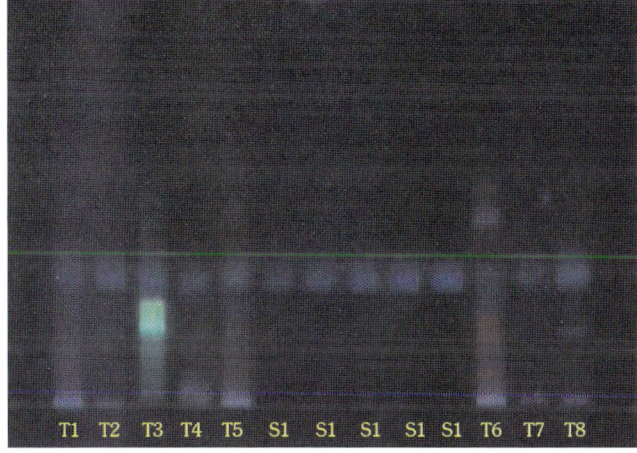

(b)

Image (visible)

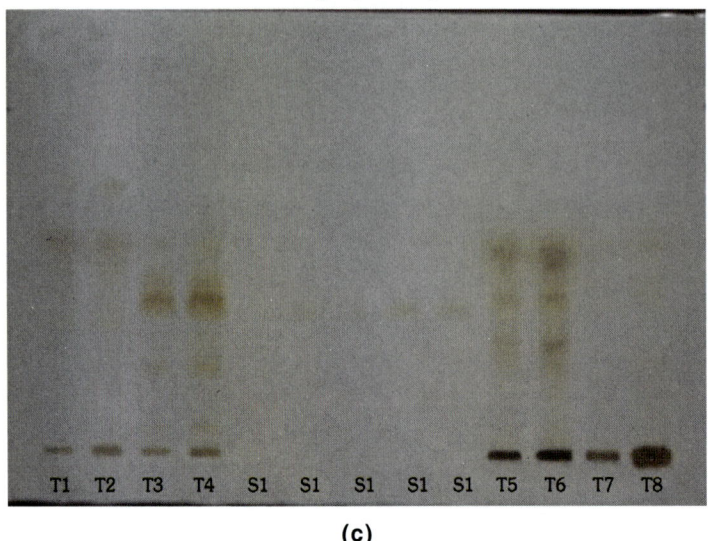

(c)

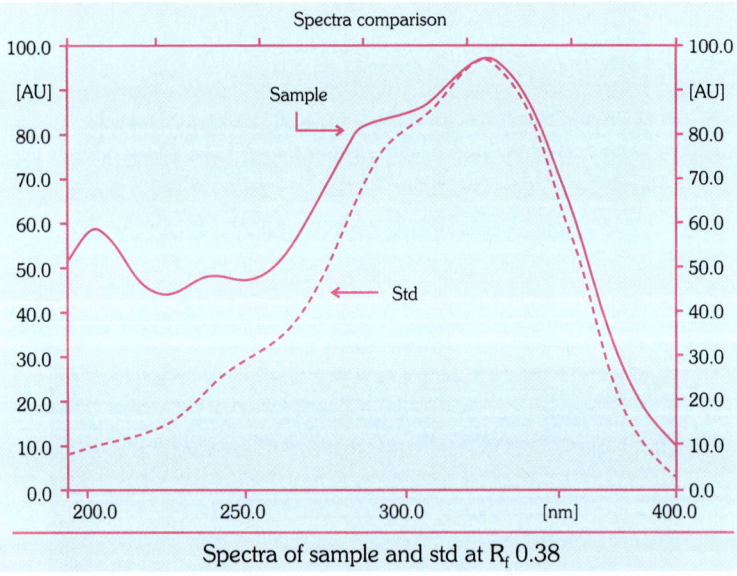

Spectra of sample and std at R_f 0.38

(d)

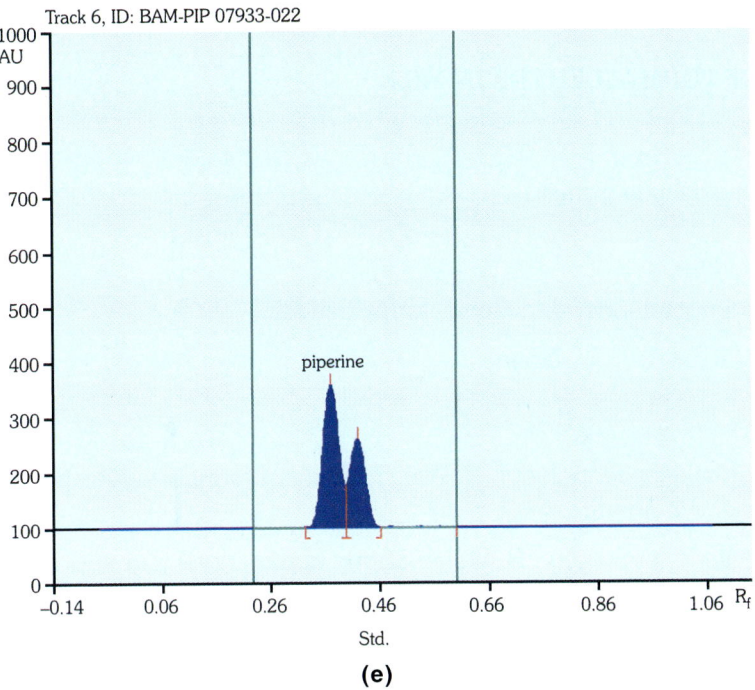

Std.

(e)

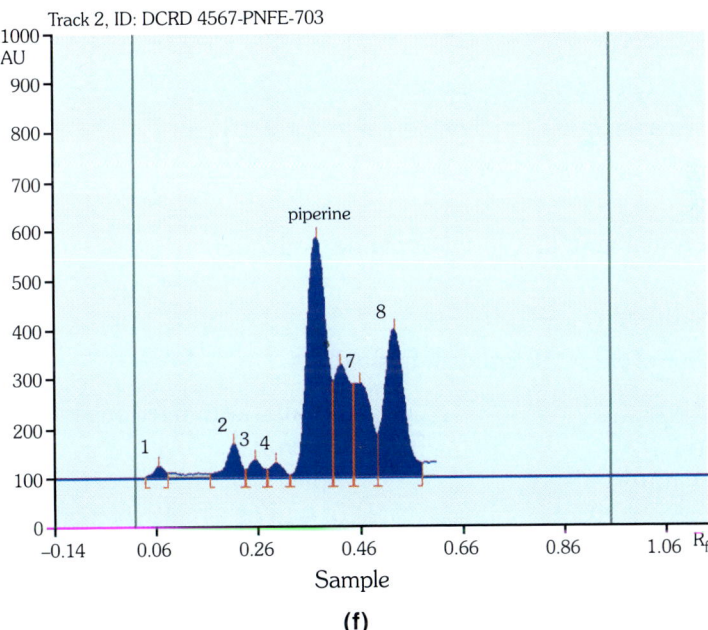

Sample

(f)

Fig. 17.10 a–f: Quantification of piperine

DETECTION OF FORMALDEHYDE IN MILK

Aim of Analysis
Detection of formaldehyde in milk.

Chemicals and Reagents
Chloroform, dichloromethane, dimedone, formaldehyde.

Reagent preparation :   None

Standard Preparation   :

Preparation of formaldehyde: Solution A: Weigh 80 mg of formaldehyde and dissolve it in 25 ml of methanol[Conc. 3.2 µg/µl].

Preparation of dimedone: Solution B: Weigh 30 mg of dimedone and add 15 ml of methanol to it [Conc. 2 µg/µl].

Sample Preparation
Take 5 ml of milk sample, add 10 ml of methanol, shake and allow it to settle for 30 min then centrifuge for 15 min.

Stationary Phase
HPTLC Al sheets silica gel $60F_{254}$ precoated cut to 20×10 cm.

Mobile Phase
Chloroform: dichloromethane (3 : 7) v/v (volume = 16 ml)

Development Chamber
Twin-trough chamber of 20×10 cm with s. s. lid.

Chamber saturation	: 20 min. with filter paper.
Plate equilibrium	: 5 min.
Sample/standard application	: Apply with the help of Camag ATS-4 or Linomat 5, 4 µl of sample solution. and 4 µl each of soln A and soln B and 4 µl of solution B (Dimedone) overspotted on 4 µl of solution A (formaldehyde). 4 µl of sample overspotted on 4 µl of solution A and 4 µl of solution B on precoated layer 8 mm from the bottom edge. Band length 8 mm. Distance from the sides 15 mm.
Development distance	: 80 mm from bottom edge.

Visualisation	: Observe under UV cabinet at 254 nm.
Post-chromatographic derivatisation	: N.A.
Photodocumentation	: At 254 nm.
Measurement mode	: UV absorbance/reflectance.

Scanning

For Quantification

Using Camag TLC Scanner 3 with winCATS software, slit-micro, 4 × 0.3 mm, scan at 254 nm.

For Identification

Record and match spectra between 190 to 400 nm (R_f = 0.73).

Image (254 nm)

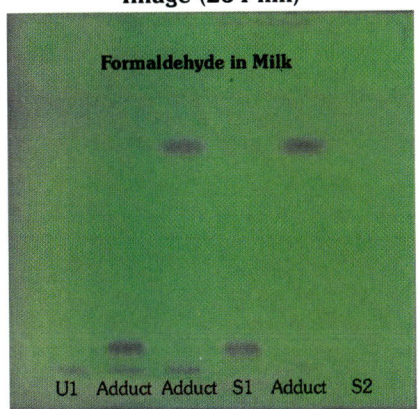

(a)

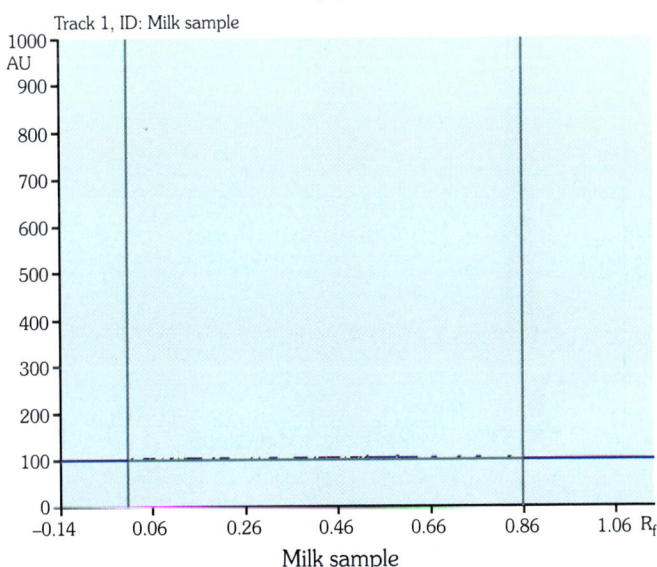

Milk sample

(b)

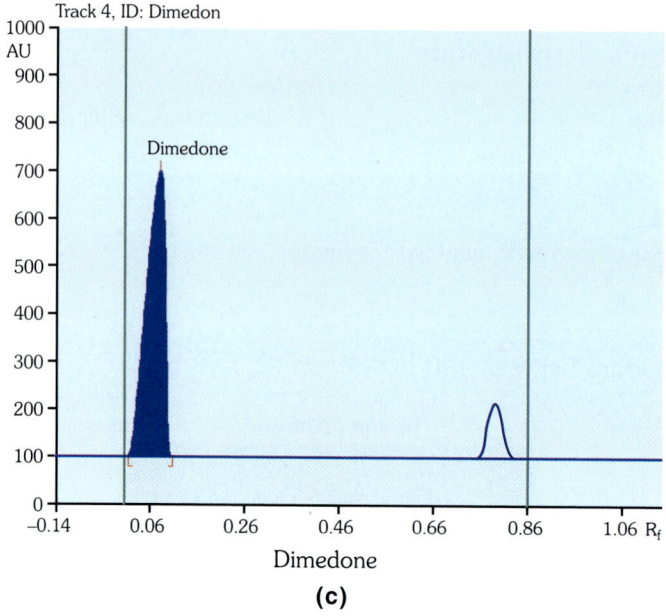

Dimedone

(c)

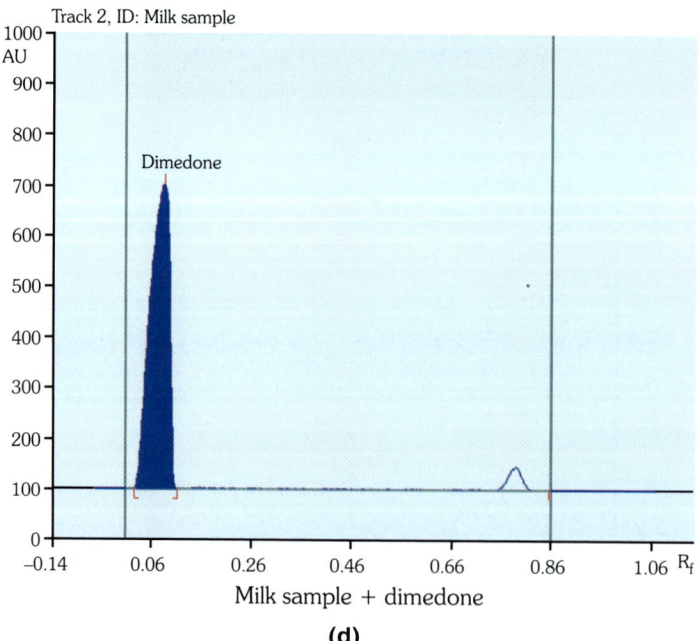

Milk sample + dimedone

(d)

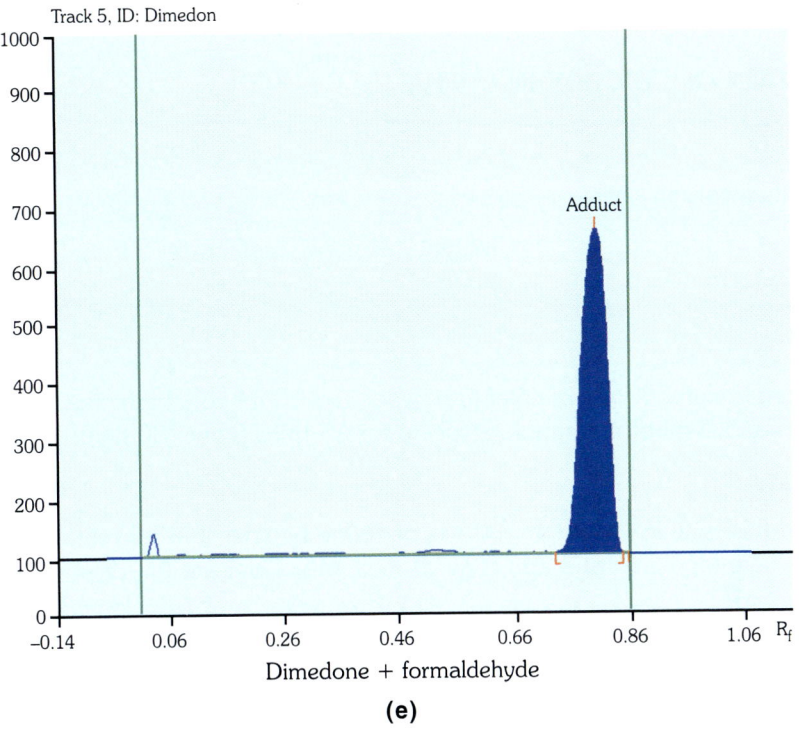

Dimedone + formaldehyde

(e)

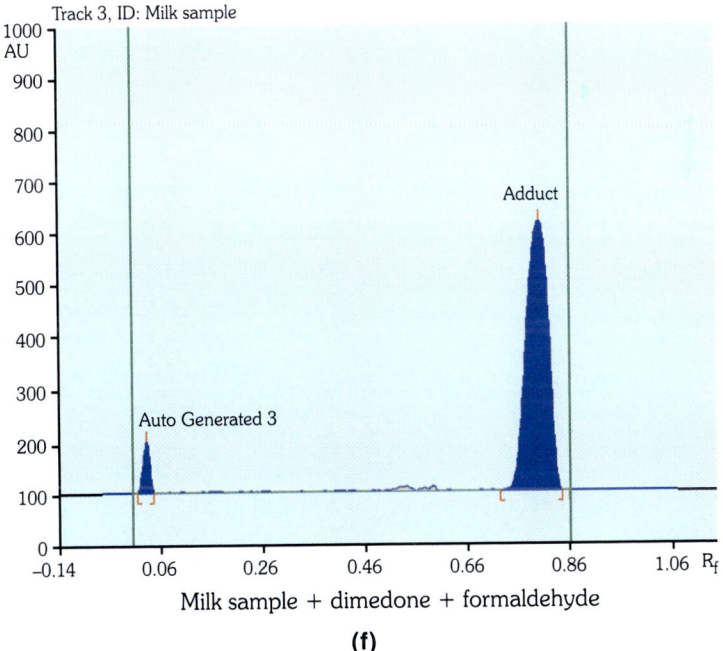

Milk sample + dimedone + formaldehyde

(f)

Fig. 17.11 a–f: Formaldehyde in milk

METHOD: 17.12

IDENTIFICATION OF CHICORY IN COFFEE

Aim of Analysis

Identification of chicory in coffee.

Chemicals and reagents	:	Toluene, acetone, methanol
Reagent preparation	:	None

Standard Preparation

Chicory solution: Weigh 500 mg of chicory powder in a 10 ml volumetric flask add 1 ml of water shake and then add 9 ml of methanol. Sonicate for 15 mins [conc. 50 $\mu g/\mu l$].

Sample Preparation

Coffee solution (test solution): Weigh 500 mg of coffee powder in a 10 ml volumetric flask. Add 1ml of water, shake for 1 min and add 9 ml of methanol. Sonicate for 15 min [conc. 50 $\mu g/\mu l$].

Stationary Phase

TLC Al sheets silica gel 60F$_{254}$ precoated cut to 10 × 10 cm.

Mobile Phase

Toluene: acetone (7 : 3) v/v (Volume = 16 ml)

Development Chamber

Camag twin-trough chamber of 10 × 10 cm with s. s. lid.

Chamber saturation	:	20 minutes with filter paper.
Layer saturation	:	None
Sample/std application	:	Apply with the help of Camag ATS-4 or Linomat 5, 5 µl of test solution (250 µg), 5 µl of chicory solution (250 µg) on precoated layer 8 mm from the bottom edge. Band length 8 mm. Distance from the side 15 mm.
Development distance	:	70 mm
Visualisation	:	Observe under UV cabinet at 254 nm.
Post-chromatographic derivatisation	:	Not required.
Photodocumentation	:	At 254 nm
Measurement mode	:	UV absorbance/reflectance.

Scanning

For Quantification

Using Camag TLC scanner Scan at 275 nm to detect the marker peak.

For Identification

Record spectra of marker (R_f = 0.30) between 190-400 nm and confirm match with standard.

Image (254 nm)

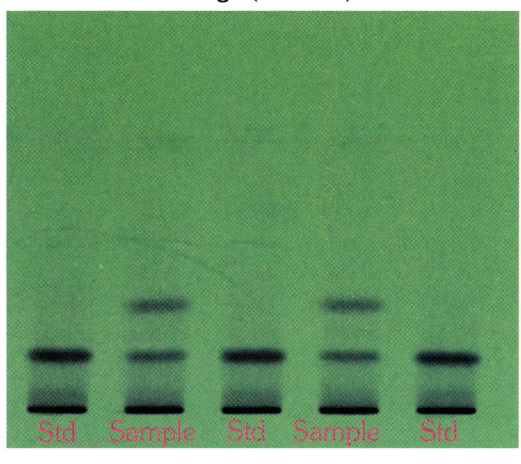

(a)

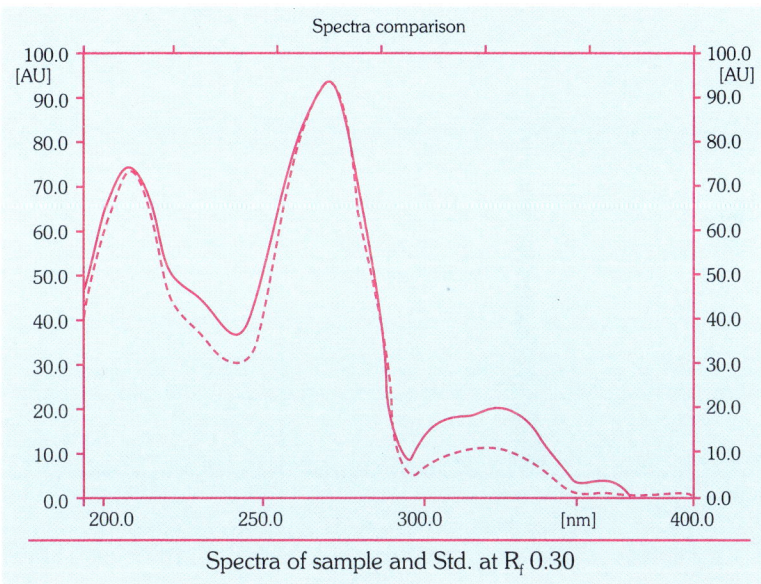

Spectra of sample and Std. at R_f 0.30

(b)

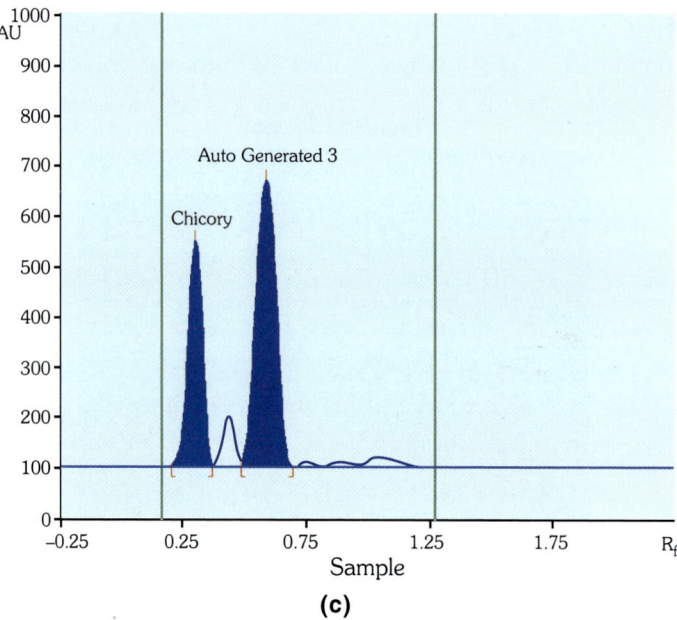

(c)

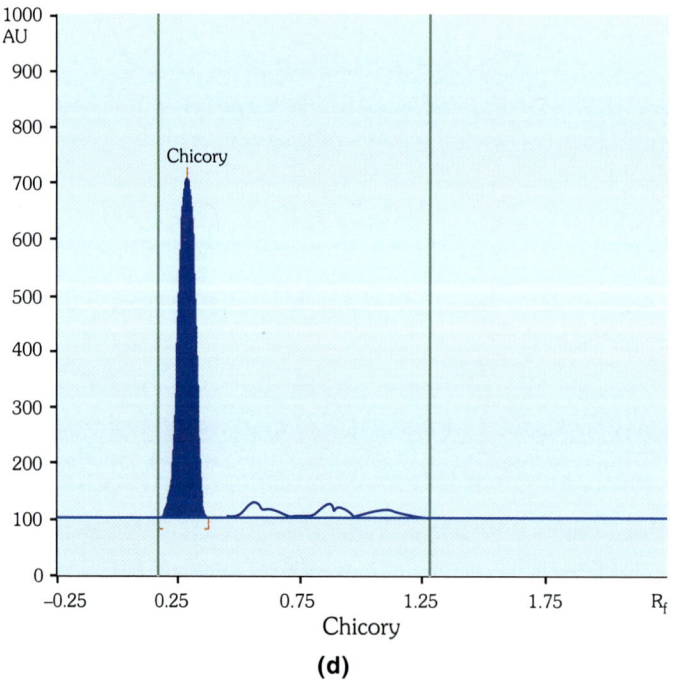

(d)

Fig. 17.12 a–d: Detection of chicory in coffee

METHOD: 17.13

IDENTIFICATION OF PRESERVATIVES IN TOMATO SAUCE

Aim of Analysis
Identification of preservatives, i.e. methyl paraben, propyl paraben and sodium benzoate in tomato sauce.

Chemicals and Reagents
n-pentane, diethylether, formic acid, methanol.

Reagent Preparation
Solvent mixture: 95 ml methanol + 5 ml water.

Standard Solution
Solution A: Weigh 10 mg of methyl paraben and add 10 ml of solvent mixture [Conc. 1 $\mu g/\mu l$].

Solution B: Take 1 ml of solution A and add 9 ml of solvent mixture [Conc.0.1 $\mu g/\mu l$].

Solution C: Weigh 10 mg of propyl paraben and dissolve it in 10 ml solvent mixture [Conc. 1 $\mu g/\mu l$].

Solution D: Take 1 ml of solution C and add 9 ml of solvent mixture [Conc. 0.1 $\mu g/\mu l$].

Solution E: Weigh 10 mg of sodium benzoate and dissolve it in 10 ml solvent mixture [Conc. 1$\mu g/\mu l$].

Solution F: Take 1 ml of solution E and add 9 ml of solvent mixture [Conc. 0.1 $\mu g/\mu l$].

Sample Preparation
Tomato sauce : Weigh 1000 mg of tomato sauce and add 10 ml of solvent mixture. Sonicate for 10 min [Conc: 100 $\mu g/\mu l$.

Mixture of tomato sauce and preservatives : Weigh 1000 mg of tomato sauce and add to it 1 ml each of prepared methyl paraben, propyl paraben and sodium benzoate. Then add 7 ml solvent mixture and sonicate for 10 minutes.

Stationary Phase
TLC Al sheets RP-18 $F_{254}S$ precoated cut to 10×10 cm.

Mobile Phase
n-pentane: diethyl ether: formic acid (12 : 2 : 0.1) (v/v) (Volume = 16 ml)

Development Chamber
Twin-trough chamber of 10×10 cm with s. s. lid.

Chamber saturation : 15 min.

Plate equilibrium : None.

Sample/standard application : Apply with the help of Camag ATS-4 or Linomat 5, 10 μl of tomato sauce solution, 10 μl of tomato sauce with preservatives, 10 μl each of standard methyl paraben, standard propyl paraben and standard sodium benzoate on precoated layer, 8 mm from the bottom edge. Band length 8 mm. Distance from the side 15 mm.

Development distance : 80 mm

Visualisation : Observe under UV cabinet at 254 nm.

Post-chromatographic derivatisation : Not required

Photodocumentation : Under 254 nm.

Measurement mode : UV absorbance/ reflectance.

Scanning

For Quantification

Using Camag scanner 3 with winCATS software, with multiwavelength scan option, slit-micro, 4 × 0.3 mm, scan at 227 and 254 nm.

For Identification

Record spectra between 190 and 400 nm and confirm match with standard.

Image (254 nm)

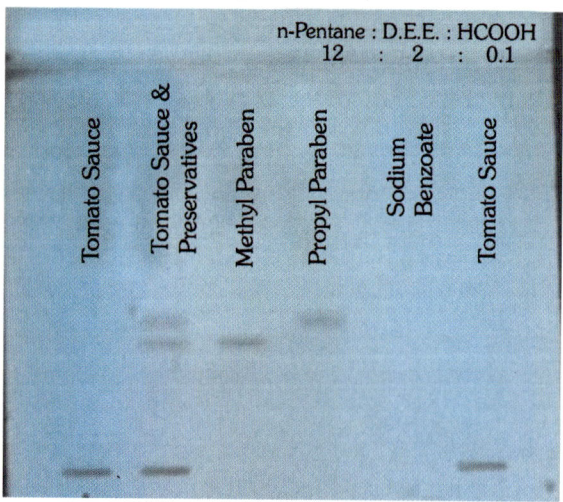

(a)

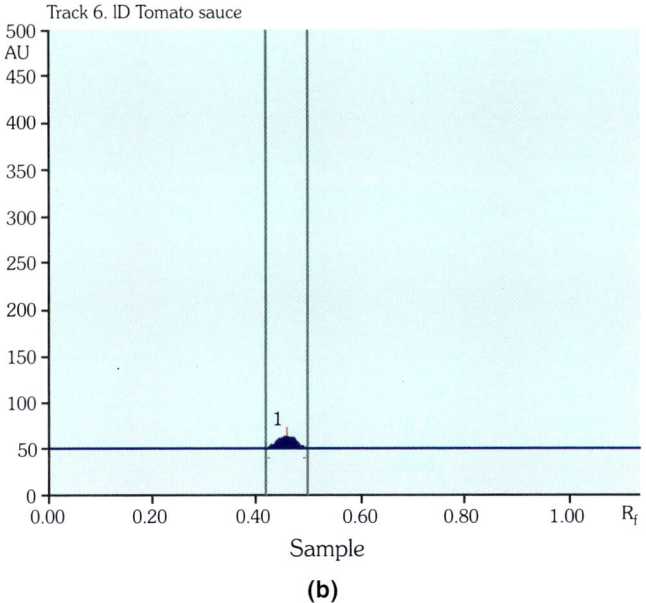

(b)

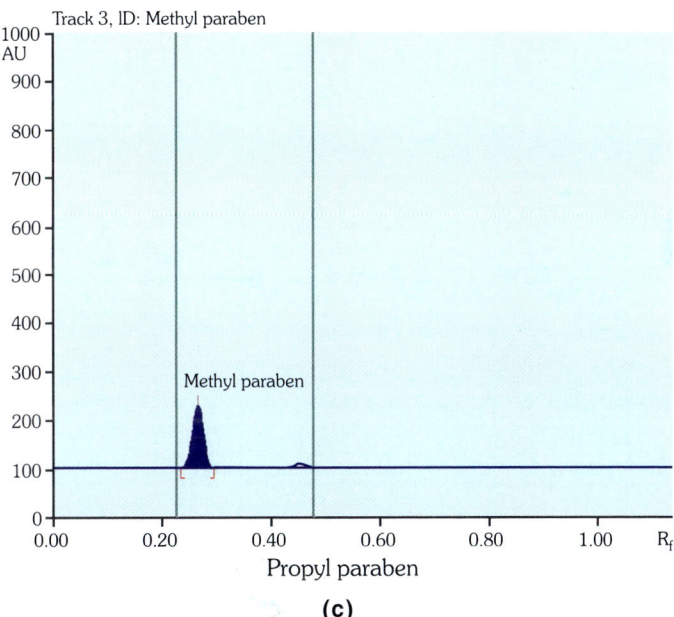

(c)

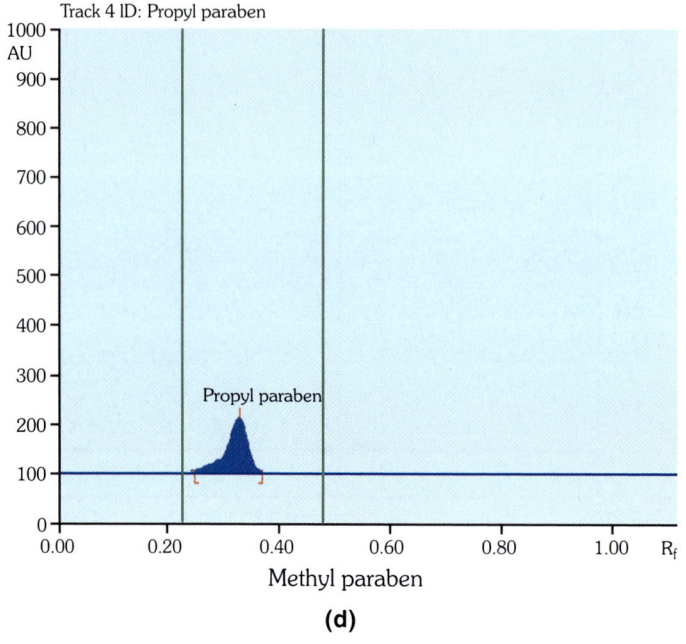

(d)

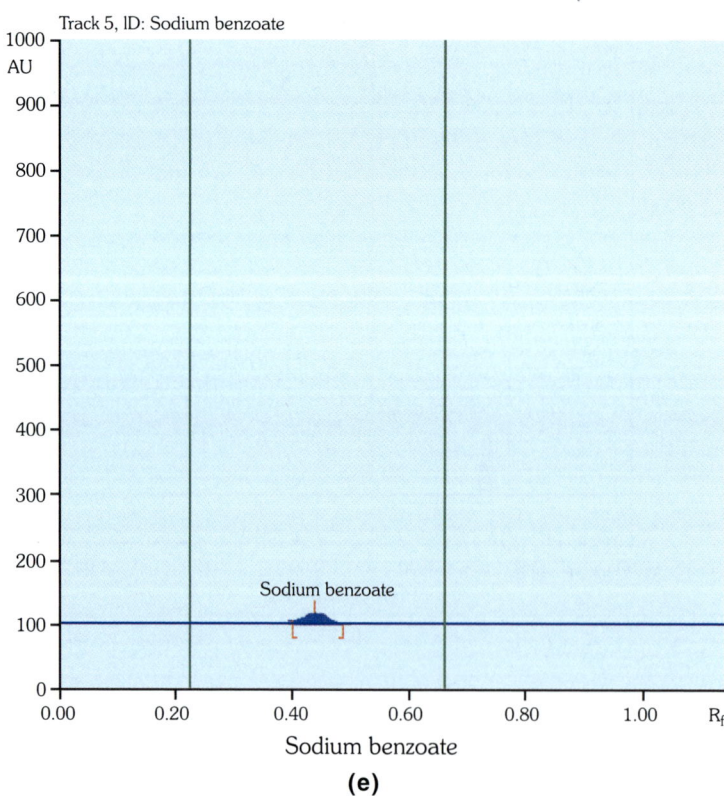

(e)

Fig. 17.13 a–e: Identification of preservatives in tomato sauce

METHOD: 17.14

QUANTITATIVE DETERMINATION OF GINGEROL IN GINGER

Aim of Analysis
Quantitative determination of gingerol in ginger.

Chemicals and Reagents
n-hexane, diethyl ether, iron chloride, potassium hexacyanoferrate (potassium ferricyanide), ethanol.

Reagent Preparation
Barton's reagent: 1g of potassium ferricyanide dissolved in 100 ml of water.

2g of iron (III) chloride dissolved in 100 ml of water.

Standard Preparation
Weigh 50 mg of 8-gingerol, add 5 ml of ethanol and shake for 1 minute (10 µg/µl)(a).

Sample Preparation
Dry ginger: Take 50 ml ethanol in a conical flask and boil in water bath. To this add 5g of dry ginger powder and heat it at 50° C on water bath for half an hour (b).

Wet ginger: Take 50 ml ethanol in a conical flask and boil. To this add 5g of wet ginger paste and heat it at 50° C on water bath for half an hour (b).

Stationary Phase
TLC Al sheets silica gel $60F_{254}$ precoated plate cut to 20×10 cm.

Mobile Phase
n-hexane: diethyl ether (3 : 7) v/v (Volume = 20 ml).

Sample/Standard Application
Apply with the help of Camag ATS-4 or Linomat 5, 2 µl of dry ginger sample, 8 µl of wet sample and 2, 5, 10, 15 and 20 µl of std. gingerol on precoated layer, 8 mm from bottom edge, band length 8 mm, distance from the sides 15 mm.

Development Chamber
Twin-trough chamber of 20×10 cm with s. s. lid.

Tank saturation	: None
Plate equilibrium	: None
Development distance	: 90 mm from bottom edge.
Visualisation	: 254 nm, 366 nm.

Post-chromatographic derivatisation : The developed plate is then derivatised with Barton's reagent (i.e. the plate is dipped in 1:1 mixture of a and b)Gingerol shows light blue coluor after derivatisation.

Photodocumentation : 254 nm, 366 nm and visible light.

Measurement mode : UV absorbance/reflectance.

Scanning

For Quantification

Using camag scanner 3 with winCATS software, slit-micro, 4 × 0.3 mm, scan at 620 nm.

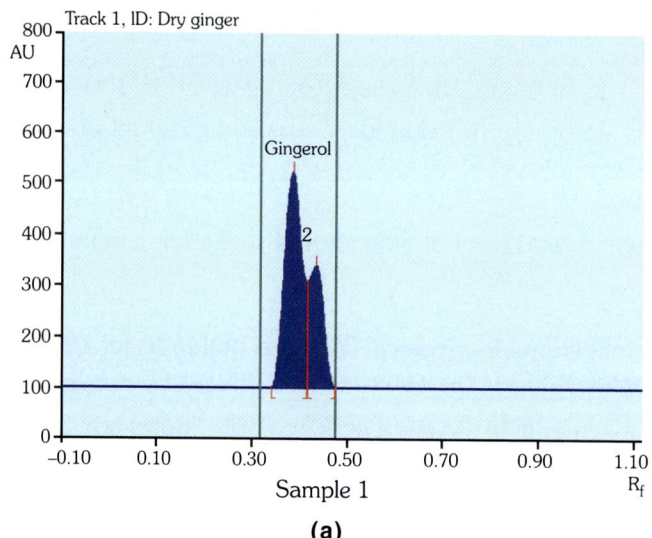

(a)

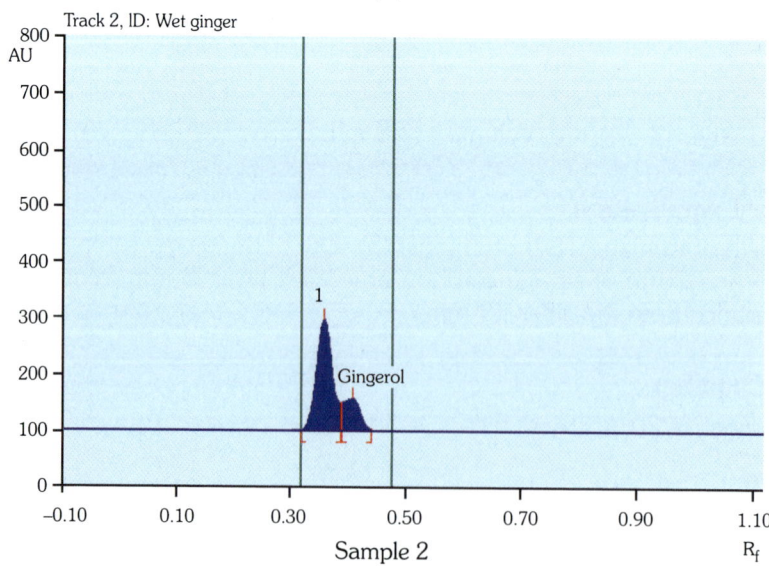

(b)

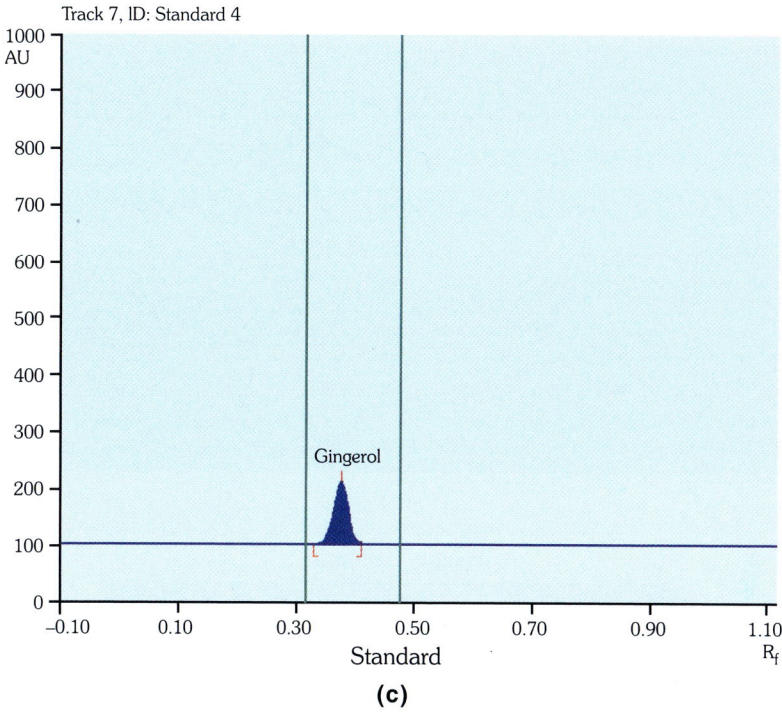

Track 7, ID: Standard 4

Gingerol

Standard

(c)

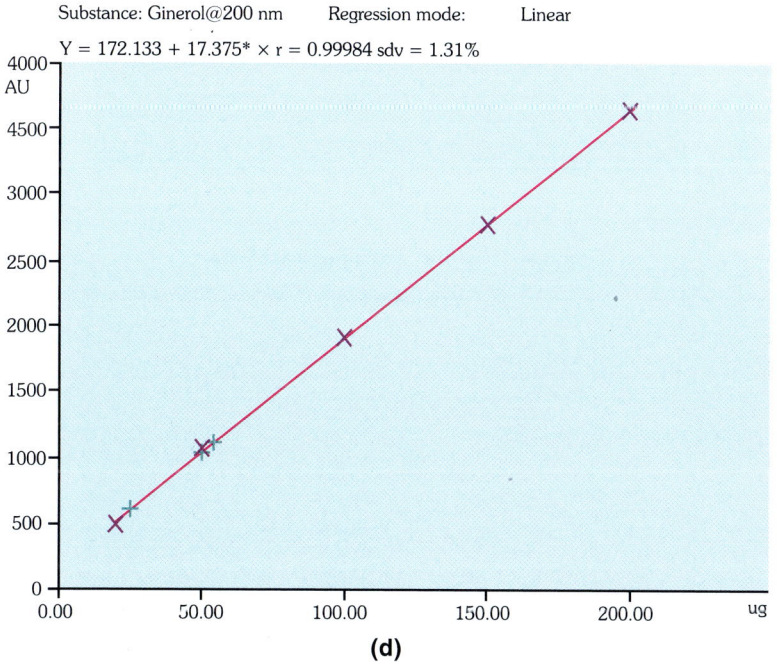

Substance: Ginerol@200 nm   Regression mode: Linear

Y = 172.133 + 17.375* × r = 0.99984 sdv = 1.31%

(d)

Image (254 nm)

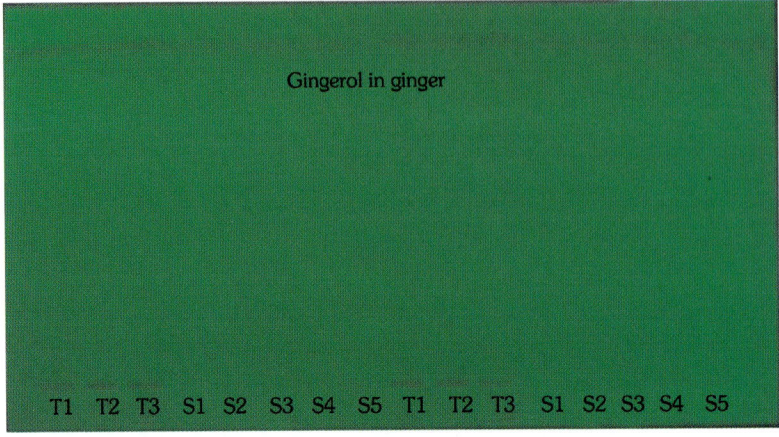

(e)

Image (366 nm)

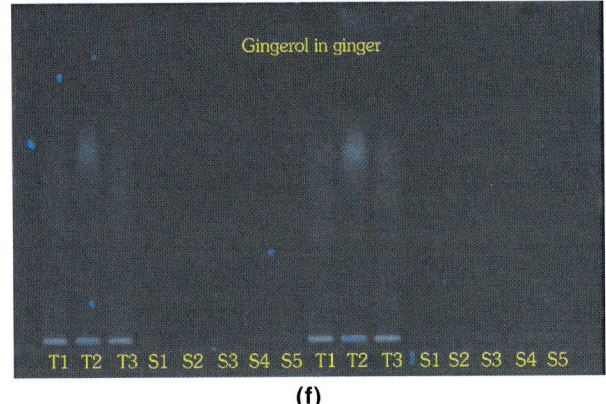

(f)

Image (Visible) after derivatisation

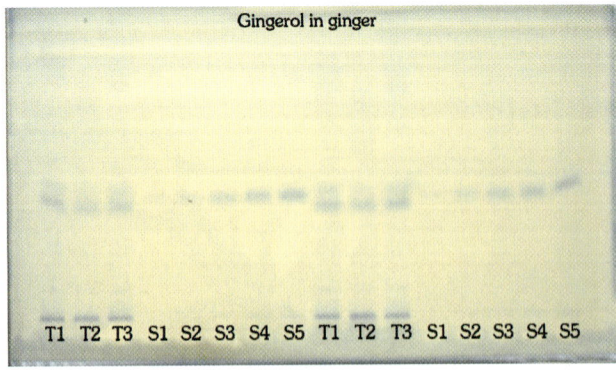

(g)

Fig. 17.14 a–g: Gingerol in ginger

DETECTION OF ARGEMONE IN EDIBLE OIL

Aim of Analysis

Detection of argemone oil in edible oil.

Chemicals and Reagents

1% Aqs. oxalic acid, iso-propanol, methanol, pet. ether.

Reagent preparation :   N. A.
Standard solution

Stock Solution (S)

Weigh 10 g of argemone seeds, crush to paste and add to 50 ml methanol. Boil it on water bath for 5 minutes. Allow to stand for about 1 hour (shake in intervals), filter this mixture and take the filtrate, evaporate this filtrate to final 5 ml volume with methanol [2000 µg/µl].

Working solution (A): Pipette out 0.1 ml from stock solution (S) and make it to 8 ml with pet ether [25 µg/µl].

Sample Preparation

Sample 1: Weigh 500 mg of sample (mustard oil), add 0.1 ml of solution (S) of standard and make the final volume to 10 ml with pet ether [50 µg/µl-mustard oil, 20 µg/µl-argemone].

Sample 2: Weigh 500 mg of sample (mustard oil) and make the final volume 10 ml using pet. ether [50 µg/µl].

Stationary Phase

TLC Al sheets silica gel 60 F_{254} precoated cut to 20 × 10 cm.

Mobile Phase

Isopropanol: 1% oxalic acid (9 : 1) v/v (vol. 16 ml)

Development Chamber

Twin-trough chamber of 20 × 10 cm with s. s. lid.

Chamber saturation	:	20 minutes with filter papers
Plate equilibrium	:	20 min.
Sample/standard application	:	Apply with the help of Camag ATS-4/Linomat 5, 4 µl of sample solution and 2 µl, 4 µl, 6 µl, 8 µl and 10 µl of standard solution A on precoated layer, 8 mm from the bottom edge. Band length 8 mm. Distance from the side 15 mm.
In situ clean-up	:	Pet ether

Development distance	:	80 mm.
Visualisation	:	Observed under UV cabinet at 366 nm. Two bright yellows spots of sanguinarine (R_f 0.66) and dihydrosanguinarine (R_f 0.77) are distinctly visible.
Post-chromatographic derivatisation	:	None
Photodocumentation	:	366 nm.
Measurement mode	:	Fluorescence/reflectance.

Scanning

For Quantification

Using Camag scanner 3 with winCATS software, slit-micro, 4 × 0.3 mm, scan 366 nm.

(Image@254 nm)

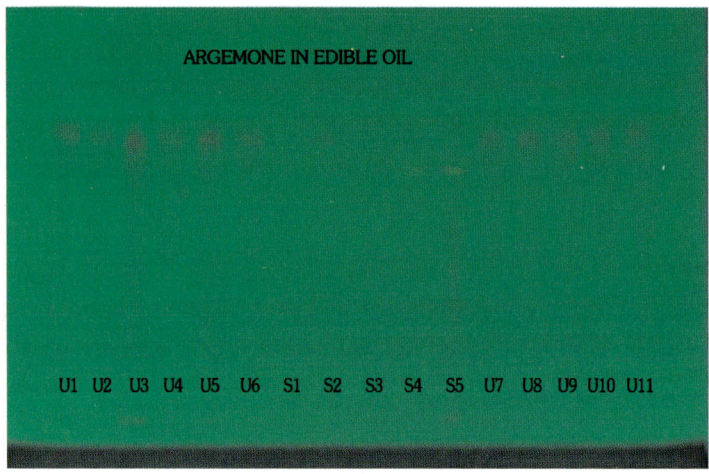

(a)

(Image@366 nm)

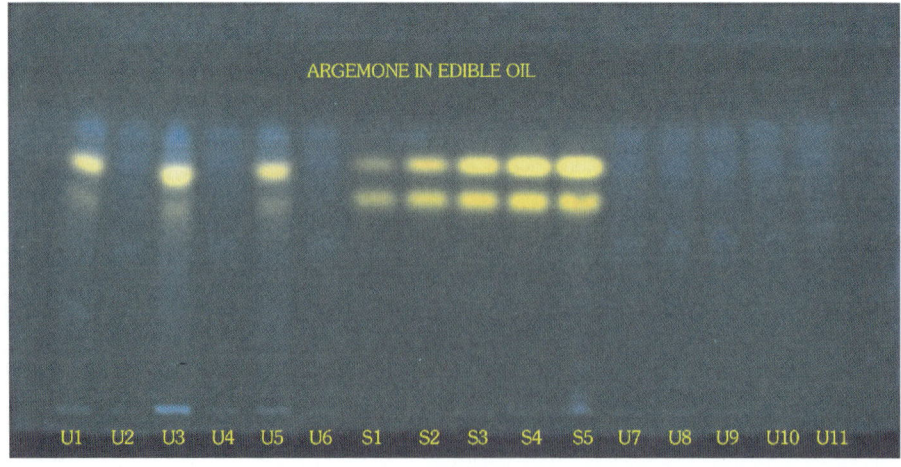

(b)

U1 – Sample 1
U2 – Sample 2
U3 – Sample 3
U4 – Sample 4
U5 – Sample 5

U6 – Sample 6
U7 – Sample 7
U8 – Sample 8
U9 – Sample 9
U10 – Sample 10
U11 – Sample 11

S1 – Argemone
S2 – Argemone
S3 – Argemone
S4 – Argemone
S5 – Argemone

Chromatograms:- @ 366 nm

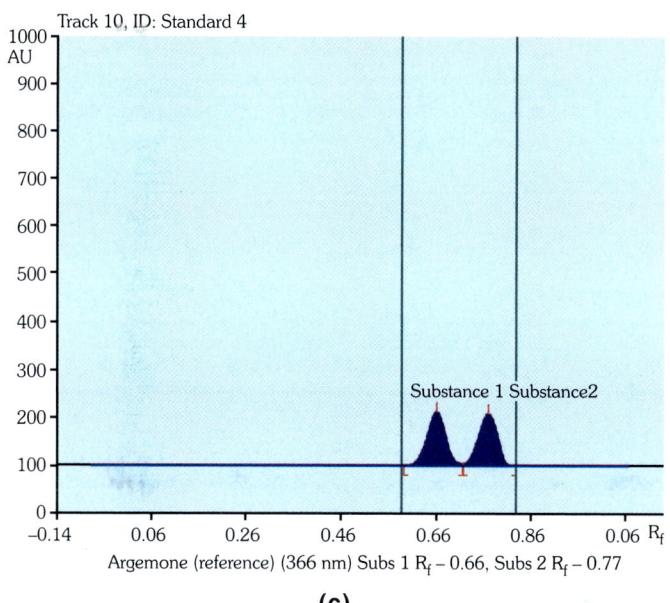

Argemone (reference) (366 nm) Subs 1 R_f – 0.66, Subs 2 R_f – 0.77

(c)

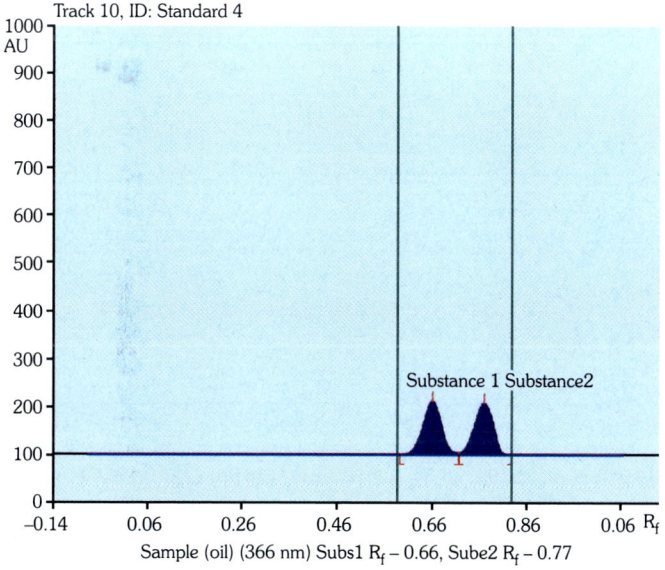

Sample (oil) (366 nm) Subs1 R_f – 0.66, Sube2 R_f – 0.77

(d)

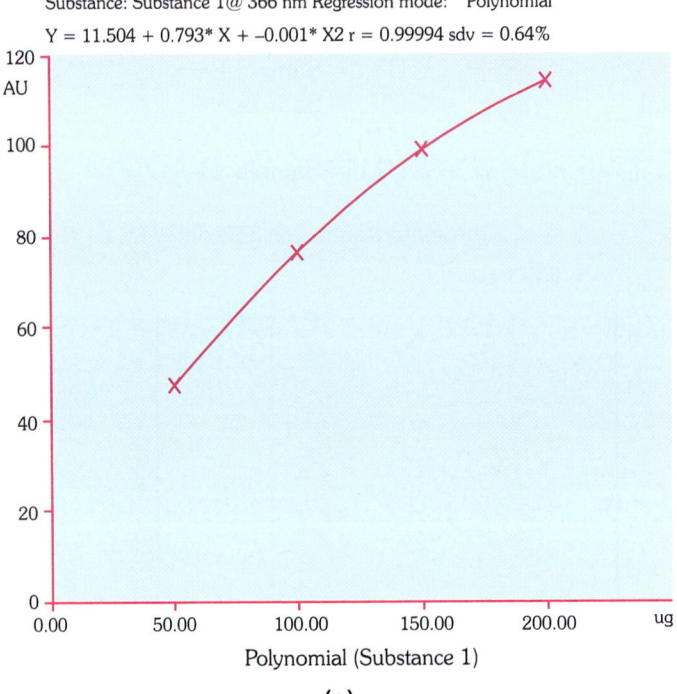

Substance: Substance 1@ 366 nm Regression mode: Polynomial

Y = 11.504 + 0.793* X + –0.001* X2 r = 0.99994 sdv = 0.64%

Polynomial (Substance 1)

(e)

Substance: Substance 2@ 366 nm Regression mode: Polynomial

Y = – 7.001 + 0.703* X + –0.001* X2 r = 0.99999 sdv = 0.05%

Polynomial (Substance 2)

(f)

Fig. 17.15 a–f: Detection of argemone in mustard oil

METHOD: 17.16 A

QUANTITATIVE DETERMINATION OF AFLATOXIN M1 IN MILK AND MILK PRODUCTS

Aim of Analysis

Quantitative determination of aflatoxin M_1 in milk and milk products.

Chemicals and Reagents

Milk, milk powder, cream, butter, cheese, fermented cheese, water, acetone, 10% lead acetate solution, hexane, chloroform, sodium sulphate, nitrogen, isopropanol.

Reagent preparation :   None

Standard Preparation

Preparation of stock aflatoxin M_1 solution: Weigh 1 mg of aflatoxin M_1 and dissolve it in 200 ml of chloroform (5 $\mu g/\mu l$).

Diluted stock solution: Pipette out 1 ml of above stock solution (5 $\mu g/\mu l$) and dilute it to 10 ml with chloroform(0.5 $\mu g/\mu l$).

Sample Preparation

Weigh 100 ml milk or 10 g milk powder or 100 ml cream, 50 g butter or 50 g cheese or 50 g fermented cheese and add 10 ml, 100 ml, 60 ml, 90 ml, 80 ml, 65 ml of water respectively to each flask. Then add 300 ml of acetone and stir. Filter through plaited filter. Transfer 300 ml of the filtrate and add 40 ml of a 10% lead acetate solution. Filter. Transfer 300 ml of the filtrate in a separation funnel and extract lipophilic by-products with 100 ml of hexane (Repeat extraction once more for cream and cheese). Extract aqueous phase with first 100 ml and then 50 ml of chloroform. Dry extract over a column with sodium sulphate. Evaporate to dryness in a rotary evaporator (under nitrogen and not exceeding 50° C). Dissolve residue in 200 μl of chloroform.

Stationary Phase

TLC Al sheets silica gel $60F_{254}$ precoated 20 × 10 cm.

Mobile Phase

Chloroform : acetone : isopropanol(85 : 10 : 5) v/v (Volume = 16 ml).

Sample/Standard Application

Apply with the help of Camag ATS-4 or Linomat 5, 40 μl of each test solutions and 2 μl, 4 μl, 6 μl of standard solution on precoated layer 10 mm from the bottom edge. Band length 10 mm. Distance from the side 15 mm.

Development Chamber

Twin-trough chamber of 20 × 10 cm with s. s. lid.

Tank saturation	:	20 minutes with whatman filter paper.
Plate equilibrium	:	None
Development distance	:	80 mm from bottom edge.
Visualisation	:	Observe in UV cabinet at 366 nm.
Post-chromatographic derivatisation	:	Not required.
Photodocumentation	:	Under 366 nm.
Measurement mode	:	Fluorescence/reflectance.

Scanning
For quantification
Using Camag scanner 3 with win CATS software, slit-micro, 4×0.3 mm, scan at 366 nm.

METHOD: 17.16 B

DETERMINATION OF AFLATOXINS B_1, B_2, G_1, AND G_2 IN FOODSTUFFS

Aim of Analysis
Determination of aflatoxins B_1, B_2, G_1, and G_2 in foodstuffs.

Chemicals and Reagents
Hexane, diethylether, peroxide-free, dried petroleum ether, toluene, dichloromethane, chloroform, acetone, acetonitrile, methanol, water, dist., acetic acid, trifluoroacetic acid, sodium chloride, sodium sulfate, paraffin oil, RP-18 cartridge, 6 ml/1g sorbent, silica gel cartridge, 3 ml/ 0.5g sorbent, standard: aflatoxin B_1, B_2, G_1 and G_2, (Aflatoxin standard Kit, 1 mg each).

Sample Preparation for Spices
- Grind or homogenize sample and mix 5.6 g with 100 ml methanol for 3 min.
- Add 40 ml water, mix for 4 min, leave to stand for 10 min, then filter.
- Shake 20 ml filtrate with 20 ml N.Cl solution (10%) and 20 ml petroleum ether for 2 min and leave to separate for 10 min (extraction of matrix in petroleum ether).
- Shake aqueous phase with 50 ml dichloromethane for 1 min and leave to separate (extraction of aflatoxins in dichloromethane).
- Dry dichloromethane phase with 5 g sodium sulfate, filter and evaporate to dryness.
- Dissolve residue in 0.5 ml toluene–acetonitrile (98:2). Use extract (= 0.8 g sample) for application to the HPTLC layer.

For some critical matrices such as paprika, it is advisable to dissolve the residue in 2 ml toluene–acetonitrile 98:2 and perform further purification:

1. Purification of the extract on a silica gel cartridge: Rinse resp. the sorbent with 6 ml toluene–acetonitrile (98:2) (Do not let the sorbent run dry). Elute extract and rinse remaining matrix with 20 ml toluene-acetic acid (9:1) and 20 ml hexane–diethyl ether–acetonitrile (6:3:1)

(dry the sorbent between and in the end). Elute the aflatoxins fractionated with 7 and 4 ml dichloromethane–acetone (3:1) direct in a pear shape flask (dry sorbent between and in the end).

2. Evaporate eluate to dryness and take up the residue in 0.5 ml methanol.
3. Purification of the extract on an RP-18 cartridge: Rinse sorbent with 2 ml methanol, dry and condition with 4 ml methanol–water (2:8) and 2 ml water (Do not let the sorbent run dry). Elute extract and rinse remaining matrix with 5 ml methanol–water (2:8), dry for 1 min. Elute the aflatoxins fractionated with 4 × 2.5 ml methanol–water (5:5) direct in a pear shape flask (dry sorbent between and in the end).
4. Shake aqueous phase for 1 min with 20 ml NaCl solution (10%) and 18 ml dichloromethane and leave to separate for 5 min (extraction of aflatoxins in dichloromethane). Separate dichloromethane phase. Repeat extraction of the aqueous phase with 2 ml dichloromethane.
5. Evaporate eluate to dryness and take up the residue in 0.5 ml toluene–acetonitrile (98:2).
6. Use extract (= 0.8 g sample) for application to the HPTLC layer.

Sample Preparation for other Commodities

Use a higher weighted amount (e.g. 80 g for nuts) if necessary and adjust the amounts of solvent, etc. accordingly.

Standard Solution

Make up a standard mixture of aflatoxins B_1, B_2, G_1, and G_2 in toluene–acetonitrile 98:2 containing 200 pg/lit each of aflatoxins B_1 and G_1 and 100 pg/lit each of G_2 and B_2.

Layer

HPTLC plates or sheets silica gel merck 60 F_{254}, 20 × 10 cm or 20 × 20 cm.

Sample Application

Apply bandwise with Camag Linomat, distance from lower edge of sheet 10 cm (for plates 6 cm), band length 8 mm, distance between tracks 4 mm, distance from left edge 15 mm = 15 applications.

Application Pattern

S1   U U U U S1 U U U U S1.
S1 = standard mixture 5 μl each, U = sample of 100 μl each

Chromatography

Double development (in opposing direction) in twin-trough chamber.

1. For the first development, which removes the matrix from the start zone, fill the chamber to a depth of 5 cm with peroxide-free, dried diethyl ether and place the sheet or plate (6 cm free side downwards) in the chamber: migration distance 50 mm (sheet) and 40 mm (plate), respectively. View sheet or plate under UV 366 nm; the fluorescent aflatoxins should have

migrated little or not at all from the start zone. Cut off the top 85-90 mm (sheet) and 25-30 mm (plate), respectively and turn the plate or sheet through 180°.

2. For the second development, which separates the aflatoxins, charge the chamber normally (to a depth of about 8 mm) with chloroform–acetone–water (140:20:0.3) and insert plate or sheet; migration distance 80 mm (sheet) and 60 mm (plate), respectively.

Densitometric Evaluation

TLC scanner with CATS software, fluorescence measurement at 366/>400 nm, single level calibration via peak height confirmed by a multilevel calibration (see Fig. 17.16,B).

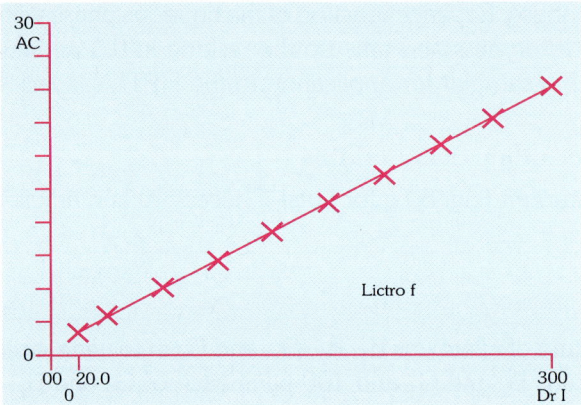

(a) Calibration function of aflatoxin B$_1$ (Peak height) after dipping in paraffin oil-n-hexane

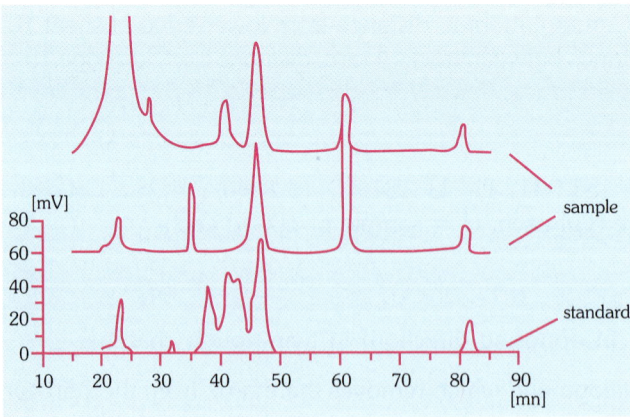

(b) TLC chromatogram with standard mixture and aflatoxin extracts from different types of paprika (extract additional purified)

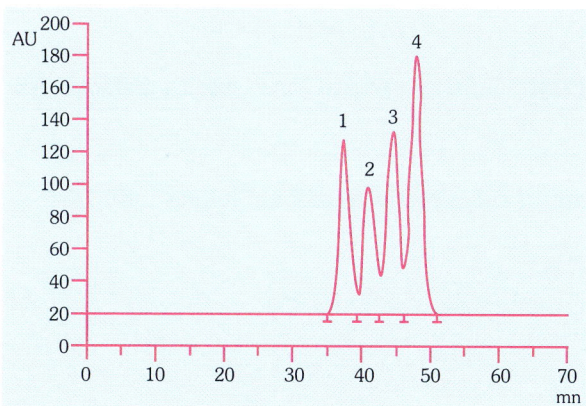

(c) HPTLC chromatogram with standard mixture (1 = G_2; 2 = G_1; 3 = B_2; 4 = B_1)

Fig. 17.16 (B) (a-c)

Discussion

Recovery is between 70 and 100%. The detection limit is 10 pg for aflatoxins B_1 and G_1 and 5 pg for B_2 and G_2. This can be improved 2 to 3-fold by dipping in paraffin oil-n-hexane (2:3).

Positive results can be confirmed by development after prechromatographic derivatization. For this purpose, additionally apply 5 ml trifluoroacetic acid to the start zones, leave for 5 min, then heat for 2 min at 35-40° C on a plate heater. The derivatives of aflatoxins B_1 and G_1 are now polar and stay behind at the start; B_2 and G_2 lie in the medium R_f range.

Precautions

- Avoid contact of aflatoxins with your skin.
- Aflatoxins are sensitive to light and oxidation. Store chromatographed HPTLC plates or sheets, standards, extracts, etc. in the dark at about 5° C.
- Verify the concentration of aflatoxin stock or standard solutions regularly by photometry.
- Place contaminated materials at least for 30 min in 6% Javel water.
- Aflatoxins are able to accumulate at synthetic material and falsify results. Avoid contact with such materials, e.g. plastic tip of an Eppendorf pipette.

References

1. SOP A9024.01B, Kantonales Laboratorium Aargau, 13.03.1997
2. AOAC Official Methods 970.44, 1990 edition (2 volumes)
3. Jork H, Funk W, Fischer W, Wimmer H. Thin-Layer Chromatography, **1a;** VCH Weinheim, 1990.
4. Nagler JM. (Natural Resources Institute, NRI, Central Avenue, Chatham, Kent, ME4 4TB), The Application of HPTLC to the Control of Aflatoxin in Philippine Copra, paper presented at the TLC-Symposium in Guildford, 3rd-5th June 1996.

METHOD: 17.17

QUANTITATIVE DETERMINATION OF SUCRALOSE IN INDIAN SWEETS (BURFI)

Aim of Analysis

Quantitative determination of sucralose in traditional Indian burfi.

Chemicals and Reagents

Methanol, distilled water, methanolic sulphuric acid

Reagent preparation :   None.

Standard Preparation

Weigh 5 mg of sucralose, add 0.5 ml water and make up volume to 24.5 ml with methanol. (conc. : 0.2 $\mu g/\mu l$).

Sample Preparation

Sample A: 1 gm of Burfi sample is taken, to it 1 ml of water and 9 ml of methanol is added (conc. : 100 $\mu g/\mu l$).

Sample B: 1 gm of Burfi sample is taken, to it 1 ml of water, 2 ml of std sucralose and 7 ml of methanol is added (conc. : 100 $\mu g/\mu l$).

Sample C: 1 gm of sample is taken, to it 1 ml of water, 3 ml of std sucralose and 6 ml of methanol is added (conc. : 100 $\mu g/\mu l$).

Stationary Phase

TLC Al sheets silica gel 60F$_{254}$ precoated 20 × 10 cm

Mobile Phase

Acetonitrile : water (9.5 : 0.5) v/v (Volume = 30 ml)

Sample/Standard Application

Apply with the help of Camag ATS-4 or Linomat 5, 10 μl of sample A, 10 μl of sample B, (2 μl, 4 μl, 6 μl, 8 μl and 10 μl) of std sucralose on precoated layer, 10 mm from bottom edge, band length 8 mm, distance from the sides 15 mm.

Development Chamber

Twin-trough chamber of 20 × 10 cm with s. s. lid.

Tank saturation	: 10 mins with filter paper.
Plate equilibrium	: 10 min.
Development distance	: 80 mm from bottom edge.
Visualisation	: None in UV cabinet.

Post-chromatographic derivatization	: Dip the plate in methanolic sulphuric acid and heat at derivatisation 120° C for 5-7 mins. Sucralose will give turquoise coloured band after heating at 120° C (R_f : 0.47).
Photodocumentation	: In UV 366 nm after derivatisation.
Measurement mode	: Fluorescence/reflectance.

Scanning

For Quantification

Using Camag scanner 3 with win CATS software, slit-micro, 6 × 0.45 mm, scan at 366 nm.

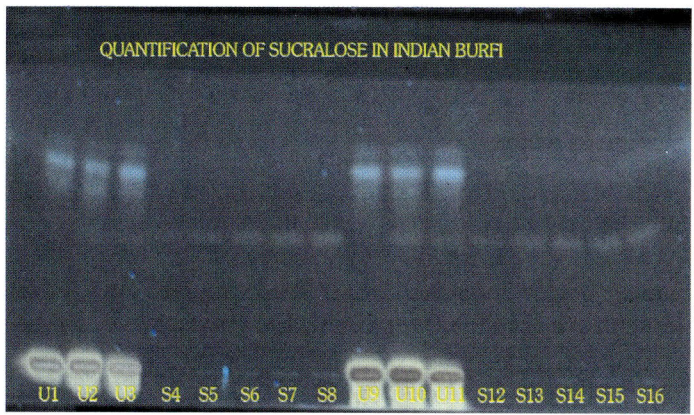

Fig. 17.17: Quantification of sucralose in Indian Burfi.

U1: Burfi sample (1000 μg)
U2: Burfi sample (1000 μg)
U3: Burfi sample (1000 μg)
S1: Sucralose standard (0.4 μg)
S2: Sucralose standard (0.8 μg)
S3: Sucralose standard (1.2 μg)
S4: Sucralose standard (1.6 μg)
S5: Sucralose standard (2.0 μg)

METHOD: 17.18

SEPARATION OF SUGARS

Aim of Analysis

Separation of glusose, galactose, mannose, xylose, sucrose, lactose.

Chemicals and Reagents

Glucose, galactose, mannose, xylose, sucrose, lactose acetone, methanol, dist. water, sodium acetate.

Reagent Preparation

Phosphomolybdic acid

Weigh 250 mg of phosphomolybdic acid and dissolve it in 50 ml of ethanol.
Sodium acetate: 0.02 M solution in distilled water.

Standard Preparation

1. Glucose (Solution A) : Weigh 20 mg of glucose and dissolve in 1 ml of dist. water and make up to 10 ml with methanol [Conc: 2 μg/μl].
   Solution B : 1 ml of solution A and add 9 ml of methanol [Conc: 0.2 μg/μl].

2. Galactose (Solution C): Weigh 20 mg of galactose and dissolve in 1 ml of dist.water and make up to 10 ml with methanol [Conc: 2 μg/μl].

   Solution D: Take 1 ml of solution C and add 9 ml of methanol [Conc: 0.2 μg/μl].

3. Mannose (Solution E) : Weigh 20 mg of mannose and dissolve in 1 ml of dist. water and make up to 10 ml with methanol [Conc: 2 μg/μl]. Solution F : Take 1 ml from solution E make it to 10 ml with methanol [Conc : 0.2 μg/μl].

4. Xylose (Solution G): Weigh 20 mg of xylose and dissolve in 1 ml of dist. water and make up to 10 ml with methanol [Conc: 2 μg/μl]. Solution H : Take 1 ml from solution G make it to 10 ml with methanol [Conc : 0.2 μg/μl].

5. Sucrose (Solution J) : Weigh 20 mg of sucrose and dissolve in 1 ml of dist. water and make up to 10 ml with methanol [Conc: 2 μg/μl]. Solution K : Take 1 ml from solution J make it to 10 ml with methanol [Conc : 0.2 μg/μl].

6. Lactose : Solution L : Weigh 20 mg of lactose and dissolve in 1 ml of dist. water and make up to 10 ml with methanol [Conc: 2 μg/μl]. Solution M : Take 1 ml from solution L make it to 10 ml with methanol [Conc: 0.2 μg/μl].

Stationary Phase

TLC Al sheets silica gel 60F$_{254}$ precoated 20 × 10 cm. Further impregnated with sodium acetate (0.02 M) by ascending chromatographic development.

Mobile Phase

Acetone : water (9 : 1) v/v (Volume = 16 ml)

Development Chamber

Twin-trough chamber of 20 × 10 cm with s. s. lid.

Tank saturation	:	20 min with filter paper.
Plate equilibrium	:	None
Sample/standard application	:	Apply with the help of Camag ATS-4 or Linomat 5, 15 μl of solution B,D, F, H, K and M (each) on precoated layer 10 mm from the bottom edge. Band length 6 mm. Distance from the sides 15 mm.
Development distance	:	80 mm from bottom edge.
Visualisation	:	None
Post-chromatographic derivatisation	:	The plate is dipped in a phosphomolybdic acid reagent then heated to 110° C for 5 or 10 min. Sugars turn blue

Photodocumentation   :   In visible light after derivatisation.
Measurement mode   :   Visible.

Scanning

For Quantification

Using TLC scanner 3 with winCATS software, slit-micro, 4 × 0.3 mm, scan at 580 nm.

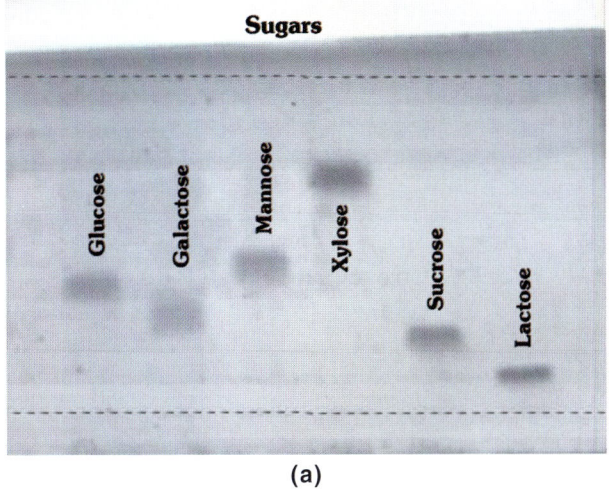

Image (Visible)

Sugars

(a)

Track 1, ID: Glucose

(b) Glucose (580 nm) at (R$_f$ 0.58)

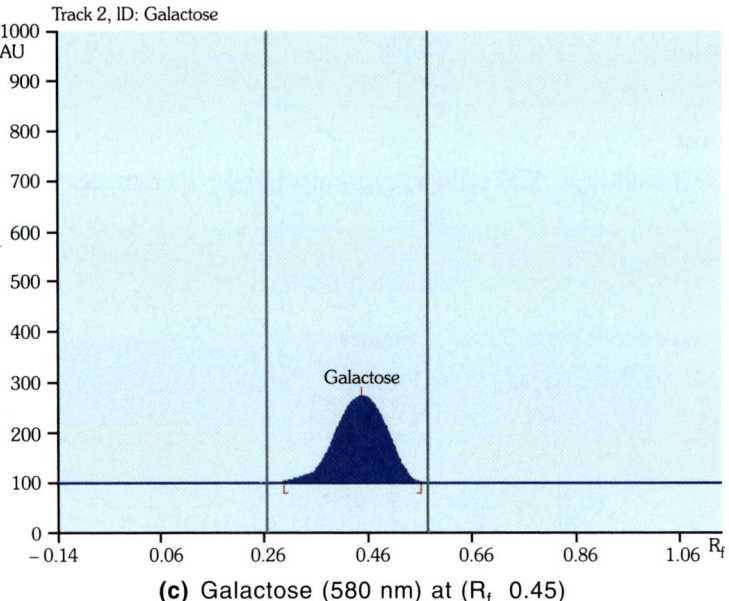

(c) Galactose (580 nm) at (R_f 0.45)

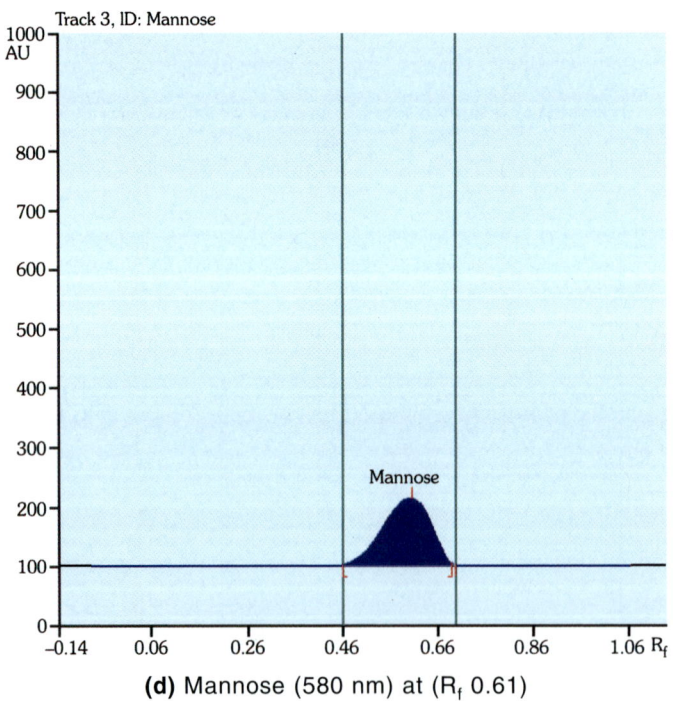

(d) Mannose (580 nm) at (R_f 0.61)

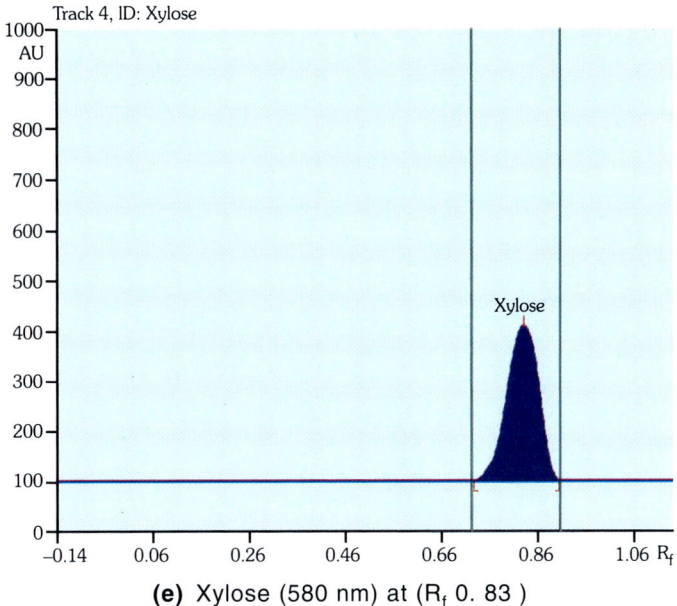

(e) Xylose (580 nm) at (R_f 0. 83)

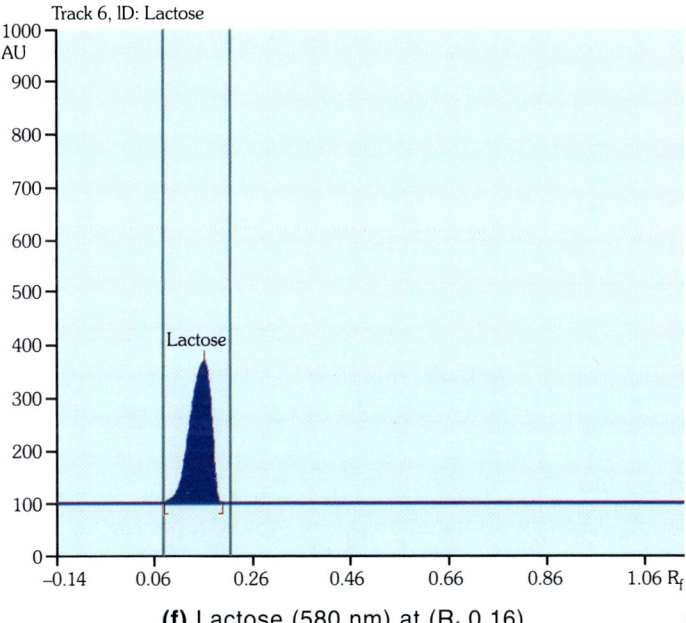

(f) Lactose (580 nm) at (R_f 0.16)

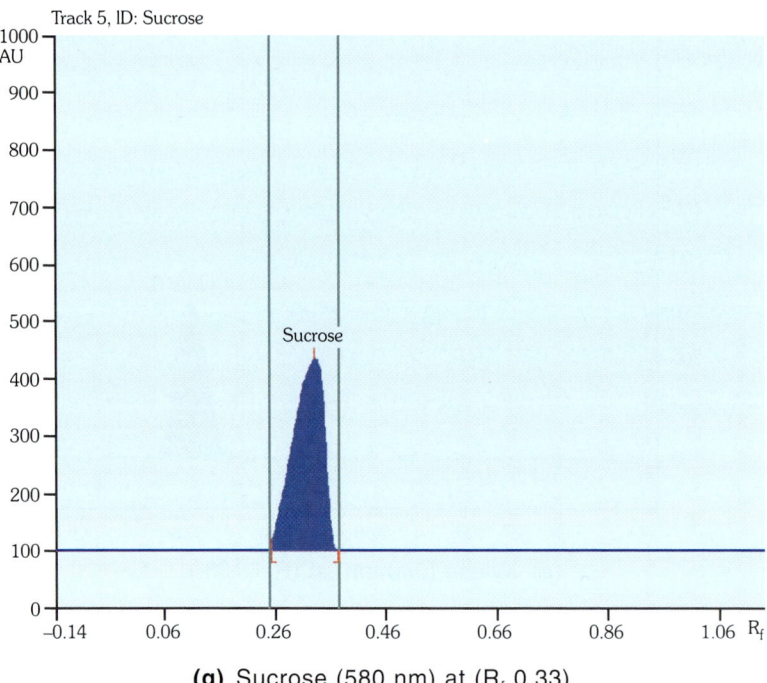

(g) Sucrose (580 nm) at (R_f 0.33)

Fig. 17.18 a–g: Separation of sugars

METHOD: 17.19

QUANTITATIVE DETERMINATION OF PESTICIDES RESIDUES IN GRAPES

Aim of Analysis

Quantitative determination of acephate, metalaxyl, dimethoate, carbaryl, monochrotophos in grapes.

Chemicals and Reagents

Ethyl acetate, methanol, water, tetrabutyl ammonium bromide, palladium (II) chloride, concentrated hydrochloric acid.

Reagent Preparation

Palladium-II-chloride reagent: Weigh 0.5 g of palladium (II) chloride and dissolve it in 100 ml of water, with 1 ml of concentrated hydrochloric acid.

Standard Solution

Weigh 10 mg of each standard dissolved in 10 ml of ethyl acetate [conc. 1 μg/μl].

Sample Preparation

Test solution: Take 75 gm of grapes (Do not wash them). Make a paste or slurry in a clean mixer (Ensure that the container does not become hot, it may cause degradation of pesticides). Add 75 ml of ethyl acetate to the contents of the mixer. Mix thoroughly for 2 min. Transfer the contents to a 250 ml of separating funnel. Wash the contents with another 75 ml of ethyl acetate and transfer it to the same separating solution. Shake vigorously the total contents for 15 min to ensure complete transfer of pesticides into ethyl acetate medium. If there is no proper separation, add ethyl acetate till two phases are clearly seen. Now transfer the ethyl acetate layer into a clean, dry conical flask through a dry funnel containing anhydrous sodium sulphate. Wash with 20 ml of ethyl acetate and collect the total ethyl acetate extract. Evaporate to dryness in a current of nitrogen at room temperature. Reconstitute with 3 ml of ethyl acetate[conc. 3 $\mu g/\mu l$].

Stationary Phase

TLC Al sheets RP-18 $F_{254}S$ precoated plate 20 × 10 cm.

Mobile Phase

Methanol: water: tetra-butyl ammonium bromide (0.001M) (9 : 6: 0.075) v/v (Volume = 16 ml).

Development Chamber

Twin-trough chamber of 20 × 10 cm with s. s. lid.

Tank saturation	:	None
Plate equilibrium	:	None
Sample/standard application	:	Apply with the help of Camag ATS-4 or Linomat 5, 10 μl of test solution and 2 μl of each standard solution on precoated layer 8 mm from the bottom edge. Band length 6 mm. Distance from the side 15 mm.
Development distance	:	80 mm from bottom edge
Visualisation	:	Observe in UV cabinet at 254 nm.
Post-chromatographic derivatisation	:	Spray with palladium (II) chloride and heat the plate at 110° C for 10 min.
Photodocumentation	:	In 254 nm and visible light.
Measurement mode	:	UV absorbance/reflectance.

Scanning

For Quantification before Derivatization

Using Camag scanner 3 with win CATS software, slit-micro, 4 × 0.3 mm, scan for acephate and metalaxyl, dimethoate at 190 nm, for carbaryl and monochrotophos at 220 nm.

PHOTODOCUMENTATION

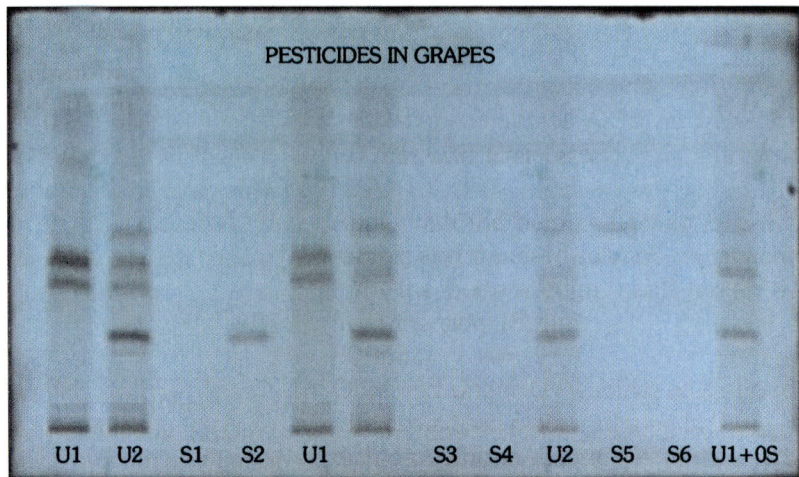

at 254 nm

(a) Before derivatisation

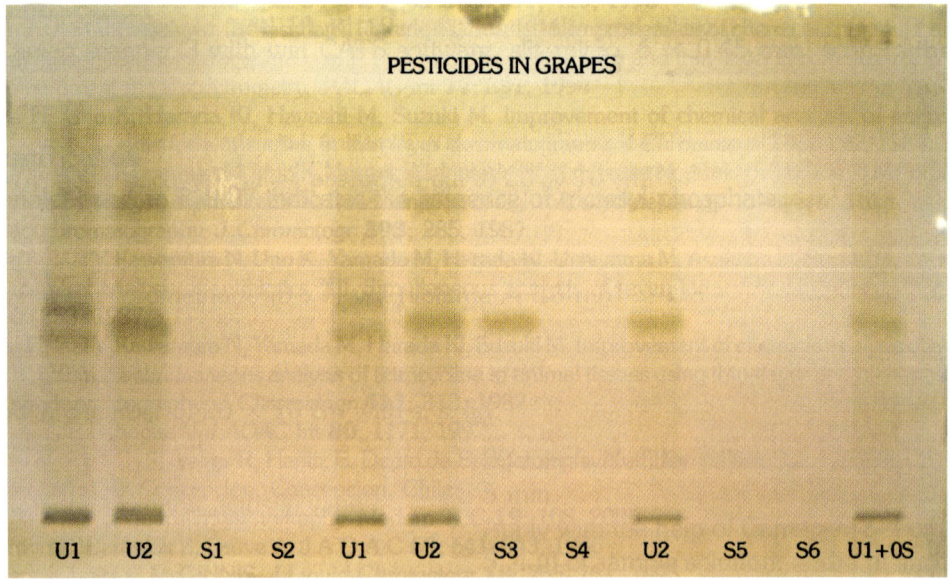

At visible

(b) Expose to iodine vapors

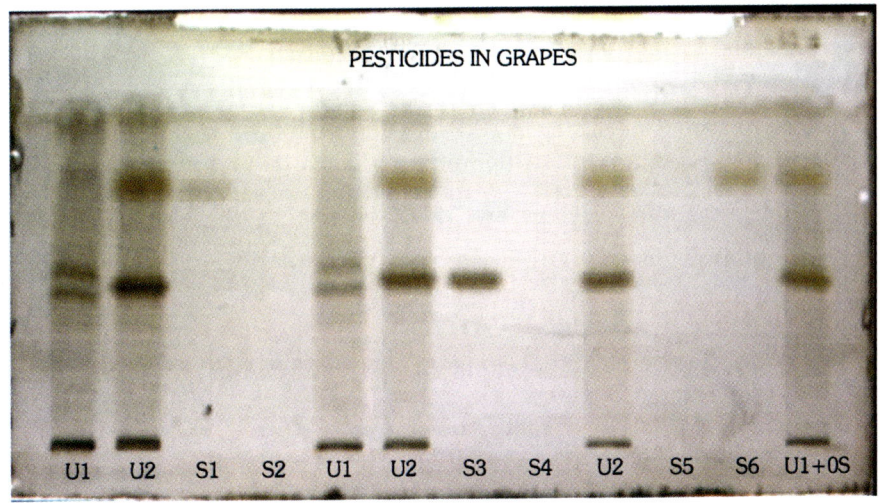

At visible

(c) After derivatisation palladium (II) chloride

Fig. 17.19 a–c

METHOD: 17.20

DETERMINATION OF CYCLOPIAZONIC ACID IN FOODSTUFFS

Aim of Analysis

Determination of cyclopiazonic acid(CPA) in foodstuffs.

Chemicals and Reagent

Petroleum ether, phosphoric acid (85%), sodium bicarbonate, hydrochloric acid conc., methanol, ethanol, chloroform, ethyl acetate, acetonitrile, ammonia(25%), 2-propanol, 4-dimethyl-aminobenzaldehyde, standard: cyclopiazonic acid.

Sample Preparation

- Mix 50 g pulverized sample (20 mesh) with 300 ml petroleum ether in a 500 ml Erlenmeyer flask. Stir gently for 4 h.

- Transfer mixture quantitatively to a suction filter lined with glass fibre paper (Whatman 93A-AH) and wash twice with 100 ml petroleum ether. Switch off vacuum when the filter cake is fairly dry, remove the filter cake and let it dry overnight.

- Mix residue with 250 ml methanol: chloroform (1:4), and 0.5 ml phosphoric acid in a 500 ml Erlenmeyer flask. Shake on a machine at moderate speed for 30 min. Then suction through a filter paper (Whatman no. 4).

- Mix a 50 ml aliquot of the extract with 50 ml 0.5N sodium bicarbonate solution in a 125 ml centrifuge vessel. Shake for 30 s, then centrifuge for 15 min at 1500 rpm.

- Decant supernatant and neutralize by adding HCl dropwise (about 7 ml). Add 25 ml chloroform, shake for 15 min, then centrifuge 10 min at 800 rpm and decant. Extract the aqueous phase with 25 ml chloroform in the same way.

- Transfer both organic phases to a 100 ml evaporating flask with dichloromethane and evaporate to dryness in a rotary evaporator at 40° C.

- If the sample is not immediately chromatographed, the residue can be stored frozen at 0° C.

- For chromatography, dissolve residue in 500 μl chloroform.

Standard Solution

Dissolve 1 mg/ml cyclopiazonic acid in acetonitrile and dilute with chloroform 1:40 (1 μl ç 25 ng CPA).

Stationary Phase

HPTLC plates silica gel merck 60 F_{254}, 20 × 10 cm.

Sample Application

With Camag Linomat as 7 mm bands, distance between tracks 3 mm, distance from left edge 15 mm, distance from lower edge 8 mm, delivery rate 5 s/μl = 17 applications per plate.

Application scheme

S1	U1	U2	U3	S2	U4	U5	U6	S3	U1	U2	U3	S4	U4	U5	U6	S5
1	10	10	10	5	10	10	10	10	10	10	10	15	10	10	10	20 μl

S1-S5 = standard in different concentrations U1-U6 = unknowns

Chromatography

In twin-trough chamber 20 × 10 cm, no conditioning required with ethyl acetate-2-propanol-ammonia (10:3:2), migration distance 60 mm.

After development, dry for 15 min at 50° C (ascertain that ammonia has been driven off).

Post-chromatographic Derivatization

By dipping for 3 s with Camag chromatogram immersion device in 80 ml Ehrlich reagent (0.5 g 4-dimethylaminobenzaldehyde in 75 ml ethanol and 25 ml conc. hydrochloric acid). CPA stains blue. Dry with a stream of hot air and evaluate after 10 min.

Densitometric Evaluation

With Camag TLC scanner and CATS evaluation software; scanning by absorbance at 546 nm with mercury lamp (offers higher signal intensity than tungsten lamp at 596 nm).

Notes

The limit of quantification is 25 ng per spot. Recovery rates of authentic matrices topped up with standards were determined 90% for peanut matrix and 85% for cereal matrix.

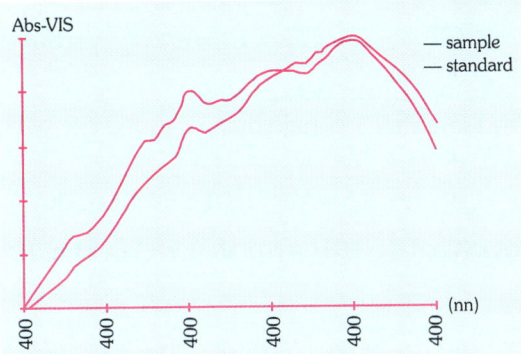

(a) Absorption spectrum of CPA-dimethylaminobenzaldehyde, absorption maximum 596 nm.

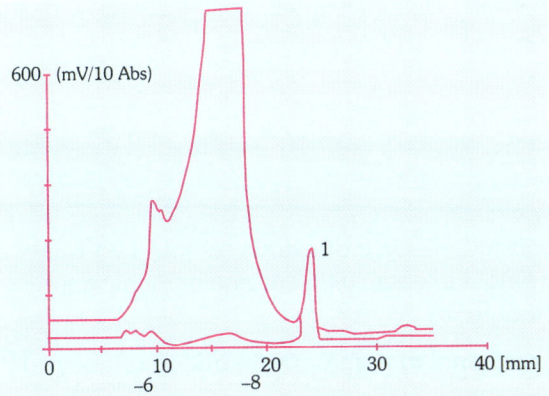

(b) Densitogram of peanut matrix and of standard toppod up (90 ng CPA (1) absolute or 90 µg/kg).

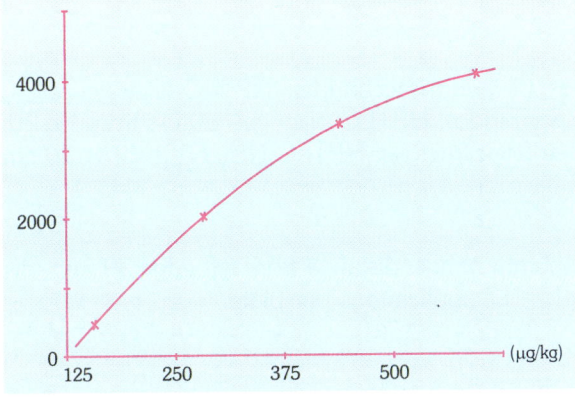

(c) Calibration function of CPA via peak area. Calibration range 125-500 ng/spot, corresponding to 125-500 µg/kg.

Fig. 17.20 a–c

References

1. Lansden JA. Assoc off Anal Chem **69**; 6, 1986.
2. Popken AM, Dose K, Fresenius Z. Anal Chem 316, 47-59, 1983.

METHOD: 17.21

DETERMINATION OF MONO-, DI-, TRI- AND POLYSACCHARIDES IN FOODSTUFFS

Aim of Analysis

Determination of mono-, di-, tri- and polysaccharides in foodstuffs.

Chemicals and Reagent

Acetone, acetonitrile, water dist., diphenylamine, aniline chloride, methanol, butanol, phosphoric acid 86%, boric acid 0.5%.

Standards

Raffinose, melezitose, lactose, maltose, sucrose, galactose, glucose, fructose.

Sample Preparation

Depending on type of foodstuff, dissolve or extract sample with water, and filter if necessary. Co-extracted proteins, which would interfere with chromatographic analysis be precipitated by adding twice the amount of acetone(cold).

Dilute the solutions with methanol to a final concentration of about 0.2 mg/ml per saccharide.

Standard Solutions

Dissolve 0.5 mg of those saccharides which shall be determined in 10 ml dist. water and dilute 1:4 with water (125 ng/μl).

Chromatogram Layers

1. To separate mono-, di- and trisaccharides: HPTLC silica gel 60F$_{254}$' 20 × 10 cm.
2. To separate polysaccharides: HPTLC silica gel 60 (without F), 20 × 10 cm.
3. To separate glucose from fructose: HPTLC NH$_2$ F$_{254}$s 20 × 10 cm.

Sample Application

With Camag Linomat as 7 mm bands, distance from left edge 20 mm, distance from lower edge 8 mm; delivery rate 15 s/μl = 17 applications per plate.

Application Scheme

S1	U1	U2	S2	U3	U4	S3	U5	U6	S4	...
2	2	2	3	2	2	4	2	2	5	...

S1-S4 standards in different concentrations; U unknowns.

Chromatography

In twin-trough chamber 20 × 10 cm with s.s. lid.

1. To separate mono-, di-, and trisaccharides, except glucose from fructose:
- 10 min conditioning with developing solvent.
- Develop three times with acetonitrile : water (85:15), migration distance 70 mm each.
- Dry with hair drier held at a distance of about 40 cm for 20 min at 40-45° C. After final run, dry in the same way or at 100° C for 2 min in an oven.

2. To separate polysaccharides:
- 10 min conditioning with developing solvent.
- Develop with butanol : methanol : water (50:25:20), migration distance 70 mm.
- Dry plate for 15 min with hair drier.

3. To separate glucose from fructose:
- Develop with acetonitrile : water : 0.5% aqueous solution of boric acid (76:24:10), migration distance 75 mm.
- Dry plate for 5 min with hair drier.

Derivatization

- By dipping for 3 s with Camag chromatogram immersion device in diphenylamine reagent (2.4g diphenylamine +2.4 g aniline chloride in 200 ml methanol, then add 20 ml 86% phosphoric acid) followed by heating at about 120° C for 10-15 min.
- Using NH_2- plates the sugars fluoresce simply by heating the plate at about 120° C for 5 min (Some sugars may require a higher temperature and time).

Note: Dipping the plate into a solution of paraffin/n-hexane (1:2) enhances the fluorescence intensity by a factor 2-3.

Densitometric Evaluation

With Camag TLC scanner and CATS evaluation software; scanning by absorbance at 620 nm with tungsten lamp or by fluorescence at 366/>400 nm with mercury lamp, slit dimension 0.2 × 4 mm, detection limit for fluorescent zones about 10 ng/zone.

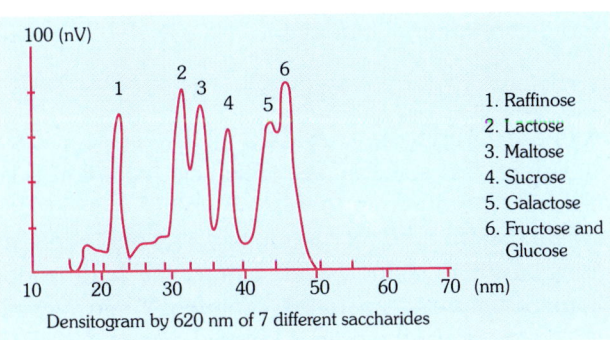

1. Raffinose
2. Lactose
3. Maltose
4. Sucrose
5. Galactose
6. Fructose and Glucose

Densitogram by 620 nm of 7 different saccharides

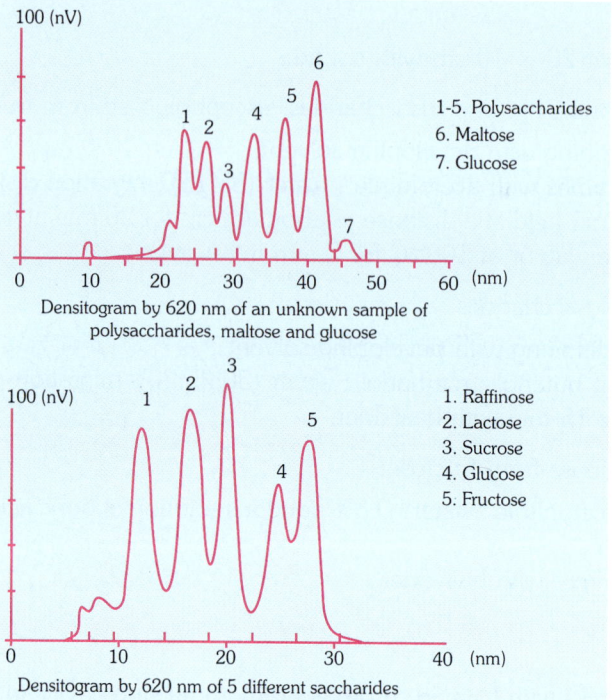

Densitogram by 620 nm of an unknown sample of
polysaccharides, maltose and glucose

Densitogram by 620 nm of 5 different saccharides

Fig. 17.21

METHOD: 17.22

QUANTITATIVE DETERMINATION OF VITAMIN C IN FRUIT JUICE

Aim of Analysis

Quantitative determination of vitamin C in fruit juice

Chemicals and Reagents

Chloroform, ethyl acetate, acetic acid, oxalic acid, (4% aqueous solution), thiourea, L-ascorbic acid [aqueous solutions = 1 mg/l (calibration standard)], 2, 4-dinitrophenylhydrazine (2% solution in 70% sulfuric acid), 2,6-dichorophenol-indophenol Na-salt (0.5% in water).

Sample Preparation

Pipette 2 ml fruit juice into a measuring flask. Add 25 ml 4% oxalic acid solution. Add 2 ml 0.5% aqueous dichlorophenol-indophenol solution. Ascorbic acid oxidation is completed within 5 min at room temperature. Add 10 mg thiourea to destroy excess of oxidation reagent. Fill up with water to 100 ml. Transfer 20 ml into a 50 ml centrifuge vessel and add 4 ml of the 2% solution of 2,4-dinitrophenylhydrazine in 70 % sulfuric acid. Extract the hydrazones with 3 × 6 ml ethyl acetate-acetic acid (98:2) for better phase separation, centrifuge 1 min after each extraction. Combine the extracts and fill up to 20 ml. Apply the orange colored extract directly on to the HPTLC layer.

Calibration Standard

L-ascorbic acid, aqueous solution 1 mg/l

Layer

HPTLC precoated plates silica gel 60 F_{254}, 20 × 10 cm, prewashed by immersing for 1 hour in isopropanol followed by 30 min drying at 120° C.

Sample Application

Bandwise with Camag automatic TLC sampler IV or Camag Linomat 5 : 6 mm bands, 8 mm from both edge, 15 mm from the side, 17 samples per plate side; delivery speed 4 sec/μl. Apply 5 μl of the unknowns and different amounts of the standard on both sides of the plate.

Recommended Application Pattern

S1   A   B   C   S2   A   B   C   S3   A   B   C   S4   A   B   C   S5
S1 = 3 μl   =   60 ng absolute
S2 = 6 μl   = 120 ng absolute
S3 = 9 μl   = 180 ng absolute
S4 = 12 μl = 240 ng absolute
S5 = 15 μl = 300 ng absolute
S   = Standard
A, B, C = Unknown 5 μl

Chromatography

In Camag horizontal developing chamber in sandwich configuration with chloroform : ethyl acetate (1:1), two runs with gentle intermediate drying. R_f of ascorbic acid derivative is about 0.4.

Densitometric Evaluation

With Camag TLC scanner 3 with winCATS software with Labdata system and CATS evaluation software. Scanning by absorbance at 510 nm with tungsten lamp, monochromator bandwidth 10 nm, slit dimensions 0.2 × 5 mm; evaluation via peak area, linear regression.

Results

The reproducibility of this method is about 1.5% relative standard deviation.

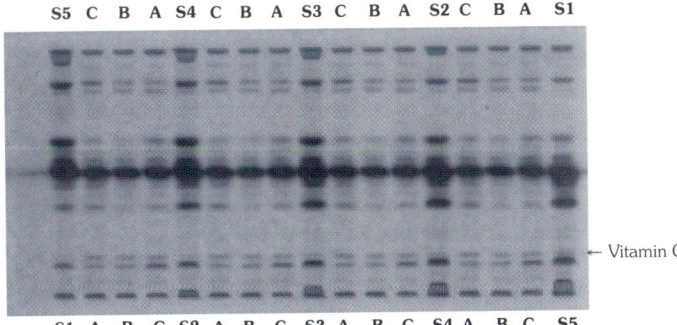

(a) Chromatogram of vitamin C derivatives; 254 nm UV

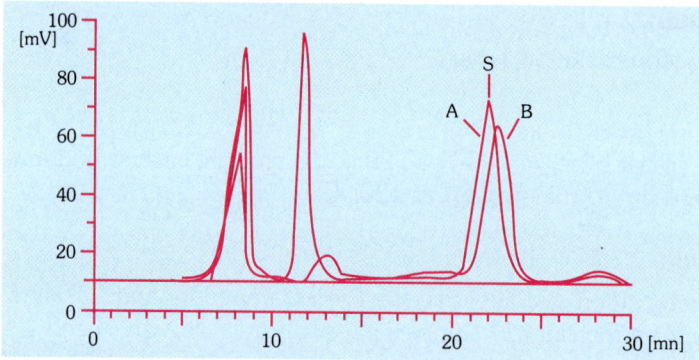

(b) Densitogram curves of two orange juice extracts (A and B) and standard (S)

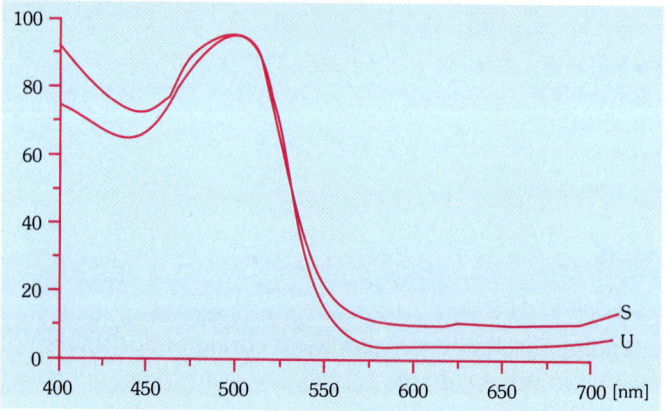

(c) In situ absorption spectra of the DNPH-dehydroascorbic acid from standard (S) and unknown (U)

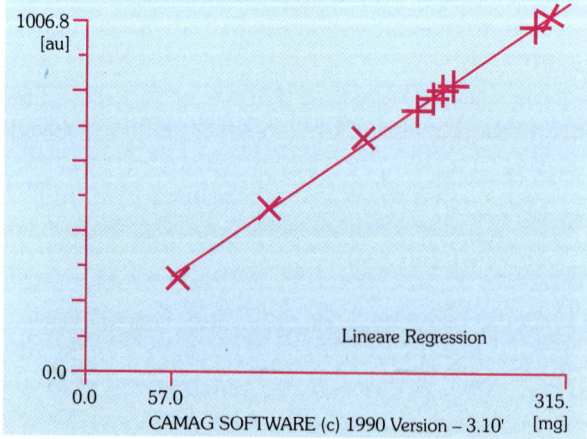

(d) Calibration function of DNPH-dehydroascorbic acid (peak area) in the range from 57-315 mg/L

Fig. 17.22 a–d

References

1. Stgrohdecker R, Henning HM. Vitamin-Bestimmungen, Verlag Chemie, Weinhein
2. Roe JH, Kuether CA. Science **95**; 77, 1942
3. Camag Application Laboratories

METHOD: 17.23

DETERMINATION OF LACTOSE, SACCHAROSE AND FRUCTOSE/GLUCOSE

Aim of Analysis

Determination of lactose, saccharose and fructose/glucose from fermentation broth.

Chemicals and Reagents

Methanol, water, 1-butanol, acetic acid, aniline, phosphoric acid (85%).

Standards: lactose, saccharose, fructose, glucose.

Sample Preparation

The culture medium (fermentation broth) is diluted with methanol: water (1:1) until the concentration (total sugar) is about 0.5%. The solution is filtered and chromatographed immediately.

Standard Solutions

Dissolve 10 mg each of lactose, saccharose, fructose or glucose* to 10 ml in methanol: water (1:1) and dilute this solution (1:2) with methanol: water (1:1) (500 ng/μl).

Application scheme:

U1	S1	U2	U3	S2	U4	U5	S3	U6	U1	S1	U2 ...	
200				600		1000			200			μl/band
100				300		500			100 ng carbohydrate each			

Stationary Phase

HPTLC plates silica gel 60 F_{254}, 20 × 10 cm.

Sample Application

Bandwise with automatic TLC sampler III, band length 4 mm, distance from side 15 mm, distance from lower edge 10 mm, 17 applications.

Chromatography

In twin-trough chamber with 1-butanol : acetic acid : water (80:100:15) with chamber saturation, migration distance 50 mm, running time about 30 min.

    After chromatography, the plate is dried for 20 min at 115° C. Then the chromatogram can be stored. The following derivatization is carried out when densitometric evaluation is feasible within one hour.

---

*In this method, fructose and glucose are not separated. If they are to be determined separately, NH_2-bonded silica must be used.

Derivatization

By dipping for 3 s with Camag chromatogram immersion device in diphenylamine reagent (3 ml aniline + 3 g diphenyl amine + 15 ml conc. phosphoric acid made up to 150 ml with methanol) followed by heating at about 115° C for about 15 min. The colours are as follows:

Lactose   : greyish blue
Saccharose   : brownish green
Fructose   : rust brown
(Glucose lies between saccharose and fructose)

Densitometric Evaluation

With TLC scanner and CATS evaluation software; scanning by absorbance at 620 nm.

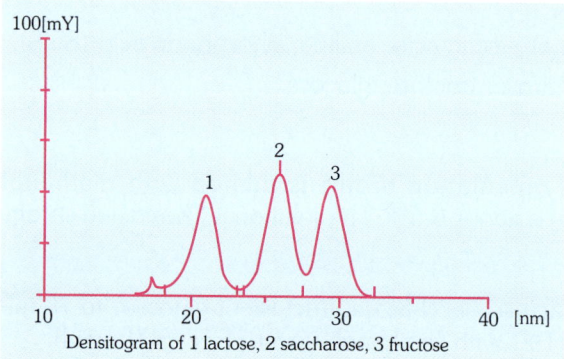

Densitogram of 1 lactose, 2 saccharose, 3 fructose

Fig. 17.23

References

1. Kreuzig F. *J Liquid Chromatog* **6**; 1227-1236, 1983
2. Camag Application Laboratories

METHOD: 17.24

IDENTIFICATION AND DETERMINATION OF P-HYDROXYBENZOIC ACID, ITS ESTERS AND 4-METHOXYCINNAMIC ACID DERIVATIVES

Aim of Analysis

Identification and determination of p-hydroxybenzoic acid, its esters and 4-methoxycinnamic acid derivatives.

Chemicals and Reagents

Acetone, ethyl acetate, pentane, dichloromethane, petrol ether b.p. 40-60° C, diethylketone, acetic acid.

Standards: p-hydroxybenzoic acid (PHB), methylparaben (MP), propylparaben (PP), parsol MCX (4-methoycinnamate) (PA), (giveaudan).

Sample Preparation

- Dissolve or suspend 1.0 g of the cosmetic preparation in 40 ml acetone : ethyl acetate (2:1) at 40° C. Cool to room temperature.
- Filter, rinse with 7 ml acetone: ethyl acetate (2:1).
- Make up to 50 ml.

Standard Solutions

Method A, determination of MP and PP in process control:

Dissolve 50 mg MP + 25 mg PP in 50 ml acetone: ethyl acetate (2:1), then dilute (1:10).

Method B, determination of PHB in stability testing:

Dissolve 5 mg PHB + 50 mg MP + 25 mg PP in 50 ml acetone: ethyl acetate (2:1), then dilute (1:10).

Method C, analysis of stabilizing and light absorbing agents:

Dissolve 50 mg MP + 25 mg PP + 500 mg PA in 50 ml acetone: ethyl acetate (2:1), then dilute (1:10).

Stationary Phase

Precoated HPTLC plates silica gel Merck 60 F_{254}, 20 × 10 cm

Sample Application

Bandwise with Linomat, 8 mm band length, distance between bands 4 mm, distance from side edge 22 mm, distance from bottom edge 8 mm = 13 applications.

Recommended application pattern

	S3	U1	U2	S3	U3	U4	S3	U5	U3	U2	U7	U3	S3
Method A	5 µL	10 µL	10 µL	5 µL	10 µL	10 µL	5 µL	10 µL	10 µL	5 µL	10 µL	10 µL	5 µL
	S1	U1	S2	U2	S3	U1	S4	U2	S5	U1	S6	U2	S7
Methods B and C	1 µL	10 µL	2 µL	10 µL	3 µL	10 µL	4 µL	10 µL	5 µL	10 µL	6 µL	10 µL	7 µL

U = unknown   S = standard

Chromatography

In twin-trough chamber 20 × 10 cm, with out pre-equilibration.

Developing solvent for parabens: pentane : dichloromethane : acetic acid (25:25:3).

Developing solvent for parabens plus methoxycinnamic acid derivatives: petrol ether : diethylketone: acetic acid (88:5:12).

Solvent migration distance about 40 mm (10 min); after chromatography, dry plate 3 min with hair dryer.

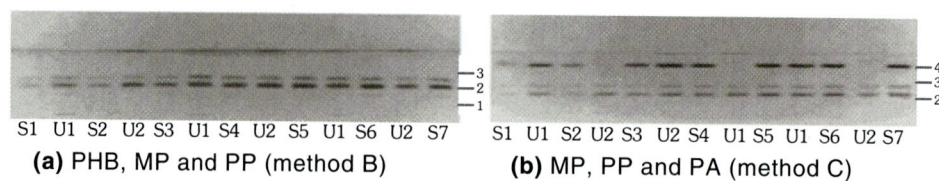

S1 U1 S2 U2 S3 U1 S4 U2 S5 U1 S6 U2 S7   S1 U1 S2 U2 S3 U2 S4 U1 S5 U1 S6 U2 S7

(a) PHB, MP and PP (method B)   **(b)** MP, PP and PA (method C)

Fig. 17.24 a, b

Densitometric Evaluation

Scanner with CATS software, deuterium lamp, scanning by absorbance at 255 nm for PHB, MP and PP, at 310 nm for PA; monochromator bandwidth 10 nm, slit dimensions 0.3 × 5 mm.

R$_f$-values and Absorption Maxima

	Substance		UV_{Max}	$A(R_F)$	$B(R_F)$	$C(R_F)$
1	p-Hydroxybenzoic acid	PHB	258	---	0.32	---
2	Methylparaben	MP	257	0.45	0.45	0.32
3	Propylparaben	PP	257	0.58	0.58	0.46
4	Parsol MCX	PA	309	---	---	0.78

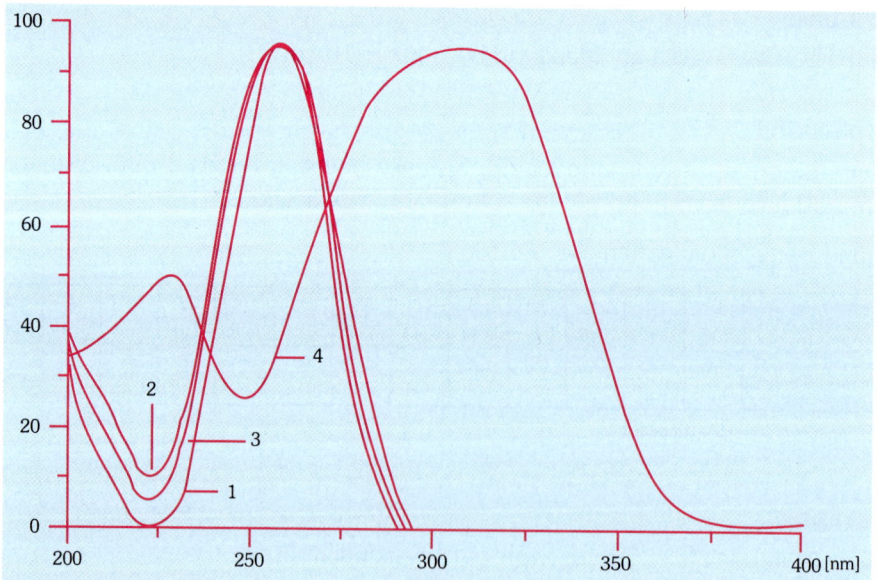

(c) UV spectra of free p-hydroxybenzoic acid PHB (1), methylparaben MP (2), propylparaben PP (3)and Parsol PA (4); the spectra 1-3 are typical for PHB esters but differences between the individual ones are very small. The individual PHB esters are identified by their R$_f$s.

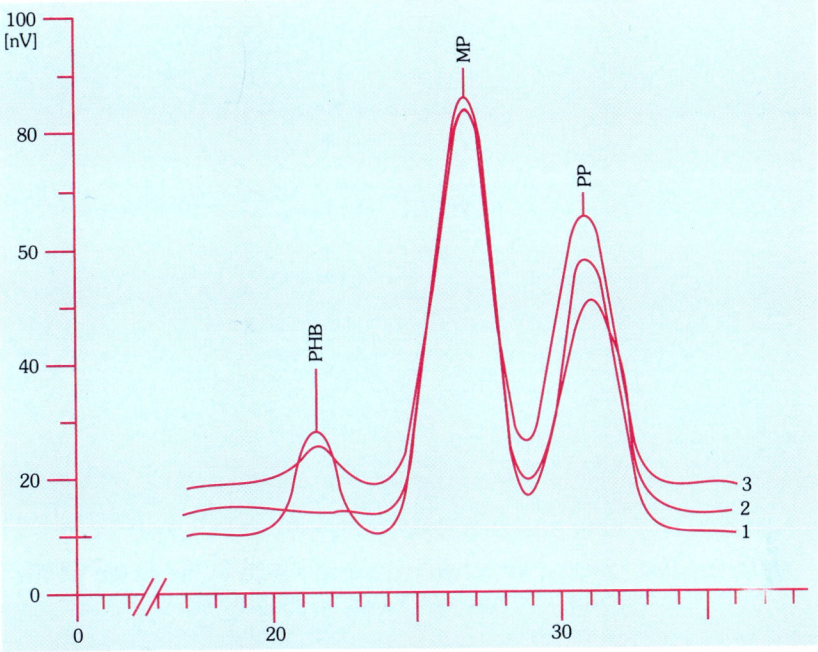

(d) Densitogram curves according to methods A and B: (1) Standard solution. (2) extract from a night cream, (3) extract from a body lotion.

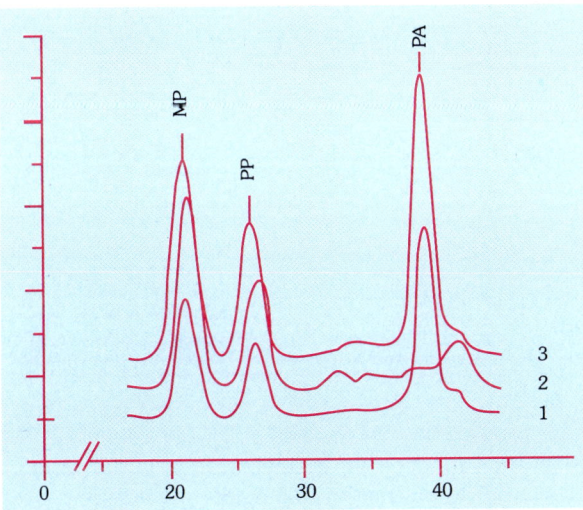

(e) Densitogram curves of standard solution C(1) of an extract from a night cream (2), and of a tanning lotion containing Parsol MCX (3)

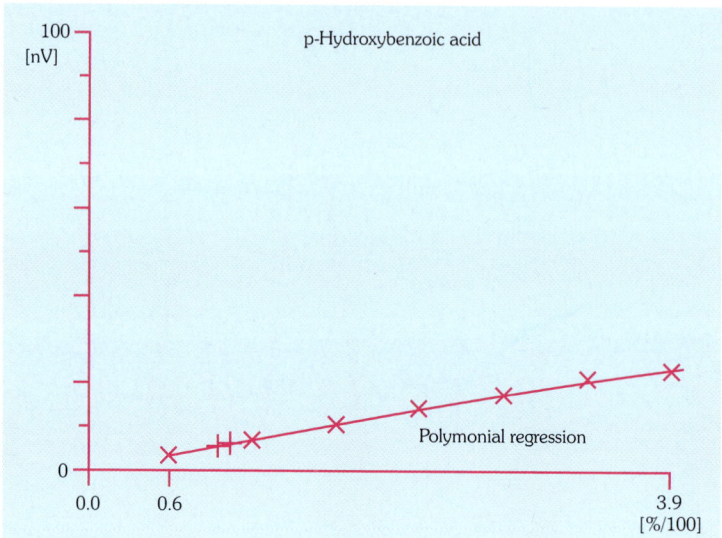

(f) Calibration function for p-hydroxybenzoic acid in the range from 0.5 to 3.5 mg/g (0.05-0.35%) method B

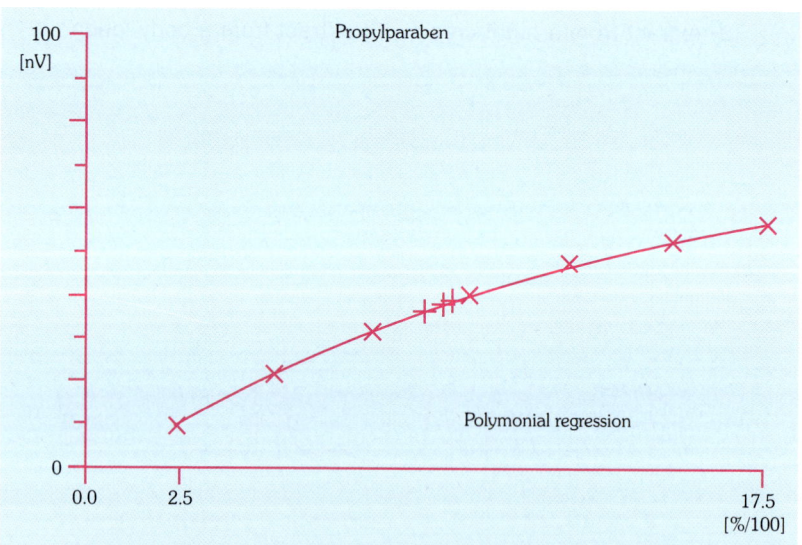

(g) Calibration function for methylparaben in the range from 5.0 to 35.0 mg/g (0.5-3.5%) method B

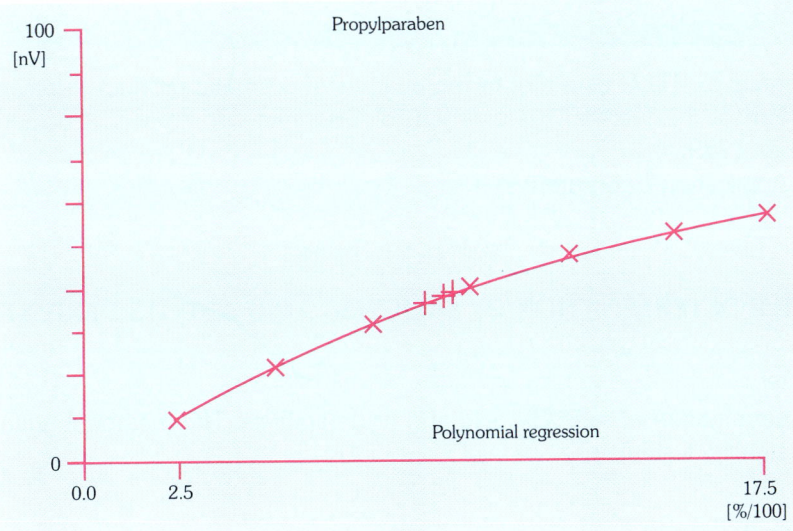

(h): Calibration function for propylparaben in the range from 2.5 to 17.5 mg/g (0.25 - 1.75%) method B

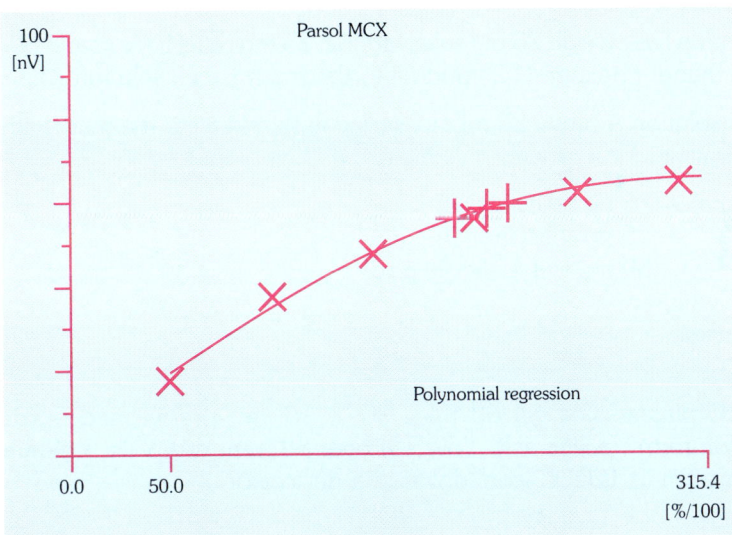

(i) Calibration function for Parsol MCX in the range from 50 to 350 mg/g (5.0-35.0%) method C

Fig. 17.24 c–i

Remark

Method precision: depending on kind of sample and component (2-6%)
Recovery: depending on kind of sample and component (87-100%)

References

1. Zimmermann G. (TLC as a tool in the analysis of cosmetic products) presented at the 8th Symposium of the German Association for Scientific and Applied Cosmetics, Hamburg, November 1989
2. Camag Application Laboratories

METHOD: 17.25

QUANTITATIVE DETERMINATION OF SALICYLIC ACID AND ITS DERIVATIVES

Aim of Analysis

Quantitative determination of salicyclic acid and its derivatives. The method is suitable for quality assurance and stability testing.

Chemicals and Reagents

Hexane, acetic acid, ethanol, cyclohexane, chloroform, salicylic acid, salicylamide, acetylsalicylic acid.

Standard Preparation

In a 100 ml measuring flask weigh 25 mg salicylamide, 25 mg salicylic acid and 50 mg acetylsalicylic acid; dissolve in ethanol : hexane (1:1) and fill to the mark (stock solution A).

Pipette 2 ml stock solution A into a 20 ml measuring flask and fill to the mark with ethanol: hexane (1:1) (solution B).

$1\mu l$ solution B contains 25 ng salicylic acid
25 ng salicylamide
50 ng acetylsalicylic acid

Sample Preparation
Stationary Phase

Precoated HPTLC plates silica gel 60 F_{254}, 20 × 10 cm, pre-washed by development with cyclohexane : chloroform : acetic acid (60:5:5) over 50 mm, dry plate 5 minutes with hair drier then 1 hour in an oven at 120° C; cool down in a desiccator.

Sample Application

Apply samples in narrow bands using a Linomat IV; 6 mm band length, distance between bands 4 mm (= track distance for scanner parameters 10 mm), distance from side edge 35 mm, 13 separation tracks, distance from bottom edge 8 mm, sample delivery speed 4 sec/μl.

Recommended Application Patterns

S1 U1 U2 S2 U1 U2 S3 U1 U2 S4 U1 U2 S5
S1   = 3 μl solution B

S2   = 4 μl solution B
S3   = 5 μl solution B
S4   = 6 μl solution B
S5   = 7 μl solution B
U1, U2 = unknown (for simulated unknowns apply 5 μl solution B)

Chromatography

In twin-trough chamber 20 × 10 cm, pre-saturated with solvent vapors; developing solvent cyclohexane : chloroform : acetic acid (60:5:5), separation distance 50 mm (running time about 11 min).

R_f's   :   Salicylamide 0.07
                    Salicylic acid 0.15
                    Acetylsalicylic acid 0.21

Densitometric Evaluation

Scanner with lab data station and CATS software. Deuterium lamp, scanning by absorbance at 200 nm, evaluation via peak height, polynominal regression.

Note 1: Salicylic acid, salicylamide and acetylsalicylic acid have absorption maxima at 200 nm and can thus be measured simultaneously at this wavelength. Salicylic acid and salicylamide can also be measured at 306 nm where both have a maximum. Acetylsalicylic acid shows practically no response at 306 nm.

Note 2: It has been observed that the absorbance response of salicylic acid at 200 nm is significantly reduced in the presence of nitrogen. No explanation could be found so far. However, in this case it is recommended to measure at 200 nm without flushing the scanner II with nitrogen.

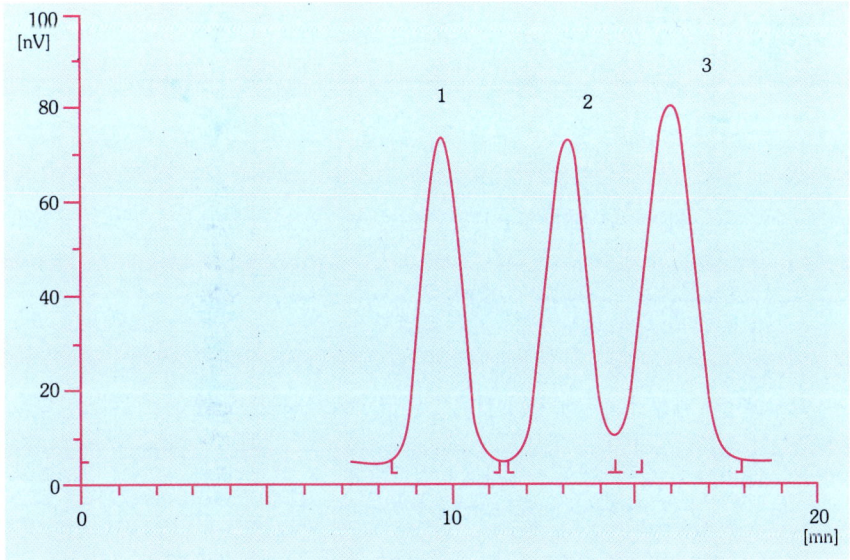

(a) Analog curve scanning by absorbance at 200 nm; (1) salicylamide, (2) acetylsalicylic acid, (3) salicylic acid

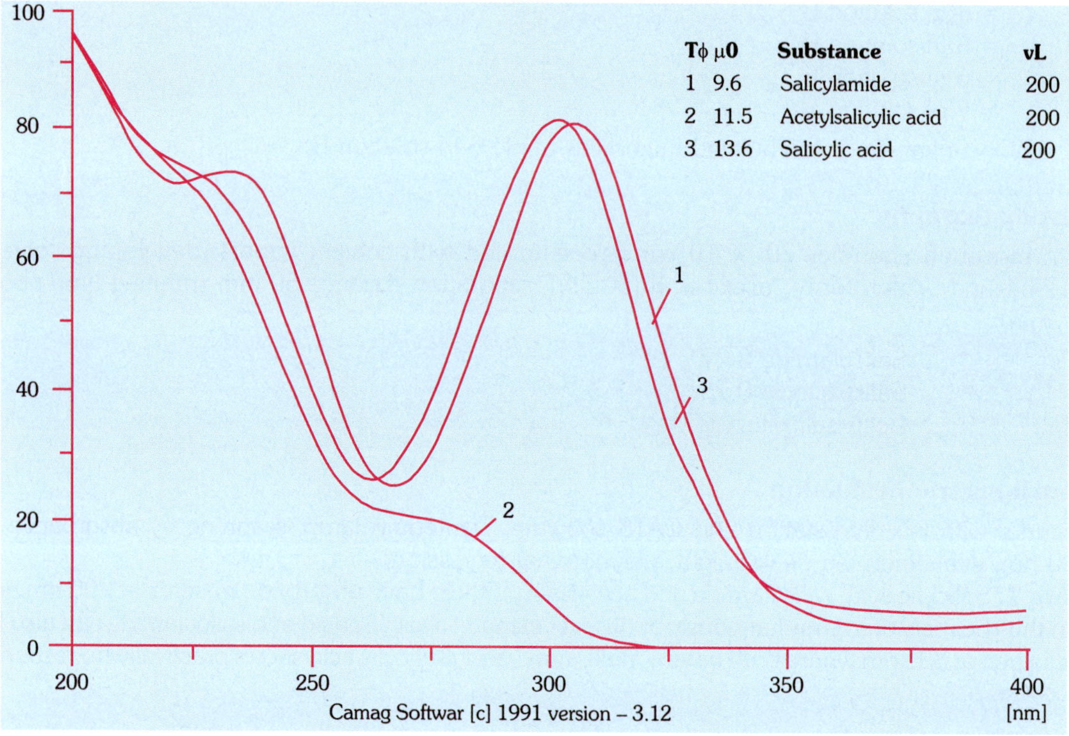

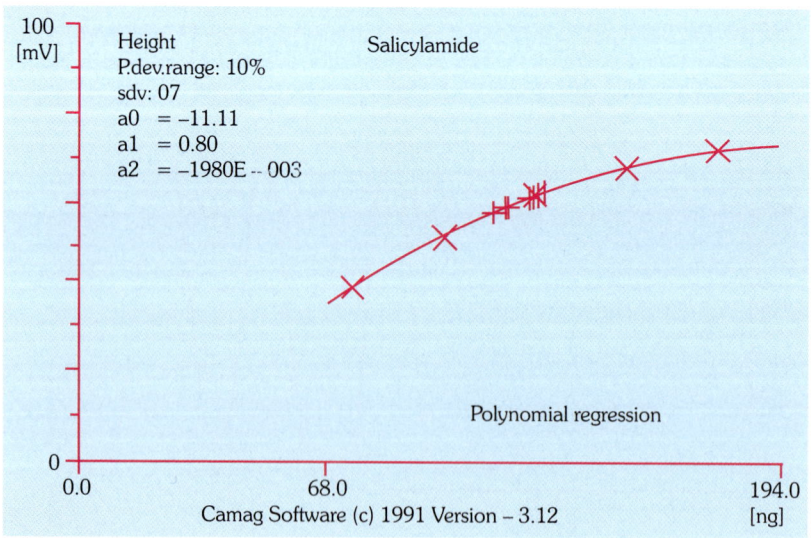

100
80
60
40
20
0
Tϕ μ0 Substance vL
1 9.6 Salicylamide 200
2 11.5 Acetylsalicylic acid 200
3 13.6 Salicylic acid 200
1
2
3
200
250
300
350
400
[nm]
Camag Softwar [c] 1991 version – 3.12

(b) Absorption spectra of (1) salicylamide, (2) acetylsalicylic acid, (3) salicylic acid

100
[mV]
0
Height
P.dev.range: 10%
sdv: 07
a0 = –11.11
a1 = 0.80
a2 = –1980E -- 003
Salicylamide
Polynomial regression
0.0
68.0
194.0
[ng]
Camag Software (c) 1991 Version – 3.12

(c) Calibration function of salicylamide in the range of 75 to 175 ng

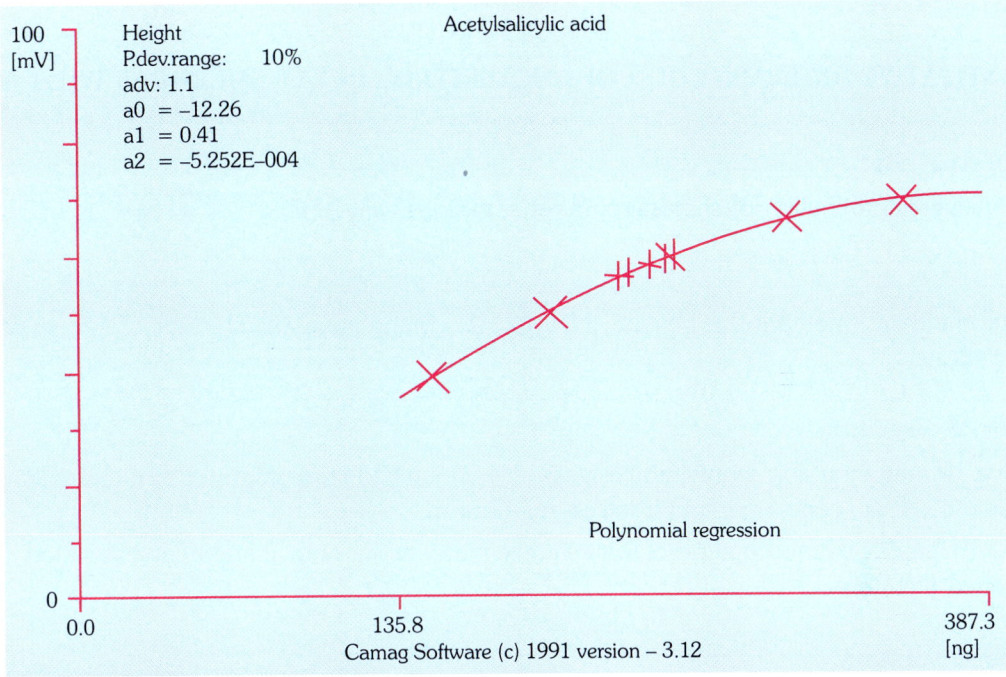

(d) Calibration function of acetylsalicylic acid in the range of 150 to 350 ng

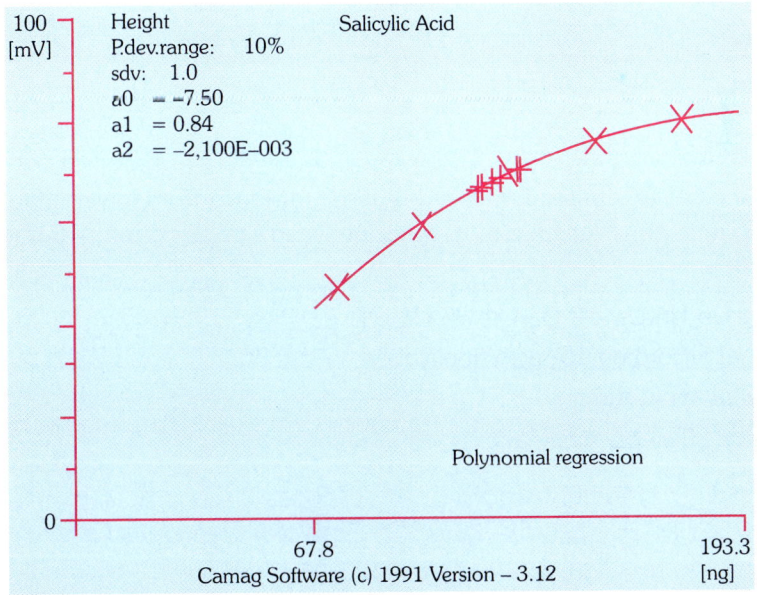

(e) Calibration function of salicylic acid in the range of 75 to 175 ng

Fig. 17.25 a–e

METHOD: 17.26

QUANTITATIVE DETERMINATION OF CHOLESTEROL IN FAT EMULSIONS WITH AMD

Aim of Analysis

Quantitative determination of cholesterol in fat emulsions with AMD.

Reagents

Dichloromethane, methanol, n-hexane, NH_3 (25%), manganese (II) chloride, conc. sulphuric acid, cholesterol.

Sample Preparation

Dissolve 100 mg emulsion (containing approx. 1% saponified cholesterol) in dichloromethane to a volume of 10 ml, centrifuge, and if necessary, decant.

For determination, extract cholesterol from sample material, saponify it and treat it as stated in the mentioned literature.

Standard Solution

One-standard calibration: Dissolve 15 mg cholesterol standard with dichloromethane to a volume of 100 ml.
1 μl contains 150 ng cholesterol.

Stationary Phase

HPTLC precoated plates silica gel 60 F, 20 × 10 cm.

Sample Application

Bandwise with Linomat IV, application speed 4 μl/sec, distance from lower edge 8 mm, distance from side edge 15 mm, band length 6 mm space between samples 4 mm, 16 tracks, according to the pattern below:

S1 U1 U2 U3 U4 U5 U6 U1 U2 U3 U4 U5 U6 S4
S = Standard; 2 μl according 15 mg/g/application
U = unknown; 2 μl/application
AMD gradient over 20 steps, schematic.

From step	1	2	6	11	16
Bottle nr.	1	2	3	4	5
Methanol	100	–	–	–	–
Dichloromethane	–	100	100	100	–
n-hexane	–	–	–	–	100
Drying time (min)	5	5	5	5	5
Wash bottle content	1N NH_3				

Post-chromatographic Derivatization

Reagent solution: dissolve 0.2 g manganese (II) chloride in 30 ml water, add 2 ml conc.sulphuric acid dropwise.

Derivatization by immersing the chromatogram (Chromatogram immersion device) into the above mentioned reagent solution for 1 sec, then heat 15 min at 120° C.

Densitometric Evaluation

TLC scanner II and suitable peripheral, measurement by absorbance at 546 nm, Hg-lamp, monochromator band width 10 nm, slit dimension 0.5 × 5 mm.

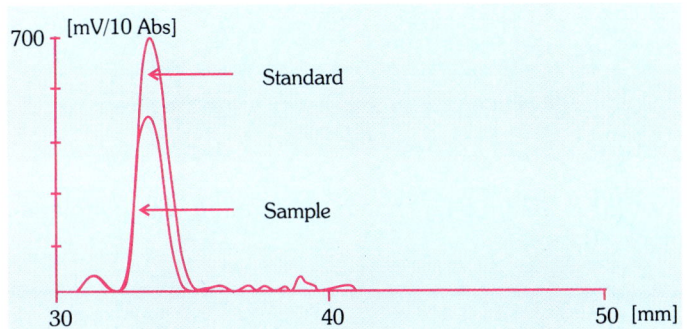

Fig. 17.26: Absorbance-scanning curves of a standard track and of an analysis track.

References

1   A.O.C.S. Official Method Ca, 6a – 40
2.   Camag Application Laboratories

METHOD: 17.27

DETERMINATION OF RAPESEED OIL IN FERMENTATION BROTH

Aim of Analysis

Determination of rapeseed oil in fermentation broth.

Chemicals

Petrol ether, hexane, methanol, manganese chloride, ethanol, sulfuric acid, dichloromethane, perchloric acid 70%, diethyl ether, paraffin oil.

Sample Preparation

Shake 10 g fermentation broth for 20 min with 50–100 ml dichloromethane in a stoppered Erlenmeyer flask 250 times/min. Transfer the lower phase by means of a pipette and press through a Ministart RS25–cartridge.

Standard Solutions

Dissolve 60 mg rapeseed oil in 10 ml dichlormethane and dilute 1:10 (= 0.6 $\mu g/\mu l$).

Stationary Phase

HPTLC precoated plate silica gel Merck 60 F_{254}, 20 × 10 cm, prewashed with methanol.

Sample Application

Apply bandwise with automatic TLC sampler 4, band length 6 mm, distance from lower edge 8 mm, distance from left edge 15 mm, delivery speed 10 $\mu l/s$ = 20 applications.

Application pattern:

S1	U1	U2	S2	U3	U4	S3	U5	U6	S4	
1	1	1	3	1	1	5	1	1	7	μl/band

S1–S4 = standards in different concentrations, U1–U6 = sample 1–6

Chromatography

In twin-trough chamber, 20 × 10 cm with chamber saturation with petroleum ether–diethyl ether (5:1), migration distance 30 mm.

Derivatization

By dipping for 1 s with Camag chromatogram immersion device in water–methanol–perchloric acid (80:20:3) followed by heating for 8 min at 120° C on a Camag plate heater (The reagent is stable for one week).

The intensity of fluorescence could be increased by a factor of 2 by dipping into a mixture of paraffin oil: hexane (1:2).

Densitometric Evaluation

TLC scanner with CATS software, scanning by absorbance at 366/>400 nm, evaluation via peak area by polynomial regression.

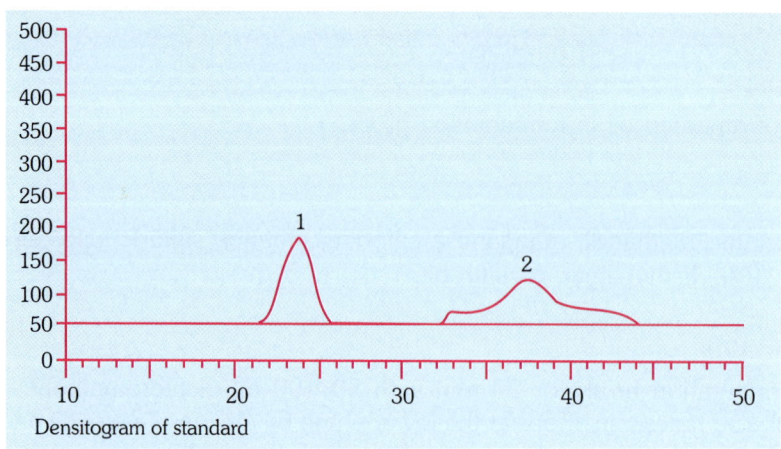

Densitogram of standard

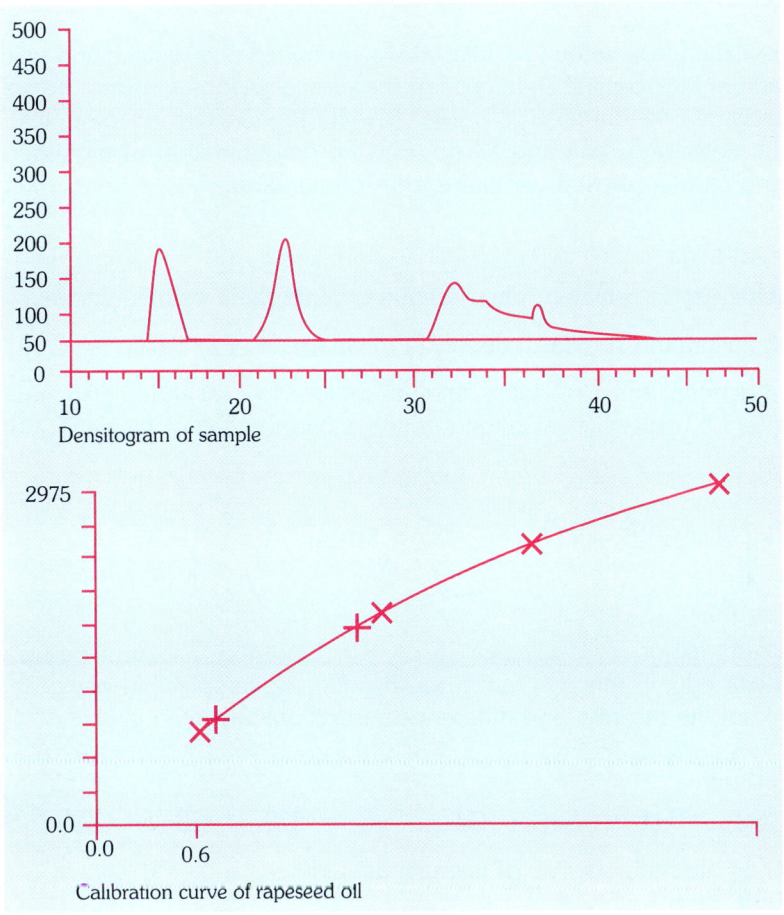

Densitogram of sample

Calibration curve of rapeseed oil

Fig. 17.27

METHOD: 17.28

QUANTITATIVE DETERMINATION OF AMINO ACIDS IN POTATOES

Aim of Analysis

Quantitative determination of amino acids in potatoes.

Reagents

Ethanol, dist. water, acetone, n–butanol, diethylether, dansylchloride, 75 mg in 30 ml acetone, sodium bicarbonate, 1% in water, sodium hydroxide 1 N, ethylenediaminetetraacetate–disodium salt (Titriplex III), paraffin : n–hexane (2:5).

Sample Preparation

Add 10 ml ethanol to 3 ml potato juice at room temperature. Filter the precipitated protein. Evaporate the light brown solution to dryness at 40° C and dissolve residue in 3 ml water.

Dansylation

Add 0.5 ml dansylchloride solution to 1 ml of the conditioned potato juice and adjust to pH 8 with the prepared sodium bicarbonate solution. For the derivatization–reaction, the solution is placed in the dark at room temperature for 16 hours. Transfer the reaction–solution with acetone: water (7:3) in a 10 ml measuring flask and fill up. For the determination of arginine, this solution is diluted (1:10) and can be applied without further preparation.

Calibration Standards

Arginine, threonine, glycin, alanine, phenylalanine, tryptophan, valine, leucine.

1 mg/ml of each calibration standard, dissolved in water.

Dansylation as described, add 1 ml dansylchloride solution to 1 ml amino acid and fill up to 100 ml with acetone–water (7:3) after derivatization reaction. Concentration of standard solution is 10 ng/μl.

Stationary Phase

Precoated HPTLC plated silica gel 60 F_{254}, 20 × 10 cm.

Sample Application

Bandwise with automation TLC sampler III or with Linomat IV; 8 mm bands, 12 mm apart, distance from lower edge 8 mm, 30 mm from the side, 12 samples per plate; delivery speed 15 sec/μl. Apply 5 μl of the samples and different standard volumes.

Application Pattern

S1   U1   S2   U1   S3   U1   S4   U2   S5   U2   S6   U2

S1 = 1 μl = 10 ng absolute U1 = 5 μl (sample diluted 1:10)
S2 = 2 μl = 20ng U2 = 5 μl
S3 = 3 μl = 30ng
S4 = 4 μl = 40ng
S5 = 5 μl = 50ng
S6 = 6 μl = 60ng
S = standard U = Sample

Chromatography

Twofold isocratic development with ethylendiaminetetraacetate disodium salt (Titriplex III): butanol: diethylether (1:2:7) in twin-trough chamber 20 × 10 cm; migration distance 80 mm.

Preparation of Developing Solvent

Weigh 5 g ethylendiaminetetraacetatedisodium salt (Titriplex III) into 100 ml measuring flask, add 50 ml dist. water; adjust to pH 9 with NaOH 1N (solution becomes clear); add 10 ml butanol, vortex, add 35 ml diethylether, vortex and use upper phase as developing solvent.

To increase the detection sensitivity, place the developed and dried plate in paraffin: n–hexane (2:5); dry with a hair dryer. This procedure enhances the measuring sensitivity by a factor of 3 to 4 compared to an untreated plate.

Densitometric Evaluation

This is done with TLC scanner II with lab data system and CATS evaluation software. Scanning by fluorescence at 313/>460 nm with mercury lamp, monochromator band width 30 nm, slit dimension 0.3×4 mm. Quantitative evaluation is done with linear regression via peak height.

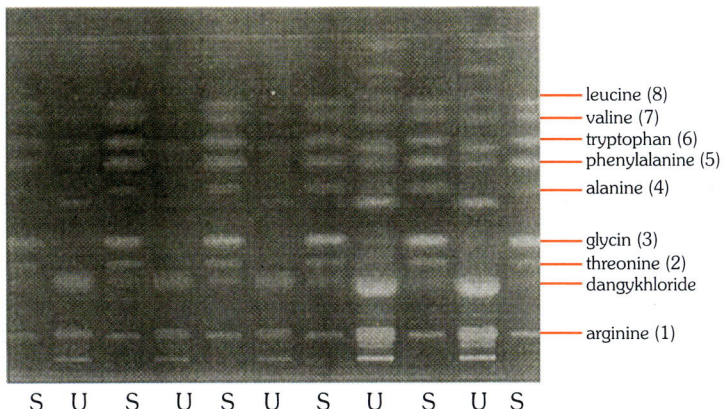

(a) Chromatogram of a potato juice sample photographed under 366 nm UV

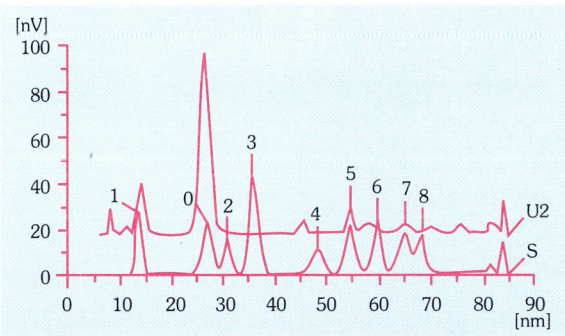

(b) Densitogram curve of a potato juice sample (U1) and a standard (S)

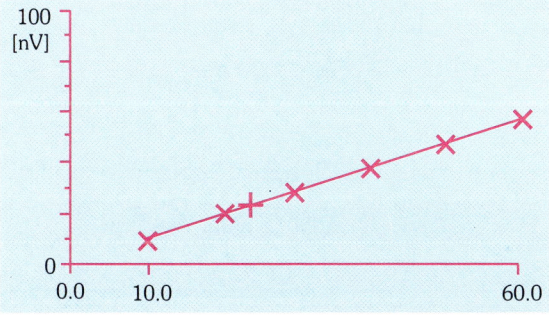

(c) Calibration curve of argining (peak height) in the range of 10 to 60 ng

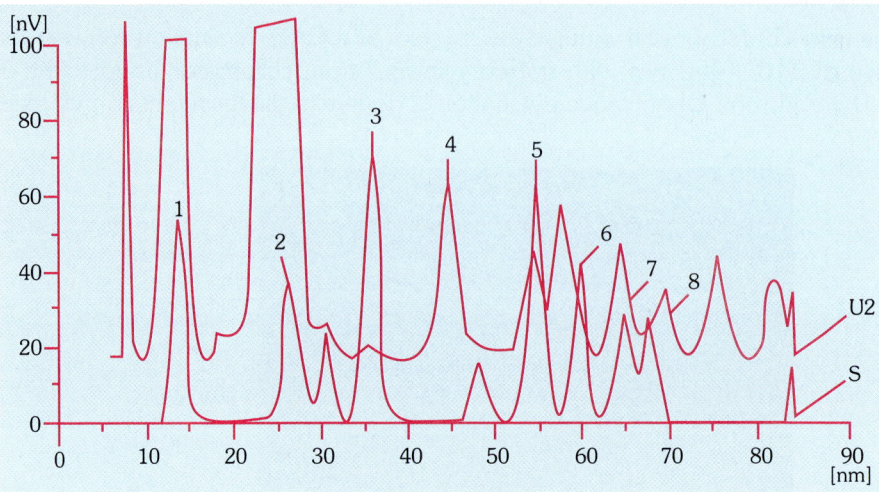

(d) Densitogram curve of potato juice sample (U2) and a standard (S)

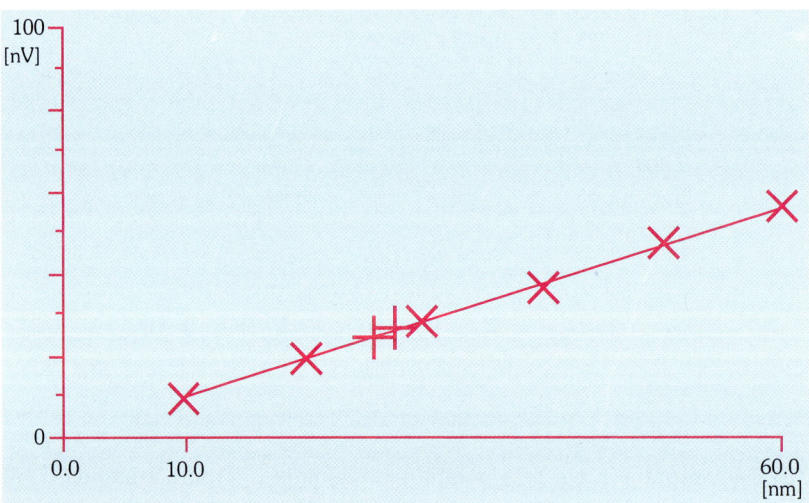

(e) Calibration curve of phenylalanine in the range of 10 to 60 ng: linear regression via peak height.

Fig. 17.28 a–e

METHOD 17.29

DETERMINATION OF PESTICIDES IN DRINKING WATER

Aim of Analysis

Determination of pesticides in drinking water by AMD.

Reagents

All solvents used as mobile phases, conditioning liquids, extraction liquids and for prewashing of the TLC plates must be free of nonvolatile, especially UV absorbing residues. Only tested lots should be used for the analysis.

Method for Testing Solvents

- 10 ml if a lot of each solvent, blow down to dryness in a stream of nitrogen.
- Take up residues separately in 100 μl of solvent and apply aliquots corresponding to 5 ml of the original solvent to a TLC plate.
- Chromatograph the TLC plate by AMD gradient A.
- Scan the plate by multi-wavelength scanning as described.
- The scan of the evaporation residue of a suitable solvent should not contain a significant peak besides the chromatographic blind

For the experiments described in this procedure, the following solvents and reagents were used.

Solvent/Reagents

Acetonitrile, ammonia, dichloromethane, formic acid, hydrochloric acid, isopropanol, methanol, n–heptane, n–hexane, tert–butyl methyl ether.

Material for Solid Phase Extraction

Bakerbond octadecyl (C18), 40 μm
Glass cartridge
PTFE frits

Calibration Standards

2, 4-D, metazachlor, pendimethalin, atrazine, metobromuron, propazin, benzanilde, metoxuron, simazine, chlortoluron, cyanazine, desethylatrazine, monuron, phenmedipham, terbutylazine, vinclozolin.

Other substances can be chosen as standards, if these have to be analyzed.

Sample Preparation

Solid phase extraction (SPE) is carried out using glass cartridges equipped with PTEE frits and filled with of C18 material; use glass tubes and glass connections only. For enforcing the appropriate through flow, arrange a peristaltic pump behind the outlet of the cartridge. A simple set-up is diagrammatically depicted in Fig. 17.29 (a)

- Condition the C-18 cartridge with 3 ml n-hexane, then 3 ml dichloromethane and finally 3 ml methanol.
- Acidify 1000 ml of the water sample with hydrochloric acid to pH 2.
- Connect the column entry to the 1000 ml water sample. Adjust a flow rate of 6 ml per minute through for two hours; pass the nitrogen through activated carbon before it enters the cartridge.
- Elute the cartridge with 2 ml dichloromethane, then with 3 ml methanol.
- Evaporate the combined elutes to dryness in a stream of clean nitrogen at 35° C.

- Dissolve the residue in 250 μl acetonitrile: n-hexane (95:5) and add 280 ng Benzanilid as internal standards.
- For obtaining the blank solution, elute a conditioned and dried cartridge through which no water was percolated and treat the residue in the same was as described for the unknowns.

Identification/Calibration Standards

For identification and calibration, 15 standards are combined in three calibration mixtures. Each mixture contains 1 ng of each individual pesticide per μl acetonitrile.

Mixture 1	Mixture 2	Mixture 3
Desethylatrazine	Metoxuron	2, 4–D
Simazine	Monuron	Atrazine
Cyanazine	Chlortoluron	Metazachlor
Terbutylazine	Propazin	Pendimethalin
Vinclozolin	Phenmedipham	Benzanilid (int. Standard)
	Metobromuron	

Extracts of unknowns, together with all three identification calibration standards, are chromatographed on two plates with two different AMD gradients (gradient A and B). Gradient A is considered as a screening run, and gradient B as confirmatory with regard to qualitative identification. However, either chromatogram confirms the result of the other. Quantification should be done with only one of the two, where separation from extraneous peaks is best. In case several pesticides are found in the unknowns, quantification of the individuals can be taken from different plates.

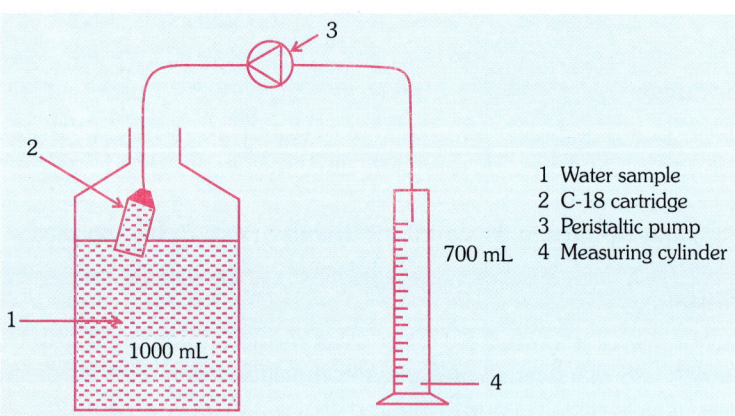

Fig. 17.29 (a): Set-up for solid phase extraction

Stationary Phase

Precoated HPTLC plates silica gel 60 F_{254}, 100 μm, 20 × 10 cm.

Immerse the plates in isopropanol for at least one hour. Then dry at 120° C for an hour.

Protect the prewashed plates suitably against contamination from the atmosphere during cooling, storage before use, and sample application.

Storage of Plates after Drying

For trace analysis in a heavily contaminated environment, place prewashed plates in a desiccator. Fill up the desiccator with nitrogen which is passed through activated carbon. Add nitrogen to pressure equilibration when the plates are cooled down to room temperature.

Before plate is removed for use, cover its layer with a clean glass plate for protection.

Sample application in a contaminated environment

For trace analysis, sample application and also AMD chromatography under a clean bench should be done. Otherwise, sample application with Linomat IV or Linomat IV–Y can be done with the inert gas blanket option.

Sample Application

Apply samples bandwise with a Linomat, preferably with a Linomat IV–Y: bands 7 mm long, 3 mm apart, distance from lower edge 8 mm, 25 mm from the side, 16 samples per plate; delivery speed should be 10 s/μl.

With a Linomat IV–Y, adjust the Y–drive at 6 mm so that rectangular sample zones 7 $\times$ 6 mm are obtained. If no Linomat IV–Y is available, apply with a regular Linomat each sample in three parallel bands, 2 mm apart in Y–direction.

Recommended application pattern for quantitative analysis:

B1	S1	S2	S3	U1	S4	S5	S6	B2	S7	S8	S9	U2	...
B		M-1		S		M-2		B		M-3		S	...

Note: B–blank; M–mixture; S–sample

Apply 20, 50 and 100 μl of each standard mixture corresponding to 20, 50 and 100 ng per pesticide and 100 μl of the unknowns (corresponding to 250 ml of the original water sample) and 100 μl of the blank on two plates.

For large numbers of samples, it is suggested to use only the 100 μl amounts of the identification calibration mixtures in the first (screening) run. On the second plate, which is used for confirmation and quantification, a complete set of three calibration standards of the mixture(s) corresponding to the suspected positives should be run.

Chromatography

Chromatographic development is performed in the AMD system using two different universal gradients.

Gradient A is most often used in a screening function. It is based on dichloromethane as the central solvent governing the selectivity. Gradient A is diagrammatically depicted in Fig. 17.29 (b). Development starts with 10 short isocratic runs with acetonitrile: dichloromethane (30:70) over 10 mm, thus extracting the active ingredients from the sample application area (away from humic acids). During steps 16 through 28, the mobile phase is acidified with small amounts of formic acid by interaction with the weakly alkaline buffer layer. This establishes a pH gradient from weakly alkaline to acidic conditions. This is important for proper chromatography of all types of pesticides in the presence of humic acids. These are prevented from moving away from the starting area under the existing conditions.

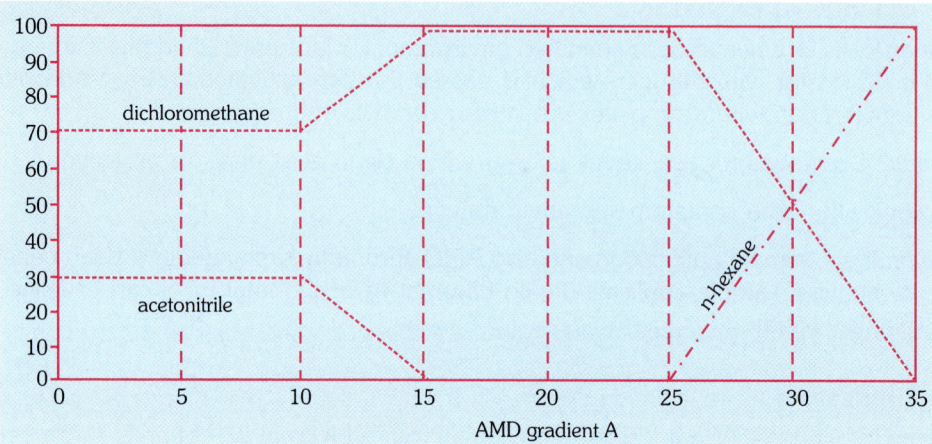

Fig. 17.29 (b): AMD gradient A

The second plate is developed with AMD gradient B, which is used as a confirmatory function. This gradient is based on tert–butyl methyl ether and exhibits a distinctly different selectivity compared to the dichloromethane based gradient A [(Fig. 17.29 (c)].

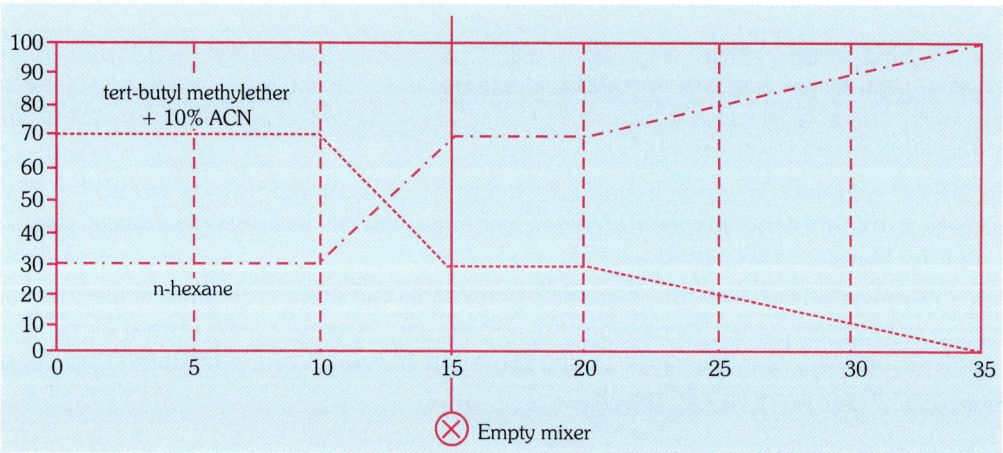

Fig. 17.29 (c): AMD gradient B

Similar to gradient A, this gradient starts with 10 isocratic steps with tert–butyl methyl ether containing 10% acetonitrile: hexane (70:30) slightly alkaline, for fixing humic acids in the starting area and extracting the active ingredients. Formation and function of the pH–gradient is similar to that of gradient A.

Densitometric Evaluation

With TLC scanner II with lab data system and CATS evaluation software. Multi-wavelength scanning is done by absorbance at 190, 200, 220, 240, 260, 280, 300 nm with deuterium lamp, monochromator bandwidth 10 nm, slit dimensions 0.2 × 3 mm, scanning speed 4 mm/sec, sens

0, span 7. In order to obtain reliable results at wavelengths 190 and 200 nm, it is proper to flush the monochromator with nitrogen (about 0.5 L/min).

The substances are evaluated with linear regression via peak height at their optimum wavelength. Benzanilide is used as internal standard for the correction of the sample volume applied to the TLC plate.

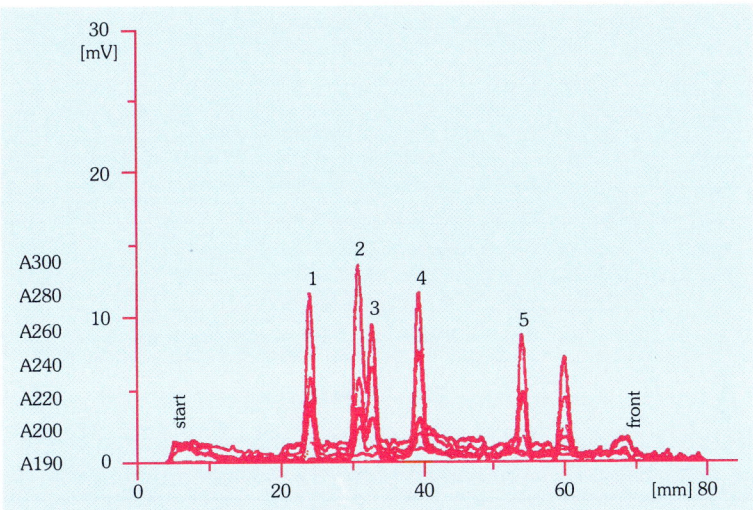

Fig. 17.29 (d): Typical chromatograms of identification/calibration mixtures

Calibration mixture 1, gradient A, 50 ng each component; 1 = desethylatrazine, 2 = simazine, 3 = cyanazine, 4 = terbutylazine, 5 = vinclozolin

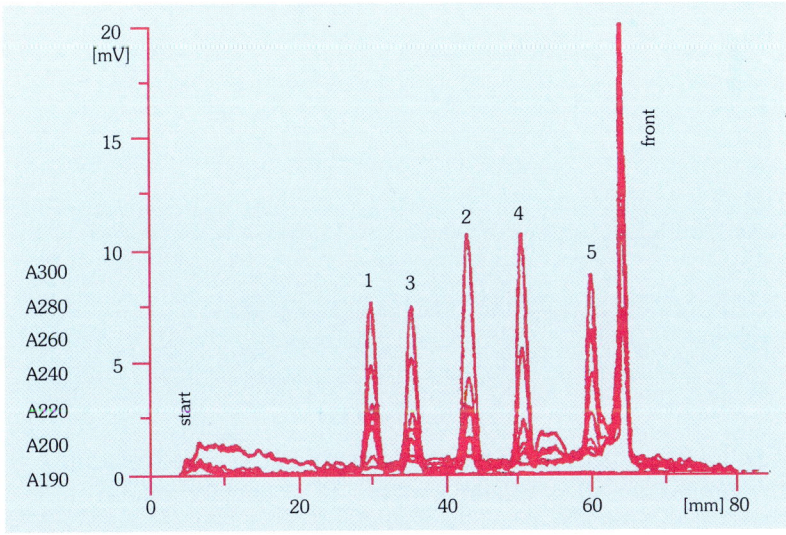

Fig. 17.29 (e)

Calibration mixture 1, gradient B, 50 ng each component; 1 = desethylatrazine, 2 = simazine, 3 = cyanazine, 4 = terbutylazine, 5 = vinclozolin

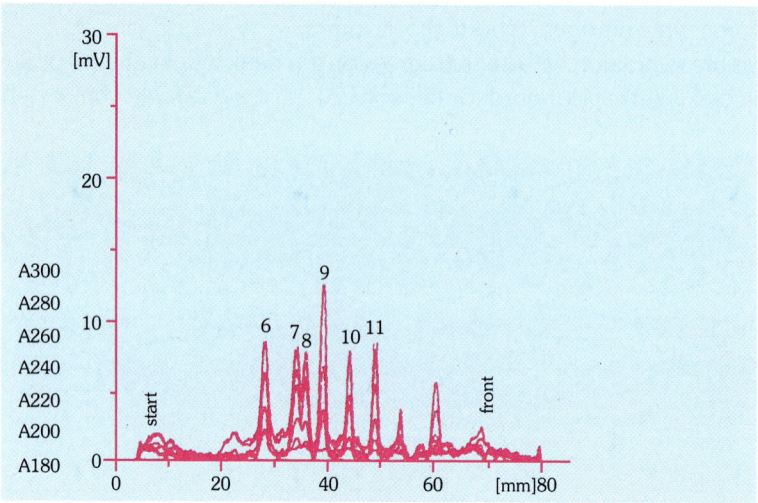

Fig. 17.29 (f)

Calibration mixture 2, gradient A, 50 ng each component; 6 = metoxuron, 7 = monuron, 8 = chlortoluron, 9 = propazine, 10 = phenmediphan, 11 = metabromuron

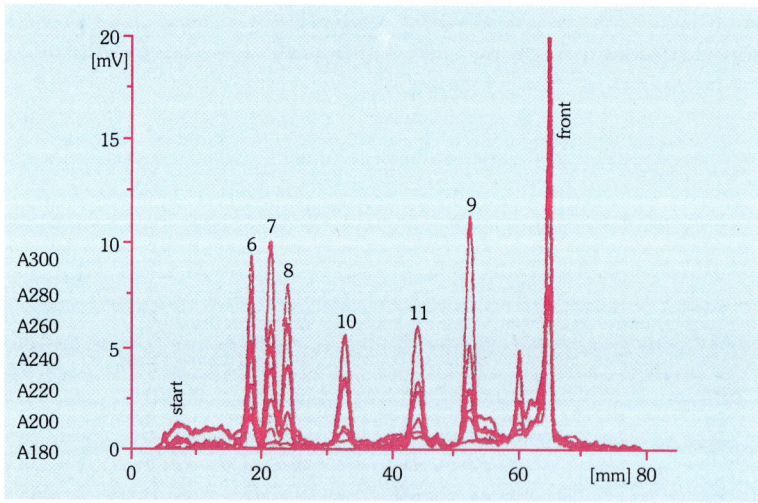

Fig. 17.29 (g)

Calibration mixture 2, gradient B, 50 ng each component; 6 = metoxuron, 7 = monuron, 8 = chlortoluron, 9 = propazine, 10 = phenmediphan, 11 = metabromuron

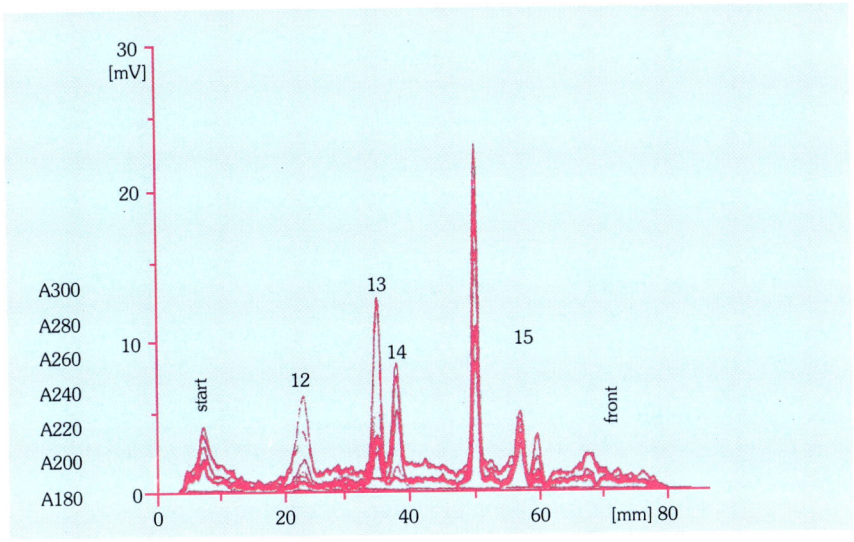

Fig. 17.29 (h)

Calibration mixture 3, gradient A, 50 ng each component; 12 = 2, 4–D, 13 = atrazine, 14 = metazachlor, i–St = benzanilid, 15 = pendimethalin

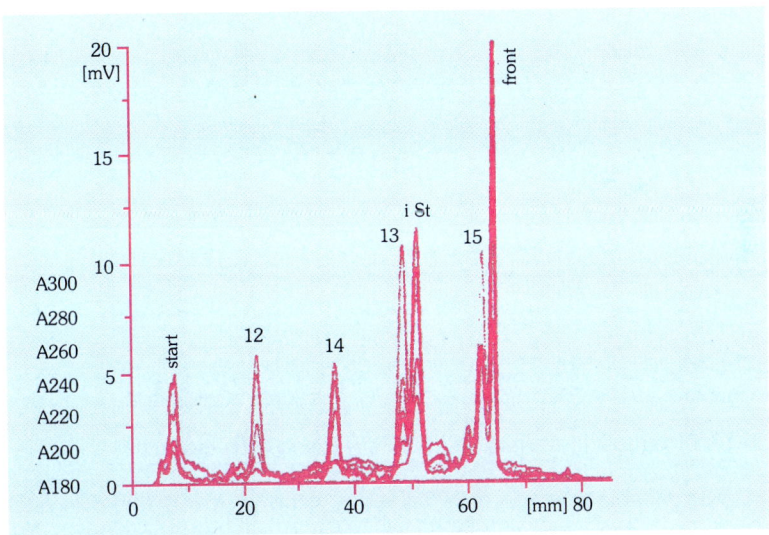

Fig. 17.29 (i)

Calibration mixture 3, gradient B, 50 ng each component; 12 = 2, 4-D, 13 = atrazine, 14 = metazachlor, i–St = benzanilid, 15 = pendimethalin

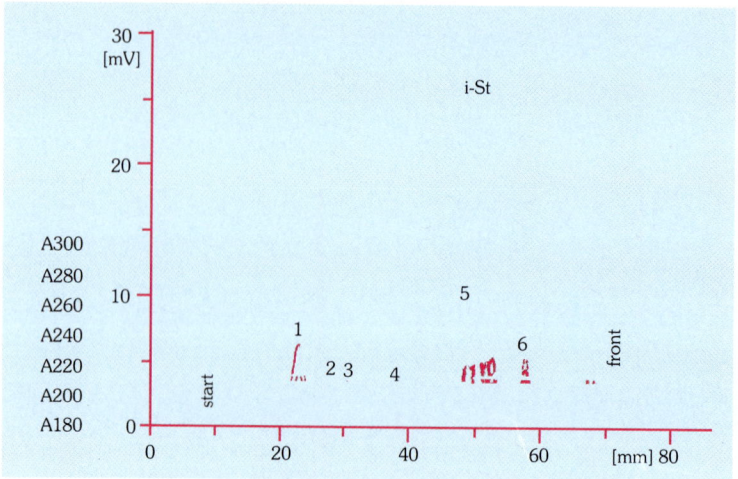

Fig. 17.29 (j): Chromatograms of a water sample

Fortified water sample, gradient A; 1 = 2, 4–D and interfering component X, 2 = metoxuron (0.08 μg/l), 3 = simazine (0.05 μg/l), 4 = metazachlor (0.06 μg/l), 5 = metobromuron (0.18 μg/l), i–St = benzanilid, 6 = pendimethalin (0.11 μg/l)

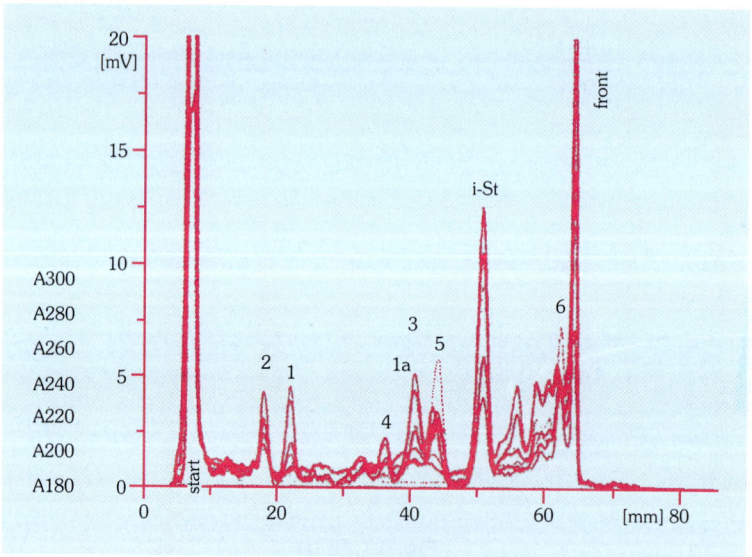

Fig. 17.29 (k)

Fortified water sample, gradient B; 1 = 2, 4–D (0, 13 μg/l), 1a = interfering component X, 2 = metoxuron (0.08 μg/l), 3 = simazine (0.06 μg/l), 4 = metazachlor (0.06 μg/l), 5 = metobromuron (0.18 μg/L), i–St = benzanilid, 6 = pendimethalin (0.10 μg/L)

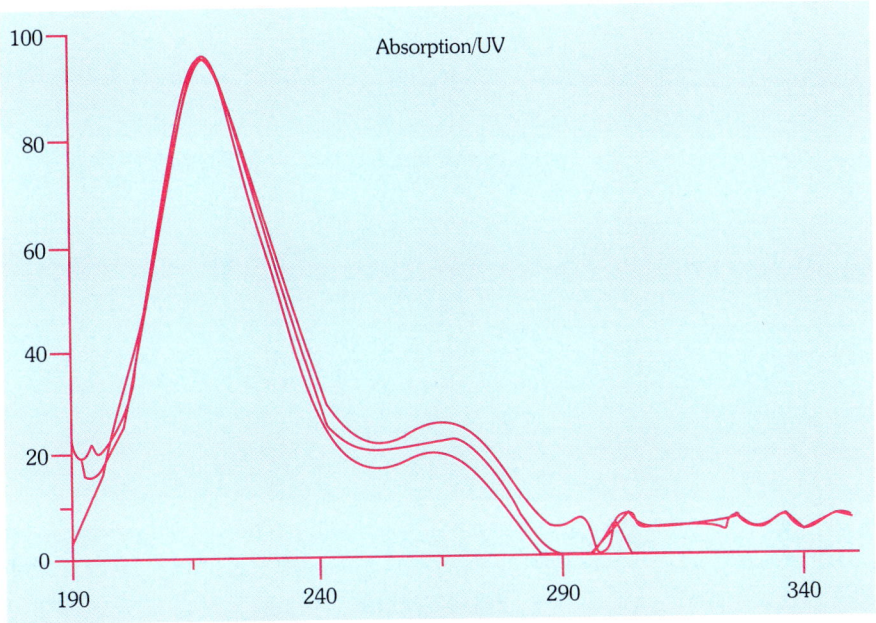

Fig. 17.29 (l): Spectra

In situ UV spectra of 20, 50 and 100 ng atrazine plotted together with the spectrum of the component in the sample, corresponding to 0.08 μg atrazine per liter. All spectra are in an acceptable accordance.

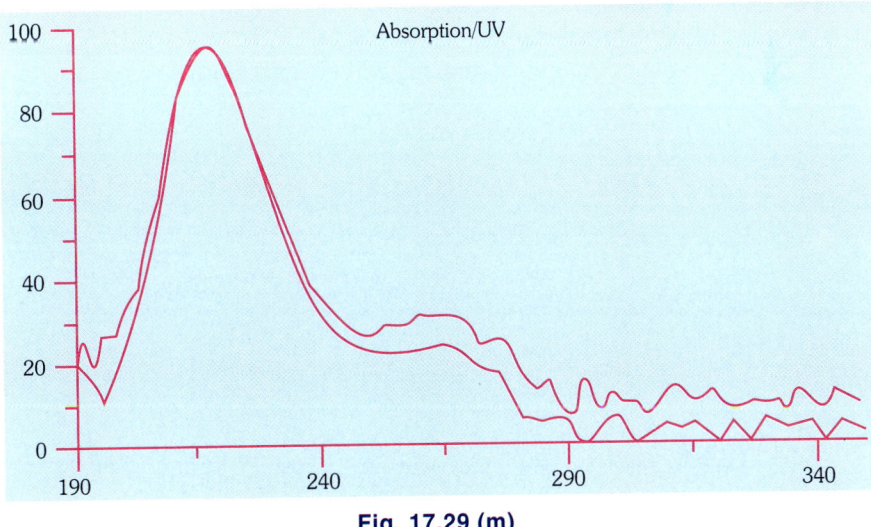

Fig. 17.29 (m)

In situ UV spectrum corresponding to 0.05 μg simazin per liter, plotted together with the spectrum of 20 ng simazin (calibration substance).

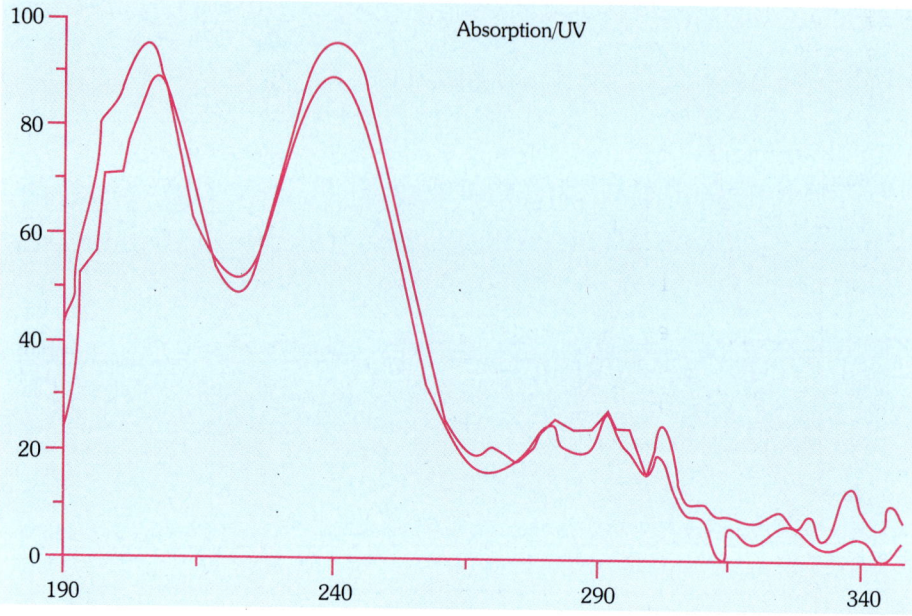

Fig. 17.29 (n)

In situ UV spectrum corresponding to 0.08 μg metoxuron per liter, plotted together with the UV spectrum of 50 ng metoxuron (calibration substance) in the range of low absorbance (>260 nm)

Recovery

The recoveries for pesticides with positive results in the samples of the conformity test are stated below. A drinking water sample was fortified with 0.2 μg of the individual pesticide, extracted and analyzed. The average of all recoveries was 80.1% ± 13.0%.
Recovery of the pesticides (% compared to the spiked concentration)

Pesticide	Recoveries observed						Mean recoveries	CV(%)
Metoxuron	102	81	92	79	79	74	84.5	12.4
2, 4–D	72	73	72	99	91	72	79.8	15.1
Metazachlor	88	71	79	68	97	88	81.8	13.6
Simazine	84	76	90	78	92	78	83.0	8.2
Metobromuron	96	77	94	76	87	70	83.3	12.7
Pendimethalin	73	82	62	65	81	96	76.5	16.4
Desethylatrazine	74	63	80	70	78	67	72.0	9.1
Atrazine	88	66	80	65	100	82	80.2	16.6

Results

Fig. 17.29 (d) shows the distinctly different separation of the three calibration mixtures in gradient A and B. Not only the distance of migration for the same pesticide is different in both gradients, even the sequence of elution is often changed.

Combining the results from both gradients is the way to distinguish between the several hundreds of pesticides on the market.

Fig. 17.29 (j and k) show the chromatograms of a fortified sample of drinking water in gradients A and B respectively. Component 1 in gradient A (Fig. 17.29j) has the same distance of migration as 2, 4–D, but the behaviour of this component in UV multi-detection differs from 2, 4–D: the peak height at 240 nm is relatively too high. In the chromatogram of the same sample in gradient B (Fig. 17.29k) component 1 is UV pure 2, 4–D and the interfering component X (1a) is separated from 2, 4–D more than half the length of the chromatogram. All other positive results from Fig. 17.29j are confirmed in Fig. 17.29k qualitatively and quantitatively.

All concentrations of pesticides stated in the legends of the chromatograms of the samples are uncorrected for recovery. In addition, confirmation of positive results is done by taking insitu UV spectra of standards and components in the unknown.

For this, some conditions are important:

1. The amount of substance in sample and calibration spot should be in the same order of magnitude; the smaller the difference of masses the better.
2. Reference spectra should be measured for each individual UV spectrum as closely as possible to the substance spot, to eliminate matrix influences.
3. Due to possible matrix interference, taking the spectra of the unknown and the identification standard from the same plate, is preferred over comparison using the spectrum library.

Operational data, recommendations, practical hints

Table 17.29.1: Program of running times per step–gradients A and B

Step	min	Step	Step min	Step min	Step min	Step min min
1–10	0.8	11	16	21	26	31
		1.0	2.9	6.0	9.6	13.9
		12	17	22	27	32
		1.3	3.4	6.6	10.5	14.8
		13	18	23	28	33
		1.6	4.0	7.3	11.2	15.6
		14	19	24	29	34
		2.0	4.6	8.1	12.1	16.6
		15	20	25	30	35
		2.5	5.3	8.7	13.0	17.5

Table 17.29.2

Step	1–10	11–15	16–20	21–25	26–28	29–35
Bottle no.	1	2	3	4	5	6
Acetonitrile	30					
Dichloromethane	70	100	100	100	50	
n-hexane					50	100
HCOOH conc.			0.1	0.1	0.1	
Washbottle: empty	10					
Drying time (min)*	1.5	1.5	1.5	1.5	1.5	1.5

- The (very short) drying time of 1.5 minutes requires that vacuum pump, all vacuum connections and the valves are in perfect condition. Selecting a drying time of 3.0 minutes would make the AMD procedure less critical, however, extends the total time by 2 hours.

Table 17.29.3 Program of AMD gradient B						
Step	1–10	11–15	16–20	21–25	26–28	29–35
Bottle no.	1	2	3	4	5	6
Tert–butylmethylether + 10% CAN	70	30	30	20	10	
n–hexane	30	70	70	80	90	100
HCOOH conc.			0.1	0.3	0.3	
Ammonia 25%	0.1					
wash bottle: empty		15				
Drying time (min)*	1.5	1.5	1.5	1.5	1.5	1.5

References

1. Burger K, Jork H, Kohler J. Application of AMD to the Determination of Crop Protection Agents in Drinking Water; Part 1: Fundamentals and Method; Journal of Planar Chromatography **3**; 504–510, 1990
2. Jork H, Kohler J, Kocher U. Application of AMD to the Determination of Crop Protecting Agents in Drinking Water; Part 2: Limitations; Journal of Planar Chromatography **5**; 246–250, 1992
3. Pfaab G, Jork H. Application of AMD to the Determination of Crop Protection Agents in Drinking Water; Part 3: Solid Phase Extraction and Affecting Factors; Acta hydrochimica et hydrobiologica, **22**; 5, 216–223, 1994
4. Burger K, Jork H, Kohler J. Application of AMD to the Determination of Crop Protection Agents in Drinking Water; part 4: Fundamentals of a confirmatory Test; Acta hydrochimica et hydrobiologica **24**; 1.6–15, 1996
5. Butz S, Stan H–J. Screening of 265 Pesticides in Water by Thin–Layer Chromatography with AMD; Anal. Chem., **67**; 220–230, 1995
6. Camag Application Labs and literature

METHOD: 17.30

DETERMINATION OF PHOSPHOLIPIDS IN FOODS

Aim of Analysis

Determination of phospholipids in foods.

Reagents

Chloroform, methanol, dichloromethane, isopropanol, n–hexane, sulfuric acid (conc.), ammonia solution 25%, manganese chloride $4H_2O$.

Standards

Sphingomyelin, triolein, cholestrol, phosphatidyl inositol, cardiolipin, cholesteryl pentadecanol, phosphatidylcholine.

Sample Preparation from Emulsifiers used in Foodstuffs

Dissolve 25 mg of the oleaginous emulsifier in 25 ml chloroform: methanol: water (70:30:5). Filter into a 50 ml measuring flask (turbidity). Rinse with chloroform and fill up 50 ml with chloroform.

Sample Preparation from Biological Tissue

Add 5 ml de-ionized water to 1 g biological matrix in a 10 ml dispersing vessel and homogenize for 1 min in an ice bath at ca. 4° C. Add 20 ml chloroform: methanol to 1 ml of the homogenized sample. Shake for 15 min and filter. Rinse dispersing vessel and filter first with 15 ml chloroform: methanol (2:1) and then with chloroform: methanol (1:2). Evaporate the collected filtrates under nitrogen. For the preparation of the lipophilic fraction, extract the residue with 1 ml chloroform, and decant (fraction A). To collect the hydrophilic components, extract afterwards the residue with 1 ml chloroform: methanol: water (4:10:2) (fraction B).

Calibration Standards

Dissolve 10 ml of each standard to 10 ml with chloroform. Pipette 1 ml of each solution into a 10 ml volumetric flask and fill up with chloroform (100 ng/μl).

Stationary Phase

HPTLC precoated plates silica gel 60F$_{254}$ "extra pure", 20 × 10 cm. The plate is prewashed by immersing it for one hour in isopropanol; dry it for 30 min at 120° C.

Sample Application

Bandwise with Linomat, band length 6 mm, 4 mm apart, 15 mm from the side, 17 samples per plate, distance from lower edge 7 mm, delivery speed 4 sec/μl. 6 μl samples of emulsifiers, calibration standards as required are applied.

Apply 5μl of fraction A and overspray with 5μl of fraction B one upon the other. For differentiation, A and B is applied separately.

Chromatography

Automated multiple development with 25 step universal gradient, based on dichloromethane, according to diagram; alkaline conditioning; 3 mm step increments; drying time between steps 6 min.

From step n =	1	2	6	11	16	21
Bottle no.	1	2	3	4	5	6
Methanol	100	70	30			
Dichloromethane		30	70	100	50	
n–hexane					50	100
Drying time (min.)	6					
Wash bottle content	NH$_3$1N					

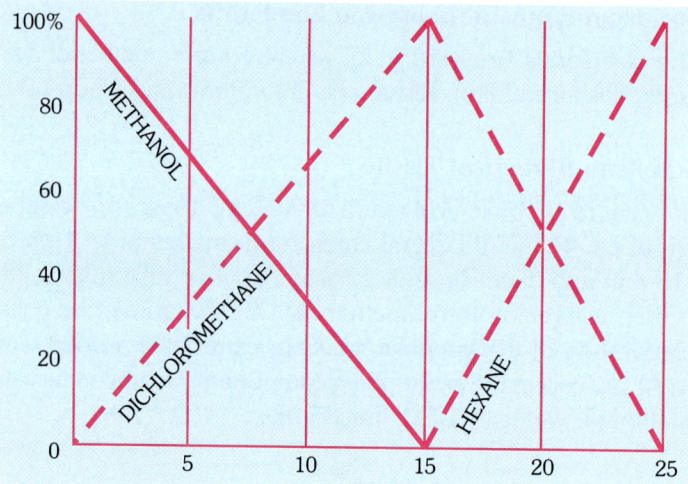

Fig. 17.30 (a): 25 step universal AMD gradient, schematic

Post-chromatographic Derivatization

Immersing solution: Dissolve 0.4 g manganese chloride. 4 H_2O in 60 ml water, add 60 ml methanol, then 4 ml conc. sulfuric acid. Allow the solution to cool to room temperature. Immerse the plate for 5 sec, then heat 20 min at 120° C.

Note: Phospholipid fractions of about ≥ 100 ng appear as brown zones. Under 366 nm UV these zones fluoresce, enabling determination limits of 10 ng could be achieved. Since in this particular case, fluorescence is less stable than the brownish colour, users are recommended to quantify sufficiently large amounts by absorbance and to resort to fluorescence scanning only for small amounts.

Densitometric Evaluation

It was performed with TLC scanner II with lab data system and CATS evaluation software. Scanning by absorbance at 55 nm tungsten or mercury lamp, monochromator bandwidth 10 nm, slit dimensions 0.2 × 5 mm; evaluation via peak height and 2nd degree polynomial. Scanning by fluorescence with mercury lamp at 366/>400 nm.

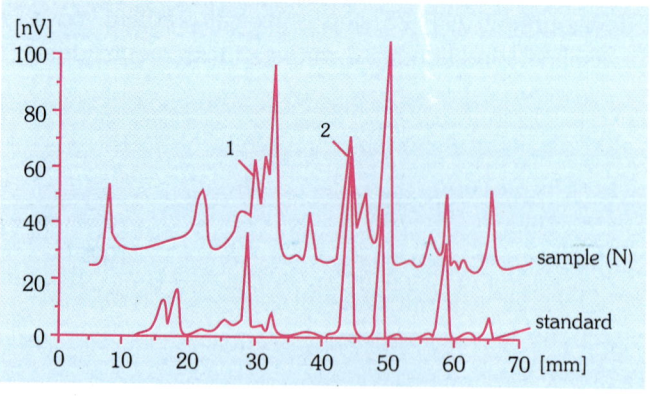

Fig. 17.30 (b)

Densitogram curves of a natural emulsifier (N) and a calibration standard. Amounts per fraction were cardiolipin (1) 266.9 mg/g and cholesterol (2) 180.0 mg/g absolute.

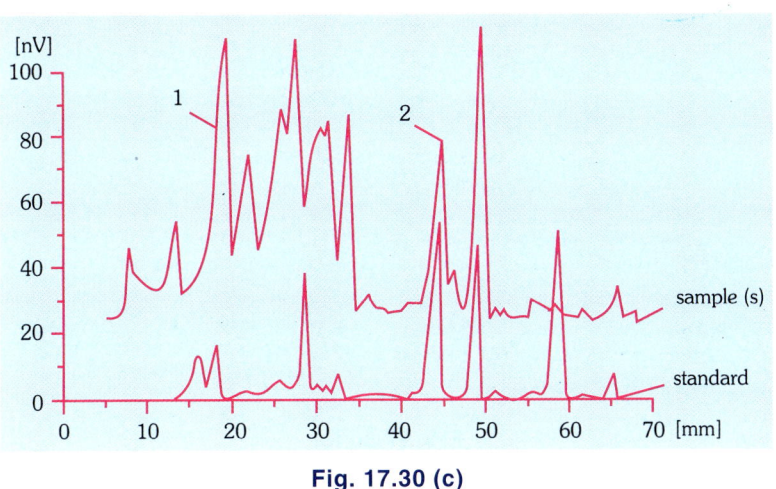

Fig. 17.30 (c)

Densitogram curves of a synthetic emulsifier (S) (without cardiolipin) containing 106.4 mg/g cholesterol (2) and 343.0 mg/g phosphatidylcholine (1)

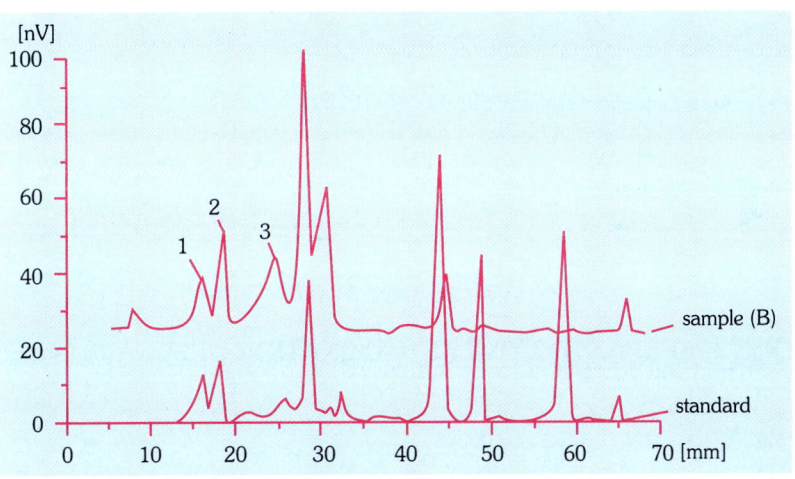

Fig. 17.30 (d)

Densitogram curves of the hydrophilic cerebral extract B and a calibration standard. Amounts per fraction were phosphatidylcholine (1) 420 ng and spingomyelin (2) 340 ng absolute.

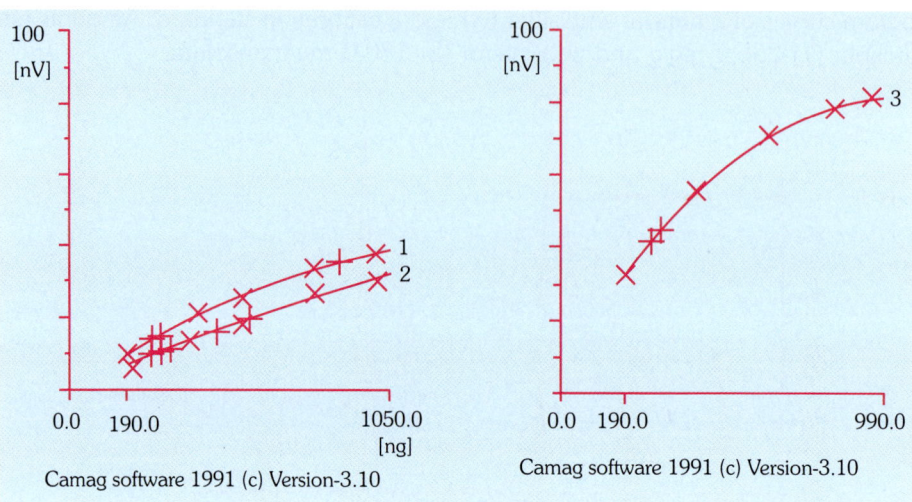

Fig. 17.30 (e)

Calibration curves (peak height, polynomial 2nd degree) of phosphatidylcholine (1)sphingomyelin (2) and cholesterol (3) calibration range 200–1000 ng absolute, corresponding to 66.67–333.35 mg/g.

References

1. Christiansen K. *Analytical Biochemistry* **60**; 93–99, 1975
2. Jork H, Funk W, Fischer W, Wimmer H. *Thin–Layer Chromatography*, Volume 1a, physical detection methoden: Fundamentals, Reagents I., VCH Verlagsges., Weinheim–Basel–Cambridge–New York, 1989
3. Camag application Laboratory

METHOD: 17.31

IDENTIFICATION AND QUANTITATIVE DETERMINATION OF SUGARS IN BEER AND WINE

Aim of Analysis

Identification and quantitative determination of sugars, maltose, glucose and fructose in beer and wine.

Chemicals and Reagents

Potassium dihydrogen phosphate, acetonitrile, acetone, maltose, glucose, fructose.

Sample Preparation

Based on expected sugar concentration, beer or wine is applied to the layer either directly or diluted with actonitrile.

Standards

Stock solutions: 10 mg of each standard sugar is dissolved in 10 ml acetone: water (1:1), For identification purposes, 1 ml of each stock solution is diluted with acetone: water (3:1) (=100 ng/μl).

Calibration Standard A (Maltose)

Dissolve 50 mg maltose (and/or malto-oligosaccharides depending on the application) in acetone: water (1:1) to 10 ml. Dilute 1 ml of this stock solution with acetone: water (3:1) to 100 ml (= 50 ng/μl).

Calibration Standard B (Glucose and Fructose)

Dissolve 50 mg glucose and 50 mg fructose in acetone: water (1:1) to 10 ml. Dilute 1 ml of this stock solution with acetone: water (3:1) to 100 ml (= 50 ng/μl).

Stationary Phase

HPTLC precoated plates silica gel NH_2 20 $\times$ 10 cm.

The plate is immersed for 15 minutes in a 0.4 M aqueous solution of potassium dihydrogen phosphate (12.52 g KH_2PO_4 in 230 ml water) with chromatogram immersion device. Dry the plate with a hair drier and leave over night. Preserve the plate under nitrogen, if not used the same day. The plate is impregnated with buffer to inhibit the formation of glycamines from reducing sugars.

Sample Application

Sample application bandwise with Linomat or automatic TLC sampler III, 9 mm bands, 6 mm apart, distance from lower edge 8 mm, 20 mm from the side = samples per plate; delivery speed 15 s/ml.

Recommended application pattern for quantitative analysis

U1 S1   U2   S2   U3   S3   U1   S4   U2   S5   U3

U = unknown – beer or wine 2 or 4 μl
S = standard A (for the determination of maltose in beer) and/or
S = standard B (for the determination of glucose and fructose in wine)
S1 = 2 μl = 100 ng
S2 = 4 μl = 200 ng
S3 = 6 μl = 300 ng
S4 = 8 μl = 400 ng
S5 = 10 μl = 500 ng

Standards and unknown shall be applied immediately before chromatogram development.

Chromatography

Automated multiple developments in the AMD system using a 15-step linear gradient is depicted in the diagram.

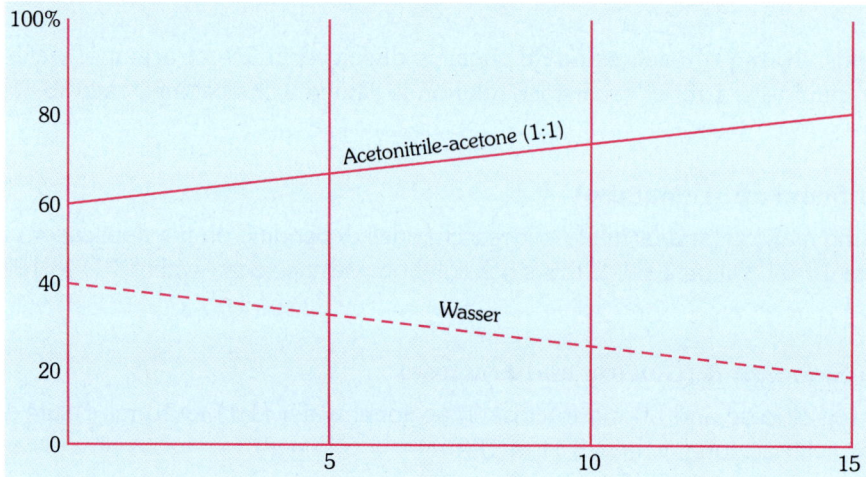

Fig.17.31 (a): 15-step AMD gradient based on acetonitrile-acetone-water, step increment 3 mm; drying time between steps 10 minutes

AMD solvent management						
From step n=	1	2	6	11		
Bottle no.	1	2	3	4	5	6
Acetonitrile	30	33	36	40		
Acetone	30	33	36	40		
Water	40	34	28	20		
Drying time(min.)	10					
Wash bottle content			empty			

AMD runtime program: [Minutes per step]							
Step	1	2	3	4	5	6	7	8
Minutes	0.8	1.0	1.5	2.0	2.6	3.5	4.5	5.5
Step	9	10	11	12	13	14	15	
Minutes	6.7	8.0	9.5	11.3	13.4	15.5	18.0	

Post-chromatographic Derivatization

It is done by thermal in situ reaction of the amino groups in the layer. After chromatography, dry plate thoroughly with a hair dryer, then place on TLC plate heater at 150° C for 3–4 minutes. Monitor the thermal reaction visually under UV 366 nm.

Densitometric Evaluation

It is done with TLC scanner with Lab data system and CATS evaluation software. Then scanning is done by fluorescence with Hg lamp at 366/>400 nm, monochromator band width 10 nm, slit dimensions 0.2 × 7 mm scanning speed 4 mm/s.

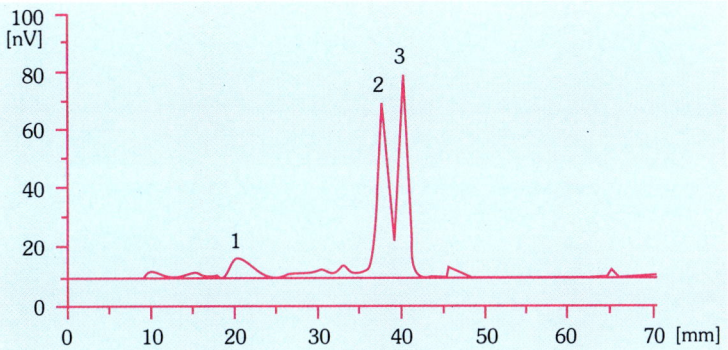

Fig. 17.31 (b): Separation of (2) glucose–264 ng and (3) fructose–290 ng in wine; Cabernet diluted 1:2 with acetonitrile, 2 µl applied.

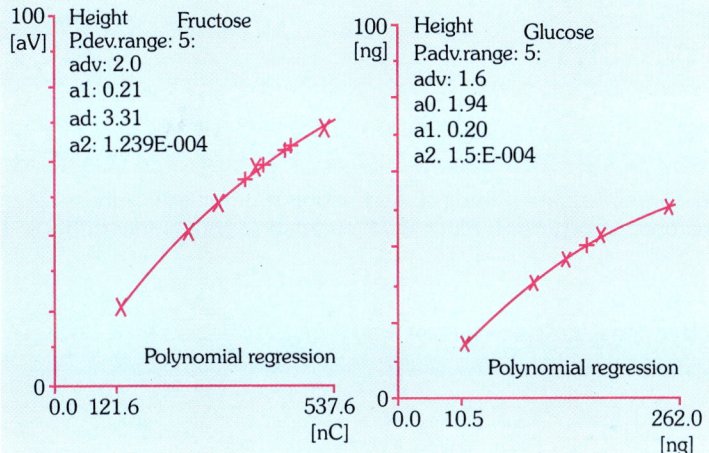

Fig. 17.31 (c)

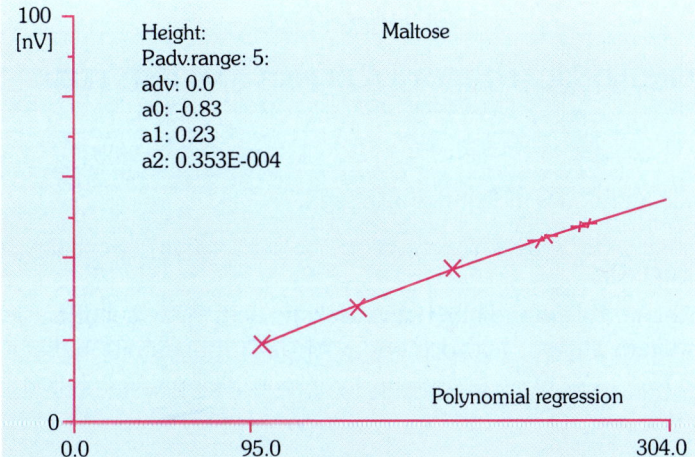

Fig. 17.31 (d): Calibration curves of fructose, glucose and maltose in the range of 100–500 ng absolute

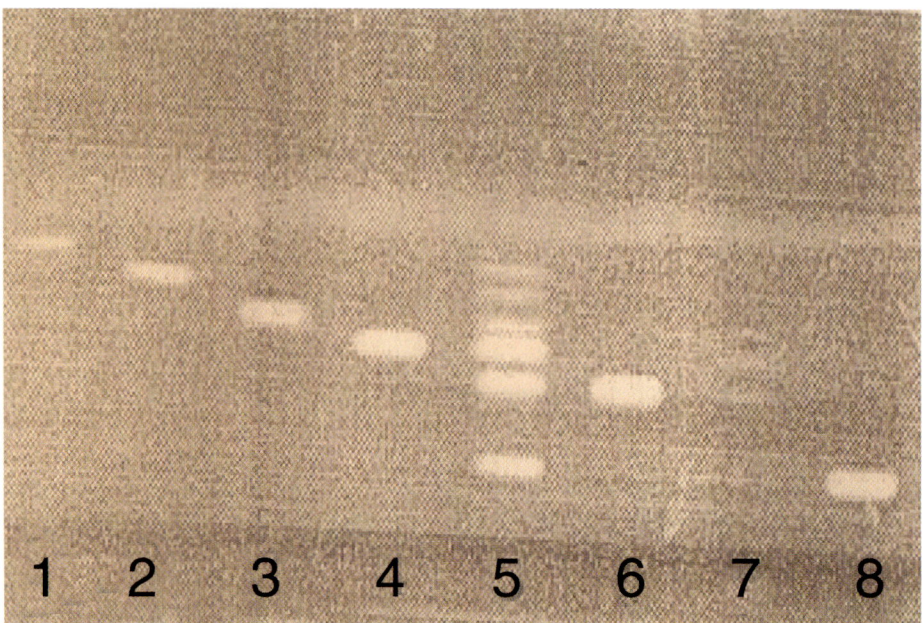

Fig. 17.31 (e): AMD chromatogram photographed under 366 nm UV: 1 ribose, 2 xylose, 3 arabinose, 4 fructose, 5 mixture, 6 glucose, 7 sucrose (partially inverted), 8 lactose

References

1. Lodi G. Universita Di Ferrara, (Analysis procedure) Dipartimento Di Chimica (private communication)
2. Klaus R, Fischer W, Hauck HE. Chromatographia **29**; 476, 1990.
3. Camag Application Laboratory

METHOD: 17.32

ANALYSIS OF QUINOLONIC ANTIBIOTICS IN FISH AND FISH FEED

Aim of Analysis

Analysis of quinolonic antibiotics in fish and fish feed.

Chemicals and Reagents

Acetonitrile, ethyl acetate, toluene, diethyl ether, formic acid 90%, sulfuric acid 95–97%, sodium hydroxide 0.1 N, sodium sulfate, hydrochloric acid 32%, phosphate buffer pH 8.0, KCl–KOH solution (KOH 0.02 M + 1% KCl), potassium hydrogen phosphate solution (0, 1 M in ethanol 50%).

Standard Substances

Flumequine, oxolinic acid, nalidixic acid

Sample Preparation

Fish feed samples: Weigh 5 g sample previously grounded in an Erlenmeyer flask and add 50 ml acetonitrile: KCl–KOH solution (7:3). Add nalidixic acid as internal standard. Shake 60 min in a mechanical shaker. Use an aliquot of supernatant for chromatography.

Fish tissue samples: Weigh about 5–10 g of fish meat in a waring blender and slurry with 10 ml phosphate buffer pH 8.0. Add nalidixic acid as internal standard (The amount of internal standard should be proportional to the expected concentration of the antibiotic). 10 g of the slurry is transferred in a capped centrifuge tube and add 12.5 ml acetonitrile: KCl–KOH solution: 0.1 N sodium hydroxide (16:4:1); sonicate for 2 min, then centrifuge at 2500 rpm. Dry 2–5 ml of supernatant, by passing it through a sodium sulfate cartridge with the help of a vacuum manifold device. Wash the cartridge with 1 ml of additional extraction mixture. Evaporate to dryness under nitrogen; residue of quinolones is taken up with 200 μl acetonitrile: KCl–KOH solution (7:3). Use this solution for chromatography.

Standard Solutions

Dissolve 500 mg of each antibiotic in 100 ml acetonitrile: KCl–KOH solution (7:3) (stock solution).

Dilute the stock solution 1:1000 with acetonitrile: KCl–KOH solution (7:3). Use this as the calibration standard solution for measuring by absorbance (S_A).

Dilute the SA–solution 1:50 with acetonitrile: KCl–KOH solution (7:3). This is used as calibration standard solution for measuring oxolinic acid by fluorescence (S_F).

Stationary Phase

HPTLC precoated plates silica gel 60 F_{254}, 20 × 10 cm; prewash with methanol by immersion and dry at room temperature then impregnate by immersion in K_2HPO_4 solution and dry the plate.

For chromatography of fish feed extracts, activate for 30 min at 120° C. Whereas for fish meat extracts, dry the plate at room temperature.

Sample Application

Bandwise with Linomat or automatic TLC sampler III. Band length 7 mm, track distance 10 mm, distance from the side 10 mm, distance from lower edge 5 mm, delivery speed 10 s/μl (100 nl/s).

Apply 2, 6, 10, 20 μl of the calibration standard solution for absorbance or fluorescence measurement respectively.

Chromatography

This is performed in the horizontal developing chamber with toluene: ethyl acetate: formic acid (6:3:1), chamber saturation with solvent vapors. When quinolones are not separated adequately in one developing run, a second run after drying needs to be performed.

Depending on the samples, a chromatographic clean-up is necessary. Non-polar matrix components can be moved to the front by a pre-run with diethyl ether. No chamber saturation is required for this clean-up run.

Post-chromatographic Derivatization (for Oxolinic Acid)

The plate is exposed to HCl gas by use of a twin-trough chamber. Into one trough pour 10 ml conc. sulfuric acid and add 2 ml conc. hydrochloric acid dropwise with care. Place the chromatogram plate in the other trough and leave it for 5–10 min.

Quantitative Evaluation

With TLC scanner, win CATS evaluation software absorption measurement is done without post-chromatographic derivatization

Mode   :   Scanning by absorbance
Wavelength :   320 nm
Calibration   :   Polynomial regression

Flumequine has absorption maxima at 320 and 246 nm, oxolinic acid at 262 and 329 nm, nalidixic acid at 319 and 253 nm. Hence 320 nm is suitable for all three. Detection limits are 8 ng/spot for flumequine and 9 ng/spot for oxolinic acid.

Fluorescence measurement of oxolinic acid is done after post-chromatographic derivatization

Mode   :   Scanning by fluorescence
Wavelength :   265/>400 nm
Calibration   :   Linear regression

The detection limit was determined at < 0.2 ng corresponding to 10 ppb in the original sample material.

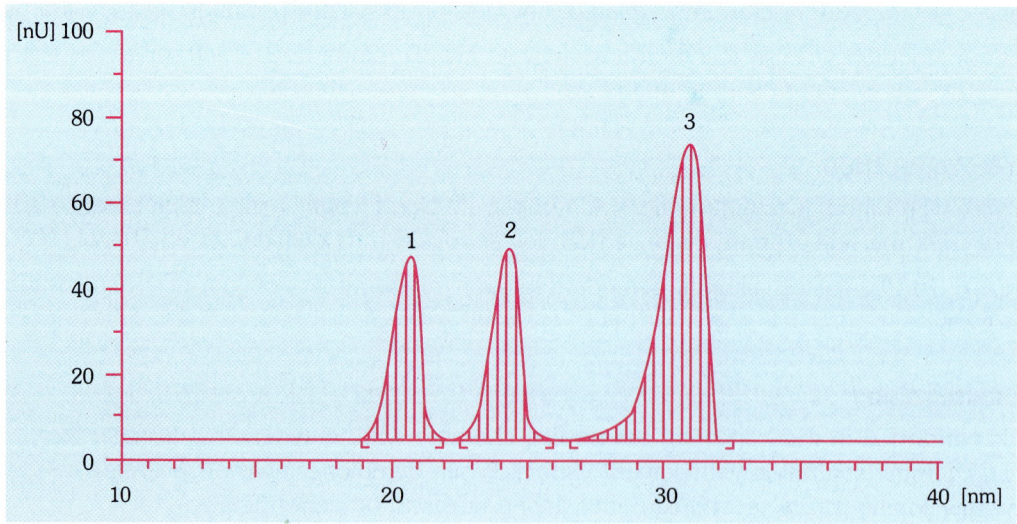

Fig. 17.32 (a): Densitogram (Absorbance at 320 nm), 1 oxolinic acid, 2 flumequin, 3 nalidixic acid, about 50 ng each

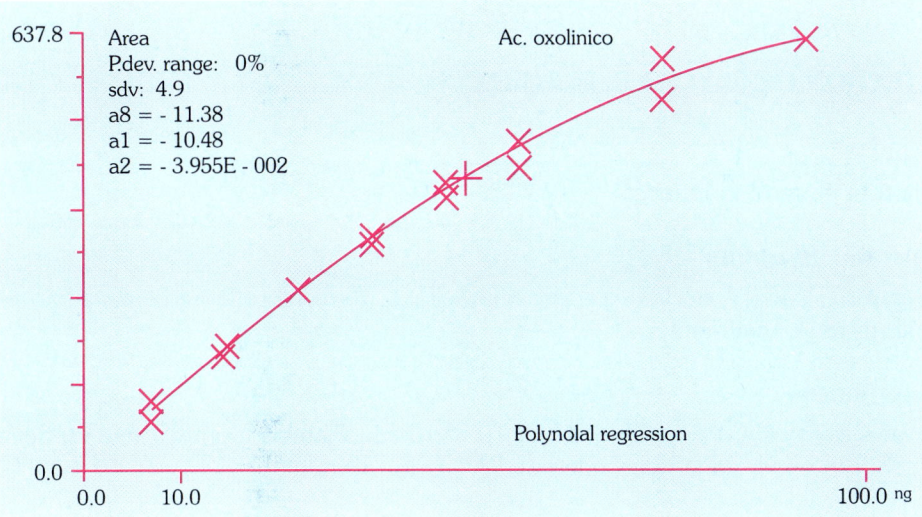

Fig. 17.32 (b): Calibration function for oxolinic acid (Absorbance at 320 nm)

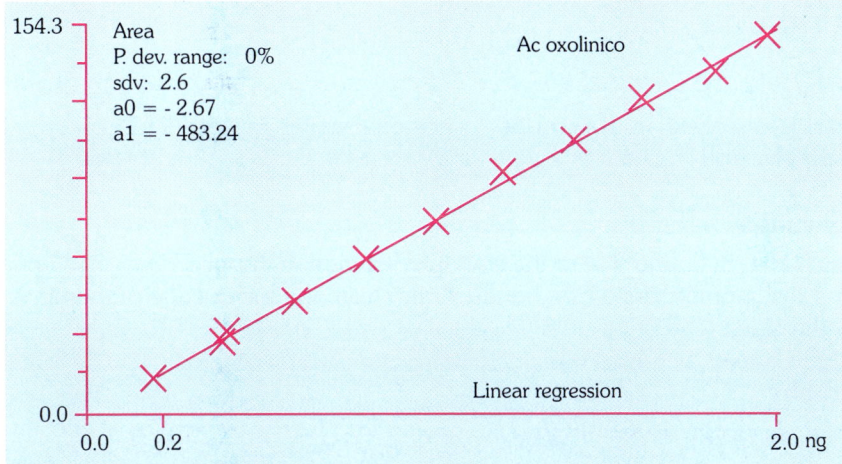

Fig. 17.32 (c): Calibration function for oxolinic acid (Fluorescence at 265/>400 nm)

References

1. Vega M, Rios G, Saelzer R, Herlitz E. Depto. De Bromatologia, Nutricion y Dietetica, Facultad de Farmacia, Universidad de Concepcion, Concepcion, Chile. P. O. Box 237, Fax 56 41 231903
2. Vega M, Garcia G, Saelzer R, Villegas RJ. Planar Chromatogr. **7**; 159–162, 1994
3. Camag Application Lab.

METHOD: 17.33

QUANTITATION OF GLYCEROL IN TOBACCO

Aim of Analysis

Quantitation of glycerol in tobacco.

Chemicals and Reagents

Acetone, glycerol, lead (IV) acetate, glacial acetic acid, 2', 7'–dichlorofluoresceine, ethanol, toluene, paraffin oil, hexane, methanol.

Stationary Phase

HPTLC plates silica gel 60 F_{254}, 20 × 10 or 10 × 10 cm. Plates are pre-washed by development with methanol to upper edge, then dried at 120° C for 30 min.

Sample

About 2 g sample is dried for 30 min at 100° C in an oven. 0.5 g of the dried sample is shaken with 6 ml water and 20 ml acetone for 5 min. 0.1 g activated carbon is added. The mixture is shaken for 5 min again and is centrifuged at 3000 rpm for 5 min. If the solution is still not clear, then filter through filter paper.

Standard

60 mg glycerol is dissolved in 100 ml of an acetone: water mixture(98:2). 1 μl of the solution contains 0.6 μg glycerol.

Sample Application

1μl of samples and 1, 3, 5, and 7 μl of the standard solution are applied with the Camag automatic TLC sampler 4 or Linomat 5 as 6 mm bands, 8 mm from the lower edge of the plate, and at least 10 mm from the sides.

Chromatography

Mobile phase: Acetone: water (98:2)

Development: 10/5 ml developing solvent is put in each trough of a 20 × 10/10 × 10 cm twin-trough chamber, containing a filter paper on one side. The lid is closed and the chamber is allowed to saturate for half an hour. Developing distance should be approx. 6 cm from lower edge of plate. The plate is dried in a stream of cold air.

Derivatization Agents

Lead–dichlorofluoresceine solution (fresh for each plate) (LDS)

Solution A: 2 g lead(IV) acetate is dissolved in 10 ml glacial acetic acid.

Solution B: 1g 2', 7'–dichlorofluoresceine is dissolved in 100 ml ethanol.

5 ml each of solution A and B are added to 190 ml toluene.

Paraffin oil–hexane solution (PHS): 40 ml paraffin oil is dissolved in 200 ml hexane.

Detection

Using the chromatogram immersion device, the plate is dipped in LDS for 6s, and then it is dried in a stream of cold air for 1 min and again dipped in PHS for 1s and dried by stream of cold air.

Results Screening

1–4: Glycerol standard, increasing volume
5: Marlboro light (Switzerland)
6–8: Aromatized water pipe tobacco(Egypt)
9: Pipe tobacco (Virginia, USA), contains no glycerol

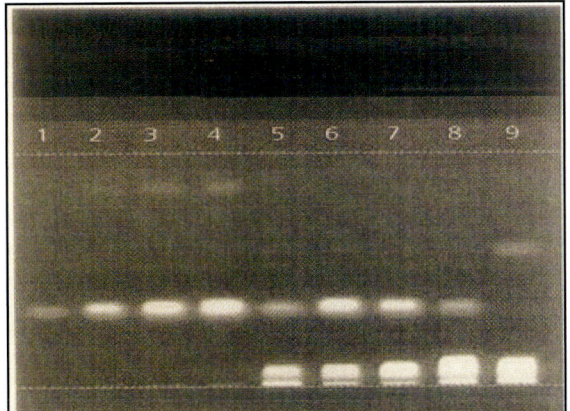

Fig.17.33 (a): Image after derivatization under UV 366 nm

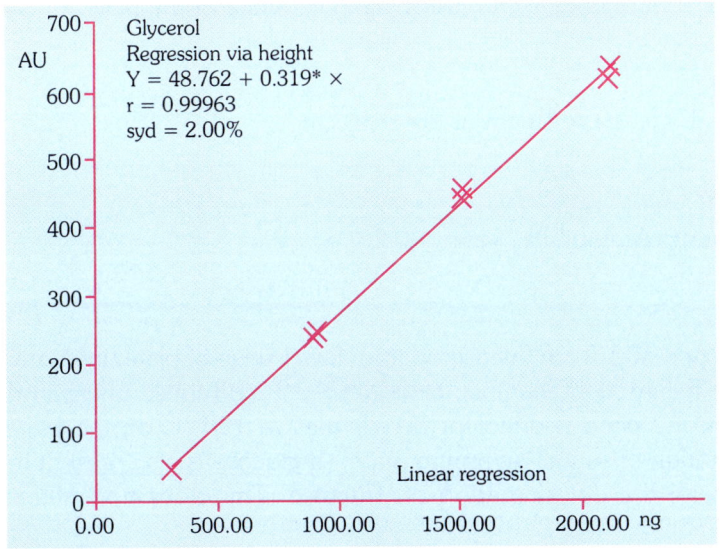

Glycerol
Regression via height
$Y = 48.762 + 0.319^* \times$
$r = 0.99963$
$syd = 2.00\%$

Linear regression

Fig. 17.33 (b)

Calibration for glycerol
Red: Standard levels (duplicates)
Blue: Samples

METHOD: 17.34

DETERMINATION OF SUCRALOSE AND FRUCTOSE IN FOOD AND BEVERAGES

Aim of Analysis

Determination of sucralose and fructose in food and beverages.

Sample

In case of carbonated beverages, 10 ml of the sample is placed in ultrasonic bath for 10 minutes. Depending on the concentration of the sweetner, the sample need to be diluted with methanol for suitable calibration range.

0.5–2g of solid sample are mixed with 2 ml of methanol: water (1:1) and placed in ultrasonic bath for 10 minutes. The mixture is filled with the solvent to 5 ml and centrifuged at 4000 rpm for 5 minutes. The supernatant is applied directly on the plate.

Standards

A solution containing 0.07 $\mu g/\mu l$ sucralose in methanol.
A solution containing 0.07 $\mu g/\mu l$ fructose in methanol.

Sample Application

2, 3, 4 and 5 μl of each standard solution and corresponding quantity of sample are applied as 8 mm bands, at least 2 mm apart, 8 mm from the lower edge of the plate.

Stationary Phase

HPTLC Si 60 NH$_2$ F$_{254}$s, 10 × 10 cm or 20 × 10 cm

Chromatography

Developing solvent: Acetonitrile: water (80:20)

Development

The plate is developed over 7 cm (measured from lower edge of plate) in an unsaturated 10 × 10 cm (or 20 × 10). Twin-trough chamber, 5 ml (resp. 10 ml) mobile phase in developing trough. After development-the plate is dried for 20 minutes at 190° C on the plate heater (thermochemical derivatization of sugars on amino plate). If glucose is also present in the sample, then it may interfere with the determination on fructose. This does not influence the sucralose peak.

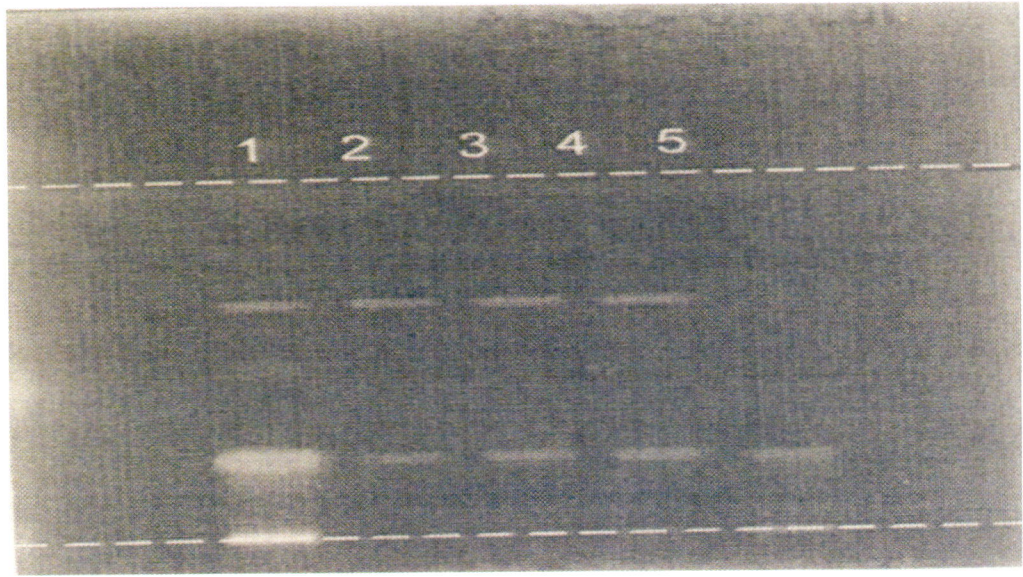

Fig . 17.34 (a): Image of a plate under UV 366 nm.

1: Neat beverage (sucralose is in the calibration range)

2-4: Fructose(R_f = 0.3) and sucralose (R_f = 0.7)

5: Diluted beverage (fructose is in the calibration range)

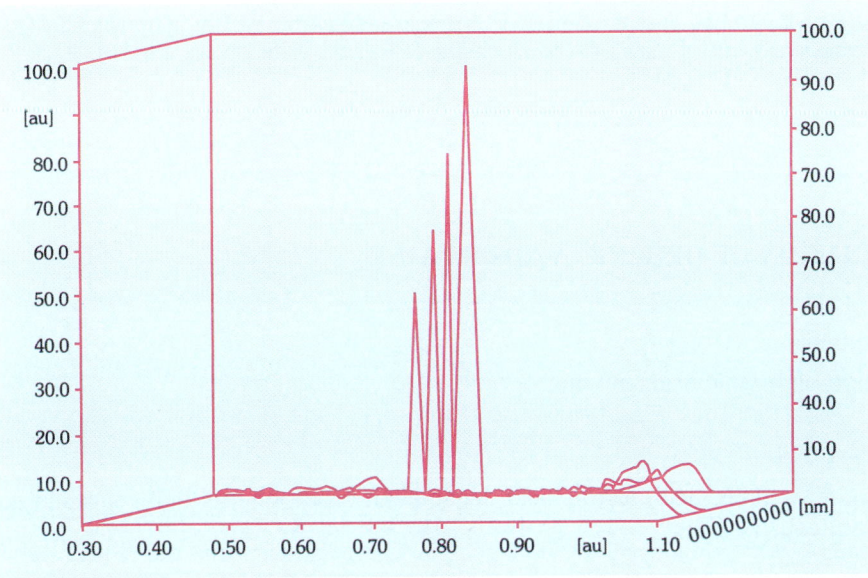

Fig. 17.34 (b): Densitogram of sucralose in a beverage (blue: 2 samples, pink: 4 standard levels)

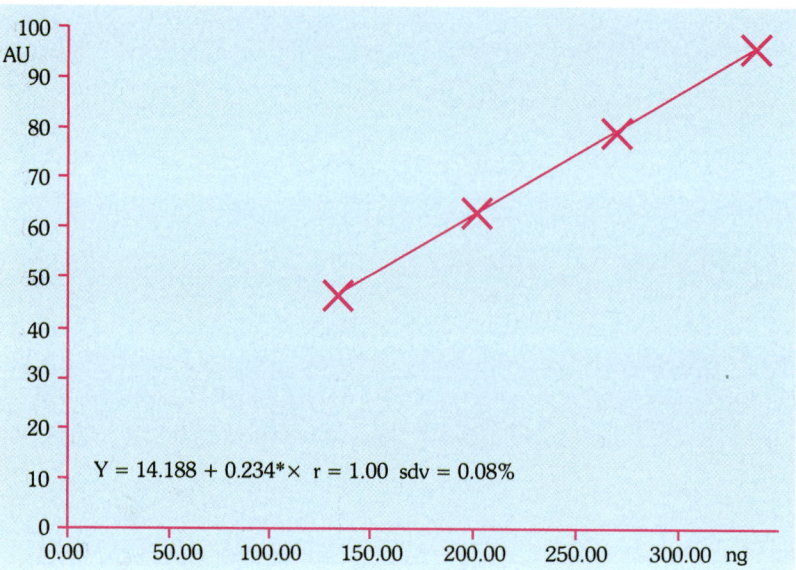

Y = 14.188 + 0.234*× r = 1.00 sdv = 0.08%

Fig. 17.34 (c): Calibration curve for sucralose in the range 140–350 ng; linear regression via peak height (blue crosses: samples)

References

1.   Spangenberg B, Stroka J, Arranz I, Anklam E. A simple and Reliable HPTLC method for the Quantification of the Sucralose. Journal of Liquid Chromatography & Related Technologies, **26(16);** 2729-2739, 2003
2.   Camag Application Laboratories

METHOD: 17.35

DETERMINATION OF ORGANIC ACIDS IN WINE

Aim of Analysis

Determination of organic acids in wine.

Chemicals and Reagents

Di-isopropyl ether, malic acid, formic acid, lactic acid, bromophenol blue, succinic acid, ethanol and 0.1 N aqueous NaOH.

Samples, Standard

Samples of red or white wine are used without any pre-treatment. Malic acid, lactic acid, and succinic acid are dissolved in water at a concentration of 1 mg/ml each.

Sample Application

5 μl of the wine samples and 3, 6, 9, 12, 15 μl of the standard solution are applied with automatic TLC sampler, 10 mm bands, and 15 mm from the side and 8 mm from lower edge of the plate, i.e. 10 applications per plate.

Stationary Phase

HPTLC plates silica gel 60 F_{254}, 20 × 10 cm.

Chromatography

Mobile phase: Di-isopropyl ether: formic acid: water (80:15:5).

Development: Twin-trough chamber 20 × 10 cm saturated for 15 min with mobile phase. 10 ml mobile phase per trough. After development, the plate is dried for 5 minutes with a hair dryer (warm air), then heated at 110° C for 15 min (plate heater or oven).

Derivatization

Reagent: 0.125 g bromophenol blue are dissolved in 250 ml denatured ethanol. 0.1 N aqueous NaOH is carefully added until the colour of the solution changes to blue.

The plate is immersed in the reagent with chromatogram immersion device. The organic acids yield yellow zones on a blue background (malic acid, lactic acid, succinic acid with increasing R_f)

Documentation, Quantitative Evaluation

Documentation with videostore under white light. Evaluation with TLC scanner 3 and CATS software in absorbance mode at 430 nm, linear regression. Detection limits are 3 μg absolute for each acid.

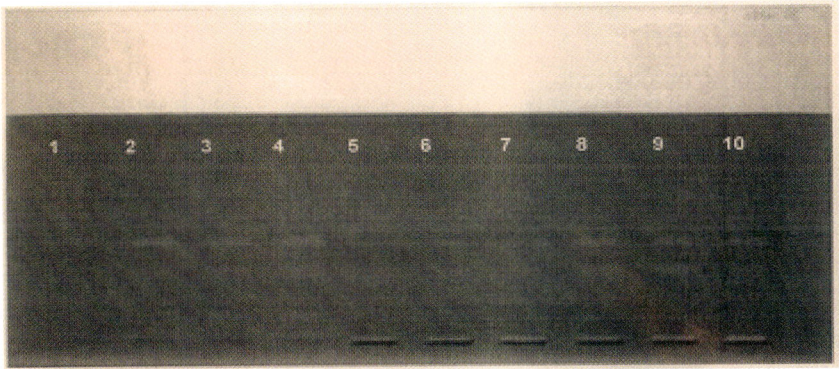

Fig. 17.35 (a)

Analysis of red wine samples, only lactic acid and succinic acid are present in the wine track assignment.

1–4: Standard levels 1–4 (3, 6, 9, 12 μg of malic, lactic, succinic acids)
5–7: Wine samples
8–10: Wine samples spiked with standard

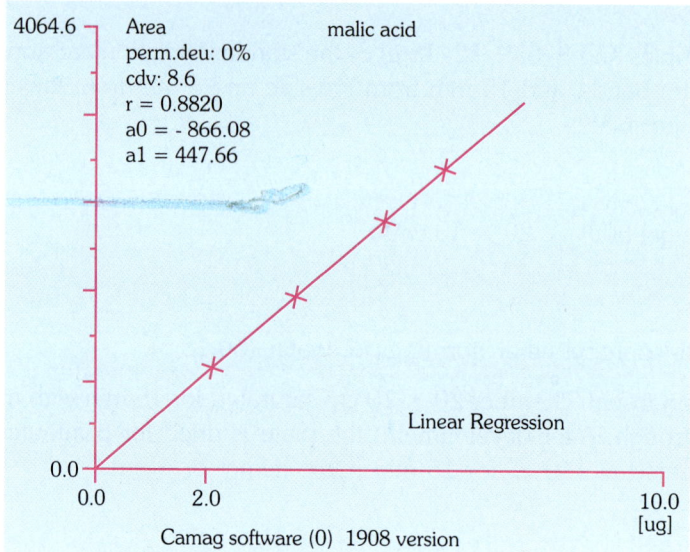

4064.6

Area   malic acid
perm.deu: 0%
cdv: 8.6
r = 0.8820
a0 = - 866.08
a1 = 447.66

Linear Regression

0.0

0.0   2.0   10.0
                                                    [ug]

Camag software (0) 1908 version

Fig. 17.35 (b): Calibration curve for malic acid

METHOD: 17.36

QUANTITATIVE DETERMINATION OF VITAMIN D$_3$

Aim of Analysis

Quantitative determination of vitamin D$_3$.

Chemicals and Reagents

n-hexane, cyclohexane, ethylacetate.

Sample Preparation

Vitamin D$_3$ in n-hexane (approx. 10 ng/μl).

Sample Application

By spray-on technique with Linomat different volumes (2–16 μl) for calibration. Band length 10 mm, distance between zone 5 mm, 11 samples per plate.

Stationary Phase

HPTLC precoated plates silica gel 60 F$_{254}$, 20 × 20 cm/20 × 10 cm.

Chromatography

In a twin-trough chamber with cyclohexane: ethylacetate (75:25), saturated with solvent vapours
Separation distance   :   7 cm

Quantitative Evaluation

Camag scanner 3. Measurement by absorbance with deuterium lamp at 280 nm.

Substance Stability

Vitamin D_3 degrades if it is irradiated by UV. Nevertheless no decrease of signal could be measured, even if the sample zones were measured several times. This can be explained by the short exposure time (1–2 sec) of the sample zones, resulting from the high scanning speed.

METHOD: 17.37

ANALYSIS OF OLIGOSACCHARIDES IN MOLASSES

Aim of Analysis

Analysis of oligosaccharides in molasses.

Chemicals and Reagents

Raffinose hydrate, 1–kestose, nystose, fructosyl nystose, acetone, 4–aminobenzoic acid, acetic acid, phosphoric acid (85%).

Samples

1 g of molasses is dissolved in 400 ml distilled water. No clean-up is required.

Standard

Raffinose hydrate, 1–kestose, nystose, and fructosyl nystose are dissolved in water: acetone (2:1) to form a solution of about 10 ng/μl each.

Sample Application

4 μl of molasses samples and 2–12 μl of the standard solution are applied with automatic TLC sampler as 5 mm bands, 20 mm from the side and 8 mm from the lower edge of the plate.

Stationary Phase

HPTLC plates silica gel DIOL 60 F $_{254}$, 20 × 10 cm.

Chromatography

AMD 2 development	:	With a nine-step gradient								
Step no.		1	2	3	4	5	6	7	8	9
Acetonitrile: acetone (1:1)		85	86	88	89	90	91	92	94	95
Water		15	14	12	11	10	9	8	6	5
Developing distance(mm)		10	17.5	25	32.5	40	47.5	55	62.5	70
Drying time(min)	:	10								
Wash bottle	:	empty								

Derivatization

Reagent: 2 g 4–aminobenzoic acids are added to a mixture of 36 ml acetic acid, 40 ml water, 2 ml phosphoric acid (85%) and 120 ml acetone.

The dried plate is immersed in the reagent using the immersion device, air dried, then heated on the plate heater or in an oven at 115° C for 15 minutes. Oligosaccharides appear as yellow-brown spots on a pale yellow background.

Documentation, Quantitative Evaluation

With video store under white light and under UV 366 nm.

Evaluation with TLC scanner 3 and CATS software, fluorescence measurement at 366/>400 nm or absorption measurement at 400 nm.

Result Screening

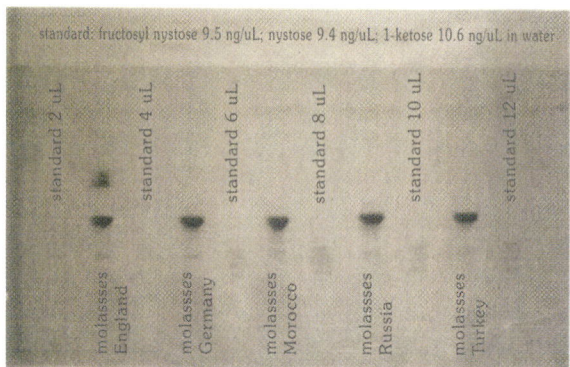

Fig. 17.37 (a): Standard: 2, 4, 6, 8, 10, 12 µl of a solution containing 9.5 ng/ µl fructosyl nystose; 9.4 ng/ µl nystose; 10.6 ng/ µl 1–ketose in water

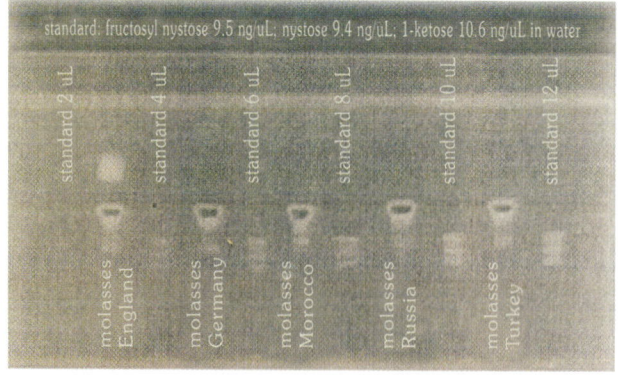

Fig. 17.37 (b): Screening of molasses samples after derivatization with 4–aminobenzoic acid, white light (top), UV 366 nm (bottom)

Results densitometry

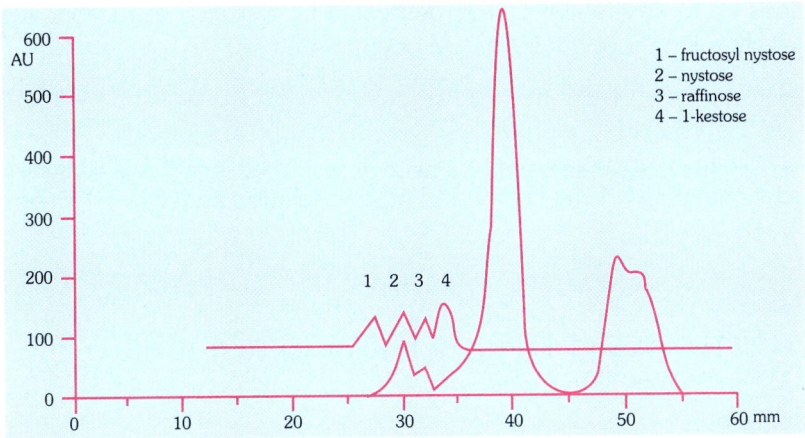

Fig. 17.37 (c): Molasses from red and standard black; wavelength 366 nm

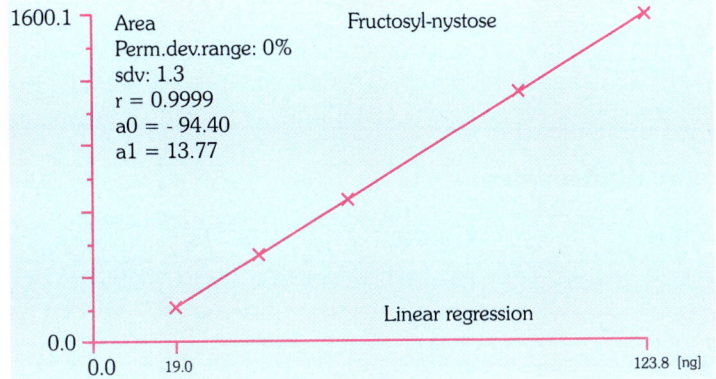

Fig. 17.37 (d): Calibration curve for fructosyl nystose/fluorescence

METHOD: 17.38

ANALYSIS OF INORGANIC AND ORGANIC MERCURY IN WATER

Aim of Analysis

Quantitative analysis of inorganic and organic mercury in water.

Chemicals and Reagents

NH_4Cl, NH_3, dithizone, chloroform, $HgCl_2$, CH_3HgCl, C_2H_5HgCl, C_6H_5HgCl.

Sample Preparation

Water: 5 ml adjusted to pH 10 with 0.1M NH_4Cl/NH_3; extracted with 10 ml 0.02 M dithizone solution in chloroform. It is evaporated to dryness in rotary evaporator and residue is dissolved in

50 μl chloroform of which 200 nl are directly spotted on the plate. (For the detection of very low Hg concentrations up to 200 ml water may be treated by repeated extraction, keeping the error below 15%, detection limit is about 0.4 ng/ml original sample).

Calibration Standards

Stock solutions of $HgCl_2$ in water and CH_3HgCl, C_2H_5HgCl and C_6H_5HgCl respectively in chloroform, each containing 0.1 mg Hg/ml; 1 μl of such solution added to 10 ml sample represents 100 ng Hg/ml.

Stationary Phase

Precoated plates HPTLC silica gel 60 F_{254}, 20 × 10 cm (or 20 × 10 cm).

Sample Application

200 nl, spotwise with nano-applicator.

Chromatography

HPTLC linear developing chamber with n-hexane: acetone (85:15) after 20 minutes preconditioning with 5M NH_4OH in HPTLC conditioning tray. (This alkaline preconditioning is essential to prevent migration of the dithizonates of other heavy metal ions, extracted simultaneously.)

R_f values of mercury dithizonates

Dithizonate	Colour	R_f	S.D.
Hg D_2	orange	0.15	0.01
CH_3 Hg D	yellow	0.32	0.02
C_2H_5Hg D	yellow	0.35	0.01
C_6H_5Hg D	yellow	0.28	0.02

(D = dithizonate)

Quantitative Evaluation

With TLC Scanner II combined with suitable peripheral unit.

Scanning is done by absorbance with Hg–lamp at 436 nm, monochromator band width 10 nm, and evaluation via peak height (Evaluation via peak height results in a better linear regression than using peak areas.)

References

1. Bruno P, Caselli M, Traini A. HRC & CC, **8**; 135–139, 1985
2. Camag Application Laboratories

METHOD: 17.39

SIMULTANEOUS QUANTITATIVE DETERMINATION OF VITAMIN D ANALOGS (PROVITAMIN D, IRRADIATION PRODUCTS OF PROVITAMIN D, VITAMINS D_2–D_3, VITAMIN D ESTERS)

Aim of Analysis

Simultaneous quantitative HPTLC determination of vitamin D analogs (provitamin D, irradiation products of provitamin D, vitamins D_2–D_3, vitamin D esters).

Reagents

Cyclohexane, cod liver oil, distilled water, diethyl ether, ethyl alcohol, n–hexane, petroleum spirit, standard D_3.

Sample Preparation

5 gm of the homogeneous cod liver oil (seven seas) was well dispersed with 25.0 ml ethanol and 25.0 ml 2% H_3PO_4 solution. 50 ml of n-hexane was added and the sample was shaken vigorously for 30 min in a 250 ml round bottomed amber glass flask. Hexane phase aliquots were evaporated to dryness under vacuum in a rotary evaporator at room temperature and dilutions were made with 5 ml petroleum spirit.

Standard Preparation

Standard vitamin D_3 100 ml/l in petroleum spirit, stock solution concentration was made 100 ng/μl.

Stationary Phase

HPTLC precoated plates, silica gel 60 F_{254}, 10 × 10 cm

Sample Application

Spray an application with Linomat IV, sample volume 4 μl; and different volumes of standard, band length 6 mm, distance between the bands 6 mm, 6 bands per plate.
Recommended application patterns:

S1	= 100 ng standard	= 1 μl
S2	= 200 ng standard	= 2 μl
S3	= 300 ng standard	= 3 μl
S4	= 400 ng standard	= 4 μl
U1	= unknown sample in duplicates	= 4 μl

Chromatography

Chromatogram was developed twice in a twin trough 10 × 10 cm chamber. First the plate was run with n-hexane only up to 5 cm without chamber saturation. Second time, the same plate was run with cyclohexane: diethyl ether (1:1) in a saturated chamber up to 7 cm, with gentle cold air drying, after each development. Chromatography development is to be carried out in a dark room and away from any UV light source.

Densitometric Evaluation

Densitometric evaluation was done by TLC scanner II with lab data system and CATS evaluation software version 3.15. Scanning was by absorbance at 268 nm with D_2 lamp, monochromator band width 10 mm, slit dimensions 0.3×5 mm. Quantitative evaluation was via peak height and linear regression.

Vitamin D_3 detection limit was 10 ng and linearity range was 100–400 ng in UV absorbance. By this method, ppb level of D_3 was quantified from cod liver oil and can be extended to any vitamin D_3 fortified oils/fats products.

Vitamin D_3 is very sensitive to UV light hence the experiment is to be carried out in dark place and away from UV light. UV scanning should be fast and the monochromator band width should be 10 nm to minimize exposure to UV.

References

1. De Luca HF, Vitamin D. Metabolism and Function, Springer Verlag, Berlin, 1979
2. Norman AW, Vitamin D. The calcium homeostatic steroid hormone. Academic press, Newyork, 1979
3. Stary E, Cruz AMC, Donomai CA, Monfardini JL, Vargas JTF. Determination of vitamins A–palmitate, A–Acetate, E–Acetate, D3 and K1 in vitamin Mix by isocratic Reverse phase HPLC, Journal of High Resolution Chromatography, **12**; June 1989
4. Camag Application L- brataries

METHOD: 17.40

DETECTION AND SEMI-QUANTITATIVE DETERMINATION OF ANABOLICA IN MEAT

Aim of Analysis

Detection and semi-quantitative determination of anabolica in meat with AMD.

Reagents

Nortestosterone (NT), estriol (E), nortestosterondecanoate (NTD), B-estradiolbenzoate (EB), 5 M HCl, acetonitrile, sodium sulfate, diethyl ether sodium carbonate solution (10% aqueous), acetone, ethanol, methanol, di-isopropyl ether, n-hexane, sulphuric acid conc.

Sample Preparation

Extract 1-5.0 g sample material in reflux with 10 ml acetonitrile and 0.2 ml 5 M HCl for 1 hour, then cool in an ice bath. Decant supernatant in a 100 ml beaker. Wash residue 1 × with 5 ml acetonitrile (contains 0.1 ml 5M HCl) and combine wash and supernatant and evaporate in a rotary evaporator under nitrogen at 50° C, adjust the cooler to 10° C. Dissolve the residue in 3 ml sodium carbonate solution and transfer it to a 15 ml stoppered glass tube. Then rinse flask with 3 ml diethyl lether. Vortex both solutions for 1 min. After phase separation, transfer the ether phase to a new glass tube over sodium sulfate, mix and centrifuge. Repeat ether extraction with 3 ml ether. Evaporate the combined ether phases under nitrogen to dryness. Dissolve residue in 0.025 ml acetone and transfer to a 3 ml glass tube, rinse with 0.25 ml acetone. Centrifuge 5 min at 6000 rpm, transfer clear supernatant to a glass tube and evaporate at 50° C under vacuum. Dissolve residue in 50 μl acetone.

Calibration Standards

Prepare dilutions of all standard substances (NT, E, NTD, EB) in the concentration range 0.1-1.0 $\mu g/kg$.

Dissolve 50 mg to a volume of 100 ml acetone, dilute 0.5 ml of this solution in acetone and make up to 100 ml and dilute again 100 μl of this solution to 100 ml acetone. 100 μl of this solution contain 250 pg steroids.

To avoid proportional–systematic errors in a quantitative determination of the steroids, add each standard substance to a blank (authentic sample material without steroids) and treat it in just the same way as the sample.

Pipette to 5 g meat blank respectively 0.2, 0.4, 0.8, 1.2, 1.6 and 2.0 ml standard solution and prepare like the sample. Apply 2 μl extract.

Stationary Phase

HPTLC precoated plates silica gel 60 F_{254}, 20 × 10 cm.

Sample Application

Bandwise with Linomat IV, application speed 4 $\mu l/sec$ distance from lower edge 8 mm, distance from side edge 15 mm, band length 6 mm, space between samples 4 mm, 16 tracks, according to the pattern below:

S1 U1 U2 S2 U3   U4 S3 U5 U1 S4   U2 U3 S5 U4 U5 S6

S1 = 2 μl = 0.1 $\mu g/kg$
S2 = 2 μl = 0.2 $\mu g/kg$
S3 = 2 μl = 0.4 $\mu g/kg$
S4 = 2 μl = 0.6 $\mu g/kg$
S5 = 2 μl = 0.8 $\mu g/kg$
S6 = 2 μl = 0.0 $\mu g/kg$

S = Standard

U = unknown 2 μl /application

AMD–gradient: automated multiple development in an AMD chamber with a relatively flat polarity gradient based on di-isopropyl ether, 20 steps.

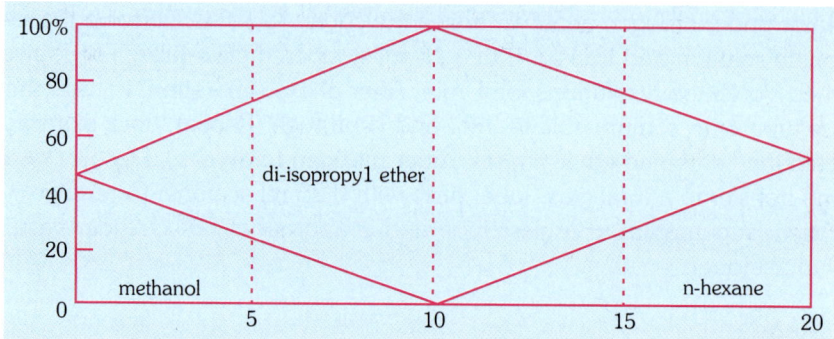

Fig. 17.40 (a)

AMD gradient over 20 steps, schematic

From step	1	2	6	11	16	21
Bottle no.	1	2	3	4	5	6
Methanol	30	15				
Di-isopropyl ether	70	85	100	75	50	
n–hexane				25	50	
Drying time (min)	5	5	4	4	4	
Wash bottle content	NH_3 1N					

Post-chromatographic Derivatization

Immerse HPTLC plate 5 sec in ethanol: conc. sulphuric acid (9:1), and then heat at 120° C for 5 min.

Densitometric Evaluation

With TLC scanner and suitable peripheral (integrator or computer) scanning by fluorescence with mercury lamp, excitation by 366 nm, 400 nm sharp cut filter, monochromator band width 10 nm, slit dimensions 0.2 × 5 mm.

A densitometric evaluation can be done by absorbance at 225 nm without derivatization. The detection limit is lower than with measurement by fluorescence.

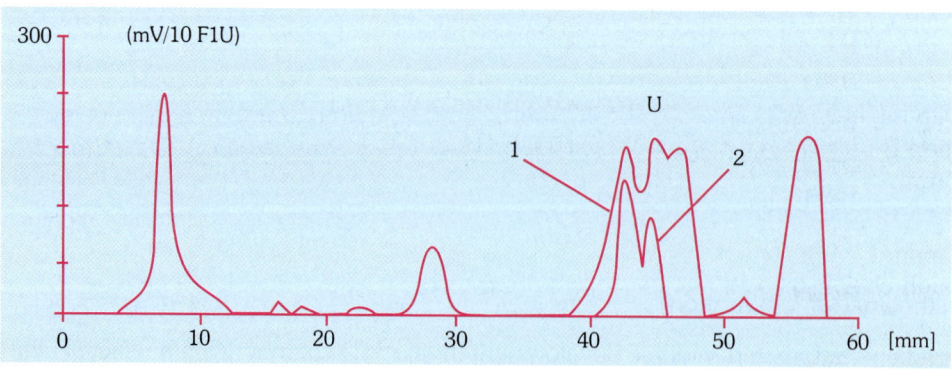

Fig. 17.40 (b)

Evaluation of a chromatogram with computer controlled system and evaluation software, spot (1) and (2) are standard anabolica NTD and EB, (U) is a meat sample containing NTD and EB.

References

1. De Ruig WG, Hooijerink H, Weseman JM, Fresenius Z. Anal. Chem, **320**; 749-752,
2. Camag Application Laboratories

METHOD: 17.41

ANALYSIS OF CURCUMA LONGA FOR CURCUMIN

Aim of Analysis
Analysis of curcuma longa for curcumin.

Chemicals and Reagents
5% sodium dihydrogen phosphate, chloroform, ethanol, curcumin.

Standard Preparation
Solution A: Weigh 5 mg of curcumin, add 5 ml chloroform (conc. 1 $\mu g/\mu l$).
Solution B: Pipette out 1 ml of the above soln. and dilute it to 10 ml chloroform (conc. 0.1 $\mu g /\mu l$).

Sample Preparation
Extract 500 mg of powdered curcuma longa sample by shaking with 10 ml methanol for 10 min with slight warming. Filter and use the filtrate.

Stationary Phase
TLC Al sheets silica gel 60F$_{254}$ precoated plate 20 × 10 cm. The plate is first impregnated with 5% sodium dihydrogen phosphate.

Mobile Phase
Chloroform: ethanol (2.5:0.1) v/v

Sample/Standard Application
Apply with the help of Linomat 5, 2 and 5 μl of sample solution, and 2 μl, 4 μl, 6 μl, 8 μl, 10 μl of std solution B on precoated layer, 6 mm from bottom edge, band length 8 mm, distance from the sides 15 mm.

Development Chamber
Twin-trough chamber of 20 × 10 cm with s.s.lid

Tank saturation	:	None
Plate equilibrium	:	None
Development distance	:	80 mm

Visualization	:	254 nm, 366 nm in UV cabinet
Post-chromatographic derivatization	:	None
Photodocumentation	:	254 nm, 366 nm, visible

Scanning for Quantification

Using scanner 3 with win CATS software, slit–micro, 4×0.3 mm, scan at 254, 366 and 440 nm.

Scanning for Spectral Match

Record the spectra between 200 to 600 nm.

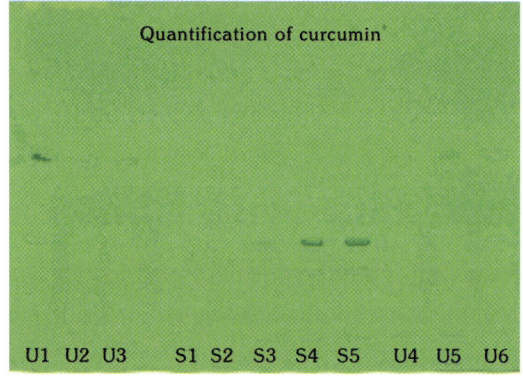

Fig. 17.41 (a): Quantification of curcumin in sample

Fig. 17.41 (b): Quantification of curcumin

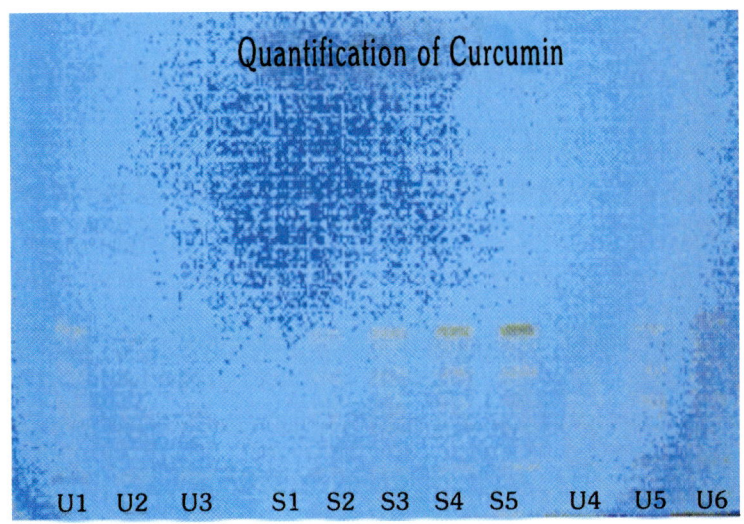

Fig. 17.41 (c)

S1: Standard # 1 ug
S2: Standard # 2 ug
S3: Standard # 3 ug
S4: Standard # 4 ug
S5: Standard # 5 ug
U1: Market sample # 50 ug
U2: Market sample # 30 ug
U3: Market sample # 20 ug
U4: Market sample # 20 ug
U5: Market sample # 30 ug
U6: Market sample # 50 ug

METHOD: 17.42

DETERMINATION OF GLYCIN, ALANINE AND LEUCINE AMINO ACIDS IN FISH SCALES

Aim of Analysis

Determination of glycin, alanine and leucine amino acids in fish scales.

Chemicals and Reagents

Ethanol, hydrochloric acid, ninhydrin reagent, acetic acid, n–butanol, barium hydroxide.

Reagent Preparation

Ninhydrin reagent (NIH): 30 mg ninhydrin is dissolved in 10 ml n-butanol, followed by 0.3 ml 98% acetic acid.

Standard Preparation

1. Glycin (Solution A): Weigh 10 mg of glycin and dissolve in 10 ml of ethanol: water (9:1). [Conc.1 $\mu g/\mu l$]

   Solution B: 1 ml of solution A, makes it to 10 ml with ethanol: water (9:1) [Conc.0.1 $\mu g/\mu l$].

2. Alanine (Solution C): Weigh 10 mg of alanine and dissolve in 10 ml of ethanol: water (9:1) [Conc.1 $\mu g/\mu l$].

   Solution D: Take 1 ml of solution C, makes it to 10 ml with ethanol: water (9:1) [Conc. 0.1 $\mu g/\mu l$].

3. Leucine (Solution E): Weigh 10 mg of leucine and dissolve in 10 ml of ethanol: water (9:1) [Conc.1 $\mu g/\mu l$].

   Solution F: Take 1 ml of solution E, makes it to 10 ml with ethanol: water (9:1) [Conc.0.1 $\mu g/\mu l$].

Sample Preparation

Weigh 1 gm of dry powdered fish scales in 250 ml of conical flask. Add 20 ml of 3N HCl, reflux it for 5 hrs. at 110° C. Cool it. Remove the HCl by adding solid barium hydroxide. Remove the precipitate by filtration. Evaporate the filtrate under vacuum; dissolve the residue in 10% isopropanol solution to 25 ml (40 $\mu g/\mu l$).

Stationary Phase

TLC Al sheets silica gel $60F_{254}$ precoated plate 20×10 cm.

Mobile Phase

n–butanol: acetic acid: water (3:1:1).

Sample/Standard Application

Apply with the help of ATS–4 or Linomat 5, 5 μl of solution B, D and F, and 5 μl of sample on precoated layer, 10 mm from the bottom edge. Band length 8 mm distance between tracks 14 mm. Distance from the side 15 mm.

Development Chamber

Twin-trough chamber of 20×10 cm with s.s.lid

Tank saturation	:	20 min with filter paper
Plate equilibrium	:	None
Development distance	:	80 mm
Post-chromatographic derivatization	:	The plate is dipped in a ninhydrin reagent then heated to 110° C for 5 or 10 min.
Photodocumentation	:	At 366 nm and in visible light after derivatization.

Scanning

Using TLC scanner 3 with win CATS software, slit–micro, 6×0.45 mm, scan at 510 nm.

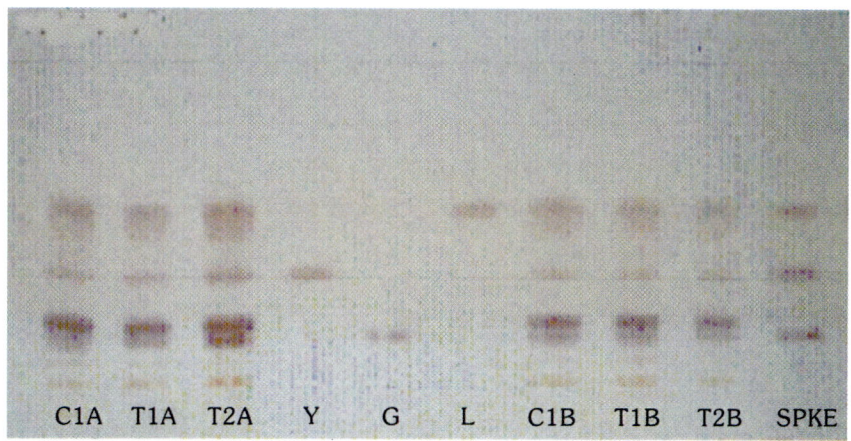

C1A   T1A   T2A   Y   G   L   C1B   T1B   T2B   SPKE

Fig. 17.42: Amino acid after derivatisation

Image@visible

V : Alanine standard
G : Glycine standard
L : Leucine standard

METHOD: 17.43

DETECTION OF ARABINOSE AND XYLOSE IN GUMS

Aim of Analysis

Detection of arabinose and xylose in gums.

Chemicals and Reagents

Arabinose, xylose, acetone, methanol, dist. water, sodium acetate, phosphomolybdic acid, ethanol.

Reagent Preparation

Phosphomolybdic acid: Weigh 250 mg of phosphomolybdic acid and dissolve it in 50 ml of ethanol.

Standard Preparation

1. Arabinose (Solution A): Weigh 20 mg of arabinose and dissolve in 1 ml of dist. water. Then make up to 10 ml with methanol [Conc. 2 $\mu g/\mu l$].
   Solution B: 1 ml of solution A, and add 9 ml of methanol [Conc. 0.2 $\mu g/\mu l$].
2. Xylose (Solution C): Weigh 20 mg of xylose and dissolve in 1 ml of distilled water, then make up to 10 ml with methanol. [Conc. 2 $\mu g/\mu l$]
   Solution D: Take 1 ml from solution C, make it to 10 ml methanol [Conc. 0.2 $\mu g/\mu l$].

Sample Preparation

Weigh 500 mg of sample, dissolve it in ethanol: water (8:2). Make up to 25 ml with the same solvent (Conc. 20 $\mu g/\mu l$).

Stationary Phase

TLC Al sheets silica gel 60 F_{254} precoated 20×10 cm, further impregnate with sodium acetate (0.02M).

Mobile Phase

Acetone: water (9:1) v/v (Volume = 20 ml)

Development Chamber

Twin-trough chamber of 20×20 cm with s. s. lid.

Tank saturation	:	20 min with filter paper
Plate equilibrium	:	None
Sample/standard application	:	Apply with the help of ATS–4 or Linomat 5, 5 μl solution B and D, and 5 μl of sample on precoated layer, 10 mm from the bottom edge. Band length 8 mm distance between tracks 14 mm. Distance from the side 15 mm.
Development distance	:	80 mm
Visualization	:	None
Post-chromatographic derivatization	:	The plate is dipped in a phosphomolybdic acid reagent then heated to 110° C for 5 or 10 min. Sugars show blue colour after heating at 110° C.
Photodocumentation	:	In visible light after derivatization.
Measurement mode	:	Absorbance/reflectance

Scanning for Quantification

Using TLC scanner 3 with win CATS software, slit–micro, 4×0.3 mm, scan at 580 nm.

METHOD: 17.44

SEPARATION OF ERGOT ALKALOIDS

Aim of Analysis

Separation of ergot alkaloids.

Chemicals and Reagents

Acetonitrile, ethanol, toluene, chloroform, methanol

Reagent preparation –   None

Standard Preparation

Weigh 2 mg of standard ergot alkaloids. Dissolve it in methanol: chloroform (1:1). Make up to 10 ml with the same solvent (Conc. 2 μg/μl).

Sample Preparation

Weigh 2 gm of sample. Samples were made basic with 15 ml 25% ammonia solution, left for 12 h, and then extract three times with 15 ml of chloroform. Chloroform extracts were evaporated under reduced pressure and redissolved in 1 ml of chloroform for chromatographic analysis.

Stationary Phase

HPTLC Al sheets silica gel $60F_{254}$ precoated plate 20×10 cm.

Mobile Phase

Acetonitrile: ethanol: toluene (8.5 : 1 : 0.5)

Sample/Standard Application

Apply with the help of ATS–4 or Linomat 5, 5 μl of sample and 2, 4, 6, 8, 10 μl of standard on precoated layer, 8 mm from bottom edge, band length 8 mm, distance between track 11.3 mm, distance from the side 15 mm.

Development Chamber

Twin-trough chamber of 20×10 cm with s.s.lid.

Tank saturation	:	10 minutes with filter paper.
Plate equilibrium	:	None
Development distance	:	70 mm
Visualization	:	Observe in UV cabinet at 254 nm and 366 nm.
Post-chromatographic derivatization	:	None
Photodocumentation	:	At 254 nm, 366 nm.

Scanning
For Identification

Using scanner 3 with win CATS software, slit-micro, 6×0.45 mm, 300 nm.

For Spectral Match
Record and match spectra between 190 to 400 nm.

METHOD: 17.45

QUANTIFICATION OF PESTICIDES IN EGG POWDER

Aim of Analysis

Quantification of pesticides in egg powder sample.

Chemical and Reagents

Fluracil, anhydrous sodium sulphate, silver nitrate.
Solvent : Petroleum ether : dichloromethane (80:20)

Sample Preparation

Weigh 5 g of egg powder sample in a 100 ml conical flask. Add 10 ml of distilled water and shake vigorously for 30 min. Then add 40 ml of solvent and shake for further 30 min, allow to settle and transfer the solvent layer to dry conical flask containing 10 g of fluracil and 10 g of anhydrous sodium sulphate. Again add 40 ml of solvent in the conical flask containing egg powder, shake for 10 min, allow to settle and transfer the solvent layer to the same conical flask containing 10 g of fluracil and 10 g of anhydrous sodium sulphate. Again repeat with 40 ml solvent and carry out the same procedure. Now shake the same conical flask containing 10 g of fluracil and 10 g of anhydrous sodium sulphate for 10 min and filter by passing through a funnel containing 5 g of anhydrous sodium sulphate and 5 g fluracil. Wash twice with 10 ml portions of solvent and collect in the same conical flask. Evaporate the solvent just to dryness in a current of nitrogen. Reconstitute with 1 ml of iso-octane (5000 mg/ml).

Sample Application

Apply with the help of Linomat–5, 30 μl of sample with a band length of 10 mm, 10 mm from the bottom edge of the plate, 20 mm from the side of the plate and distance between the tracks 22.8 mm. Dry the plate in cold air.

Standard Preparation

S1 preparation: Mixture of 7 standards are as follows (alpha HCH, beta HCH, methoxychlor, chorothalonil, dieldrin, endrin and endosulphan sulphate).

Each of 100 μl of above standards of concentration 50 ng/μl are added and total volume made of 700 μl. Apply 10 μl (71.4 ng).

S2 preparation: Mixture of 7 standards are as follows (delta HCH, dicofol, endosulphan, gamma HCH, aldrin, 2, 4 DDT and 4, 4 DDT).

Each of 100 μl of above standards of concentration 50 ng/μ are added and total volume made to 700 μl. Apply 10 μl (71.4 ng).

Egg Powder Sample with Pesticide Preparation

Weigh 5 g of egg powder sample in a 100 ml conical flask. Add 300 μl each from S1 and S2 preparation standard pesticides into the conical flasks. Add 10 ml of distilled water and shake vigorously for 30 min. Then add 40 ml of solvent and shake for further 30 min, allow to settle and transfer the solvent layer to a dry conical flask containing 10 g of fluracil and 10 g of anhydrous sodium sulphate. Again add 40 ml of solvent in the conical flask containing egg powder, shake for 10 min, allow to settle and transfer the solvent layer to the same conical flask containing 10 g of fluracil and 10 g of anhydrous sodium sulphate. Again repeat with 40 ml solvent and carry out the same procedure. Now shake the same conical flask containing 10 g of fluracil and 10 g of anhydrous sodium sulphate for 10 min and filter by passing through a funnel containing 5 g of anhydrous sodium sulphate and 5 g fluracil. Wash twice with 10 ml portions of solvent and collect in a dry conical flask. Evaporate the solvent just to dryness in a current of nitrogen. Reconstitute with 1 ml of iso-octane (500 mg/ml).

Application

Apply with the help of Linomat–5, 30 μl of sample with a band length of 10 mm, 10 mm from the bottom edge of the plate, 20 mm from the side of the plate and distance between the tracks 22.8 mm. Dry the plate in cold air.

Stationary Phase

TLC aluminium sheets aluminium oxide 60F$_{254}$ neutral (type E) precoated 20 × 10 cm

Mobile phase	: Pet ether (20 ml)
Tank saturation	: None
Plate equilibrium	: None
Development distance	: 90 mm

Dry the plate at room temp for 10 minutes.

Derivatization

Dip the plate in the silver nitrate chromogenic reagent for 30 sec (flushing with nitrogen gas). Dry the plate and expose under 254 nm UV lamp until the colour develops.

Scan   –   At 550 nm with tungsten lamp.

Image at visible light after 1st dipping

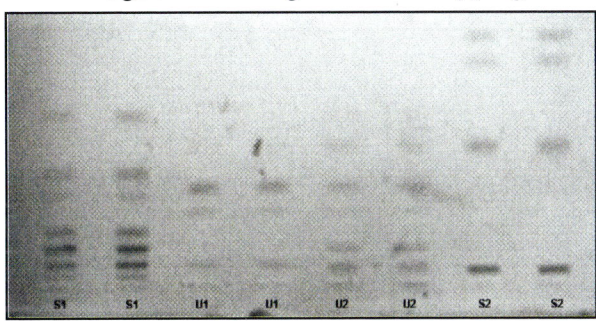

Fig. 17.45 (a): Quantification of pesticides in egg powder sample after derivatisation

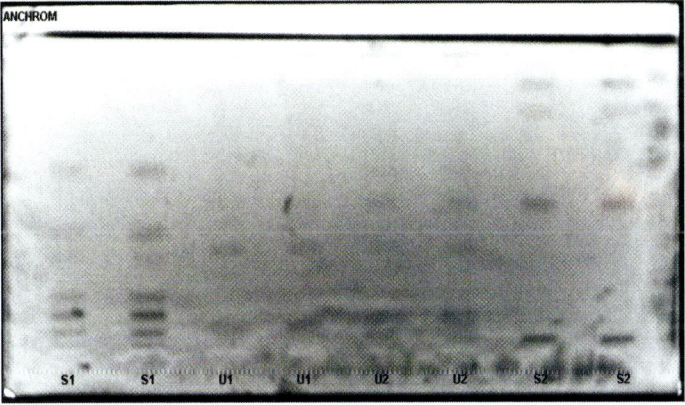

Fig. 17.45 (b): Image at visible light after 2nd dipping

S1 – 10 μl
S1 – 15 μl
U1 – 30 μl (Egg sample)
U1 – 30 μl (Egg sample)
U2 – 30 μl (Egg sample with pesticides)
U2 – 30 μl (Egg sample with pesticides)
S2 – 15 μl
S2 – 15 μl

METHOD: 17.46

DETERMINATION OF SELENIUM IN WATER

Aim of Analysis

Determination of selenium in water.

Chemicals and Reagents

HNO_3 (65%), H_2O_2 (30%), HCl (25%), EDTA solution (3.7%), NaF (1.4%), formic acid (50%), NH_3 (25%), cyclohexane, butylhydroxytoluene, chloroform, methanol, ethanol (abs.), triton X–100, 2, 3–diaminonaphthalene (DAN, recrystallized) (for preparation of DAN solution), standard :titrisol.

Sample Preparation

1. Conversion of sample and standard by weight is by chemical treatment.

An amount of 0.5 to 5.0 ml sample or 2.5 ml standard solution is evaporated to dryness at 100° C. After cooling to 20° C, the residue is dissolved in 5 ml HNO_3 and heated at 110° C for 60 min. Three 2 ml portions of H_2O_2 are added at 15 min intervals, and the reaction mixture is left at 110° C. The clear solution is then evaporated to dryness at 120° C and the colourless precipitate is taken up with 2 ml H_2O_2 and again evaporated to dryness. The residue is dissolved in 20 ml HCl and heated to 110° C for 45 min to reduce the selenate to selenite.

If this sample is left to stand for prolonged periods (30–60 min), chlorine is evolved and can oxidize the selenite back to selenate. It is desirable to remove the chlorine by purging with nitrogen.

2. Derivatization of sample and standard.

To mask interfering ions, 1 ml aqueous EDTA solution, 0.5 ml aqueous NaF solution, and 2 ml formic acid are added to the solution. The solution is adjusted to pH 1.8 with NH_3 (or HCl), and 2 ml DAN solution is added.

Preparation of DAN solution

1 g DAN is dissolved in 40 ml ethanol (abs.) at 80° C, the solution is passed through a porcelain filter, and then DAN is allowed to crystalline out. The mixture is then filtered through a glass frit (No. 4), the precipitate is washed with 50% ethanol (cold), then dried (brown crystals with a silky sheen).

0.5 g recrystallized DAN is dissolved in 100 ml 0.1 N HCl by heating, extracted three times in a separating funnel with 10 ml cyclohexane, then filtered through a glass frit (No.4). This solution must be made up fresh each time before use.

The mixture is then heated to max 40° C for a few minutes in the dark, adjusted to pH 6 with NH_3, transferred to a 250 ml (sample) or 500 ml (standard) measuring flask, and filled up to the mark with water (pH = 6).

3. **Extraction of sample and standard.**

A 20 ml portion of the derivatized sample/standard is transferred to an extrelut column and after about 12 min it is eluted with 40 ml cyclohexane (containing 0.01% butylhydroxytoluene for stabilization). The eluate (about 25 ml) is collected in a light–protected tapered–neck flask and evaporated to dryness in a rotary evaporator at max. 35° C. The residue is dissolved in 1 ml chloroform, and transferred to a curved rim beaker (this operation is repeated three times). The combined chloroform extracts are evaporated to dryness at room temperature under nitrogen, and the residue is dissolved in 500 μl chloroform.

Standard Solution

Make up one ampoule titrisol (Merck No. 9915, SeO_2) to 1 L with double distilled water. Dissolve 50 μl of this mixture in 5 ml water, and proceed as outlined in sample preparation. Make-up quantitatively to 1000 ml (1μl–5 pg Se).

Stationary Phase

HPTLC plates silica gel 60 F_{254}, 20 × 10 cm

The plates are prewashed three times in succession by blank chromatography with a mixture of methanol : chloroform (1:1), then dried for 20 min at 110° C.

Sample Application

With Camag Linomat or ATS as 7 mm bands, distance from left edge is 20 mm, distance from lower edge 8 mm.

Application Pattern

S1 U1 S2 U2 S3 U3 S4 U4 S1 U1 S2 U2 S3 U3 S4 U4
S1–S4 = 1, 3, 6, 10 μl standard solution, U1–U4 = 1 μl unknown each.

Chromatography

Horizontal developing chamber is used with chloroform (0.01% butylhydroxytoluene as antioxidant). Migration distance is 50 mm with running time 9 min, R_f = 0.51.

Derivatization (Intensification of Fluorescence)

It is performed by dipping the still wet plate three times for 2 sec each in a solution of Triton X–100: chloroform (1:4) with immersion device. Between each dip, the plate is left for 15 min at room temperature in the dark to allow the solvent to evaporate.

For quantitative evaluation, the solvent has to be evaporated completely after the final dip (absence of odour). The fluorescence intensity is increased by a factor of 90, and stabilized for about 9 hours.

Densitometric Evaluation

With TLC scanner and CATS evaluation software; scanning is done by fluorescence at 366/>560 nm.

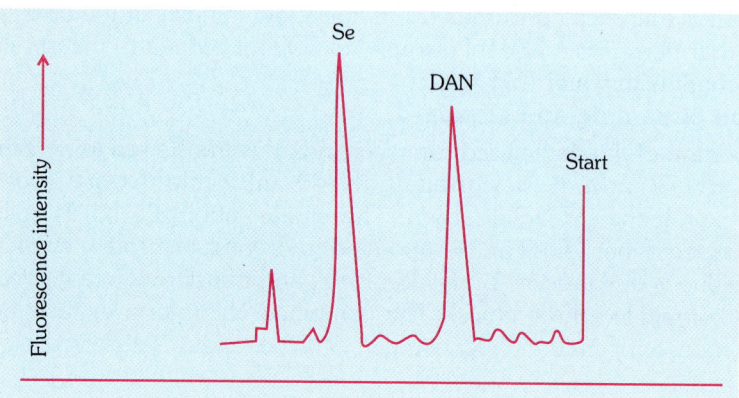

Fig. 17.46 (a):   Densitogram of the fluorescence measurement at 366/>560 nm.
Se = 10 pg 2, 1, 3–naphthoselenediazol, DAN = 2m 3–diaminonapthalene.

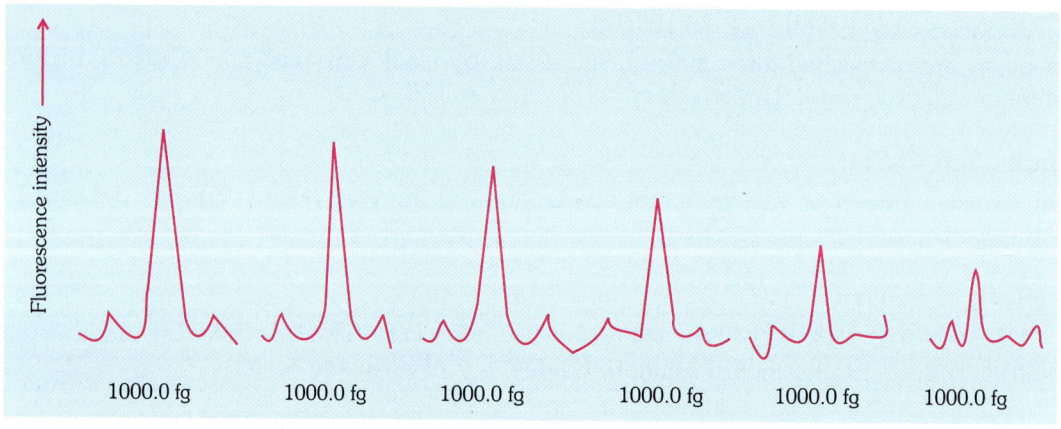

Fig. 17.46 (b):   The limit of determination for this method is about 1 pg Se/spot and the limit of detection is about 250 fg Se/spot.

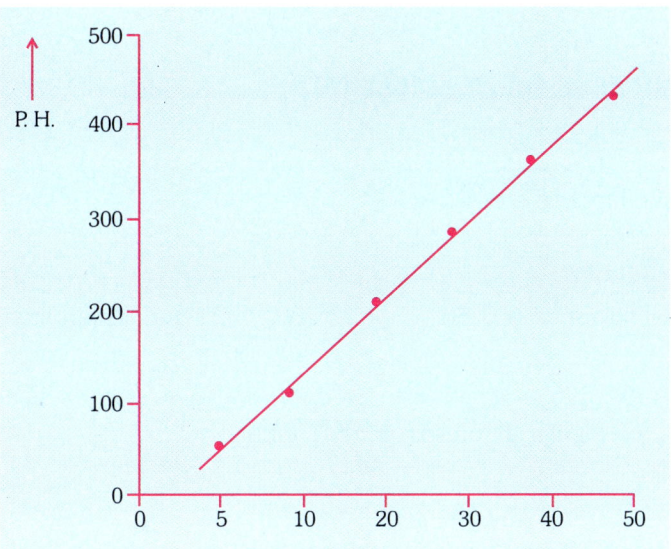

Fig. 17.46 (c)

The linear regression of 2, 1, 3–naphthoselendiazol shows that the decomposition and extraction procedure does not lead to constant–systematic errors.

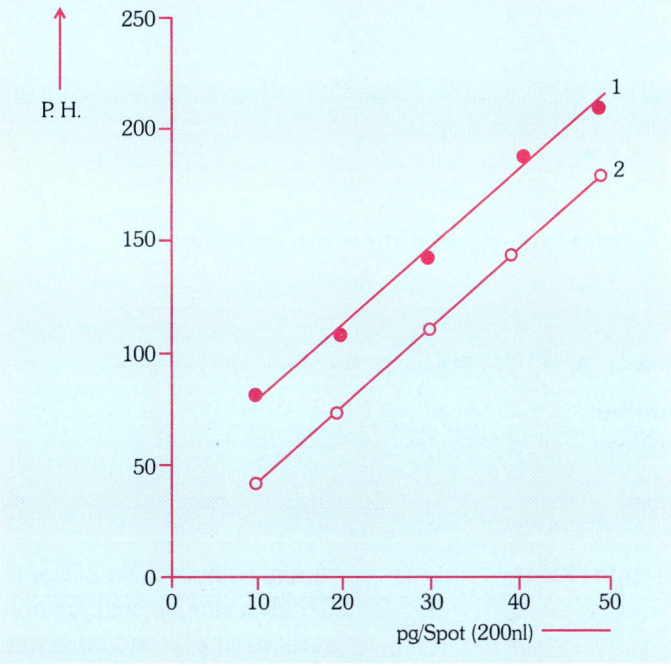

Fig. 17.46 (d)

The standard addition curve (1) is compared with the calibration curve (2) of analogously treated selenocystine standards. The recovery rate is 100.2%. A statistical check of this rate shows that this method is free from any systematic errors.

METHOD: 17.47

DETECTION OF MINERAL OIL IN EDIBLE OILS

Aim of Analysis

Detection of mineral oil in edible oils.

Chemicals and Reagents

Petroleum ether (boiling range 60°–80° C or 40°–60° C)2', 7'–Dichlorofluoresein, chloroform

Reagent Preparation

0.2% solution of 2', 7'– dichlorofluoresein in 96% ethanol.

Standard Solution

Stock solution (S): Weigh 10 mg of liquid paraffin in a 10 ml volumetric flask. Make up to volume with chloroform. (1 µg/µl)

Solution A: Take 1 ml of stock solution (s) in a 10 ml volumetric flask. Make up to volume with chloroform (0.1 µg/µl)

Sample Preparation

Sample: Weigh 1000 mg of oil sample (vegetable oil) in a 100 ml volumetric flask. Make up to volume with chloroform [10 µg/µl–vegetable oil].

Stationary Phase

TLC Al sheets silica gel 60 F_{254} precoated cut to 20 × 10 cm.

Mobile Phase

Petroleum ether (boiling range 60°–80° C or 40°–60° C)

Development chamber
Camag TTC 20 × 10 cm

Chamber saturation	:	N. A
Plate equilibrium	:	N. A
Sample/standard application	:	Apply with the help of Camag ATS–4/Linomat 5, 10 µl of sample solution and 2 µl, 4 µl, 6 µl, 8 µl and 10 µl of standard solution (A) on precoated layer 20 mm from the bottom edge. Band length 8 mm. Distance from the side 15 mm.
Development distance	:	80 mm.
Visualisation	:	Observed under UV cabinet at 366 nm.

Post-chromatographic derivatisation	:	The developed plate is dipped in a 0.2% solution of 2', 7'–dichloro-fluoresein in 96% ethanol.
Photodocumentation	:	366 nm.
Measurement mode	:	Fluorescence.

Scanning

For Quantification

Using Camag scanner 3 with winCATS software, slit–micro, 6 × 0.45 mm, scan 366 nm.

For Identification

A yellow fluorescent band on the solvent front indicates the presence of mineral oil. The vegetable oil forms a yellow streak about 2–3 cm long from the point of application.

METHOD: 17.48

DETECTION OF TRICRESYL PHOSPHATES IN EDIBLE OILS

Aim of Analysis

Detection of Tricresyl Phosphates in Edible Oils.

Chemicals and Reagents

Iso-octane, ethyl acetate, petroleum ether (40°–60° C), acetonitrile, 2.6–dichloroquinone chlorimide, ethyl alcohol.

Reagent Preparation

0.5% solution of 2, 6–dichloroquinone chlorimide (E. Merck) in absolute ethyl alcohol (Gibbs reagent). Store reagent at 10° C and use within 5 days.

Standard Solution

Stock solution (S): Weigh 10 mg of tricresyl phosphate in a 10 ml volumetric flask. Make up to volume with methanol (Conc. 1 µg/µl).

Sample Preparation

Sample: Take 10 ml oil sample containing ca. 50 µg TCP into separatory funnels; add 50 ml petroleum ether (40° to 60° C) to dissolve the oil followed by 10 ml acetonitrile previously saturated with petroleum ether. Shake contents vigorously and let stand 10 min. Collect lower aqueous layer in beaker and evaporate solvent on hot water bath. Dissolve residue in 1.0 ml ethyl or methyl alcohol.

Stationary Phase

TLC Al. sheets silica gel 60 F_{254} cut to 10 × 10 cm.

Mobile Phase
Iso-octane: ethyl acetate (90:10)

Development chamber
Camag TTC 10 × 10 cm

Chamber saturation	:	20 min with filter paper.
Plate equilibrium	:	N. A.
Sample/standard application	:	Apply with the help of Camag ATS–4/Linomat $5\mu l$ of sample and standard solution on precoated layer, 8 mm from the bottom edge, band length 8 mm. Distance from the side 15 mm.
Development distance	:	80 mm.
Visualisation	:	Observed under UV cabinet at 254 and 366 nm.
Post-chromatographic	:	The developed plate is dipped in 0.5% solution of 2, 6–dichloroquinone chlorimide and heated at 100° C on Camag TLC plate heater III for 5 min.
Photo documentation	:	254 nm, 366 nm and white light
Measurement mode	:	Absorbance

Scanning

For Quantification
Using Camag scanner 3 with win CATS software, slit–micro, 6 × 0.45 mm, scan in the visible region.

For Identification
A blue–violet band at R_f 0.27 indicates the presence of tricresyl phosphate.

Appendix A

Activity

It is the surface properties of the sorbent. It is affected by its water content which can be adjusted. Silica gel and aluminium oxide layers are activated in a drying oven at temperature $110°$ C for 30 min to remove the physically bound water. This makes the layer very active. The presence of moisture and exposure to water vapor reduce the activity of a layer.

Active Solid

A solid with sorptive properties.

Adsorption

Adsorption is the interaction between the solute and surface of the adsorbent. The forces deployed may be strong (e.g. hydrogen bonds) or weak (e.g. Van der Waals forces). In case of silica gel, the silanol group is responsible for adsorption. Functional group of the solute interacts with this group.

Adsorption Chromatography

Separation largely based on differences between the adsorption affinities of the sample components for the surface of a active solid.

Aluminium Oxide

It is a porous adsorbent with a slightly basic surface. Silica has an acidic surface.

Ambient Temperature (T_a)

It is the temperature outside the chromatographic system.

Amino Phase

Silica gel modified with aminopropyl functionality is used by and large as stationary phase in normal bonded phase chromatography. It is normally reactive for solute molecule or mobile phase additive which reacts with amines. It is a weak basic anion exchanger and the layer is totally wettable with water.

Anticircular Chromatography (Anticircular Development)

In this process, the sample as well as the mobile phase is applied at the periphery of a circle and these move towards the centre.

Ascending Chromatography (Ascending Development)

Means of operation in which the plate is in a vertical or slanted position. The mobile phase is supplied from its lower end to the upward movement based on capillary action.

Automated Multiple Development (AMD)

It is a technique which permits stepwise gradient development on an HPTLC plate. Therefore every step covers a longer developing distance than the previous and uses a solvent of lower strength. After each development, the plate is dried by vacuum automatically.

Band Broadening, Dispersion and Spreading

This refers to dilution of chromatographic band migrated on the plate. The sample is applied as a narrow band. Every separated component will elute as a narrow zone of pure compound. The measure of brand broadening is band width.

Binders

These are additives used to adhere the solid stationary phase to the inactive plate/sheet.

Bonded Phase

A stationary phase covalently bonded to the support particles.

Bonded–phase Chromatography (BPC)

A stationary phase which is chemically bonded to a support for the purpose of separation. The popular support is microparticulate silica gel and bonded phase is organosilane.

Cellulose

It is organic macromolecule of glucose units. Used as a stationary phase in its microcrystalline form.

Chamber

It is a closed apparatus/equipment in which a plate is developed. There are different types of chambers, i.e. from very simple glass tank to computerized automatic developing device.

Chamber Saturation (Saturated Development)

The expression refers to the uniform distribution of the vapour of the mobile phase in equilibrium with mobile phase throughout the chamber prior to chromatography.

Chromatogram

In planar chromatography, "chromatogram" may refer to the paper or layer with the separated zones and also to the peak data from scanning.

Chromatography

Chromatography is a physical method of separation wherein components to be separated are distributed between two phases. One is stationary phase while the other is mobile phase which moves in a definite direction.

Chromatograph when used as Noun

The assembly of equipments for conducting chromatographic separation.

Chromatograph when used as Verb

Separate by chromatography.

Clean-up

It is a process for removing the interfering matrix from a sample before conducting chromatographic analysis.

Concentration Zone

It is a 2–4 cm wide strip of sorbent layer consisting of adsorption–inactive silicon dioxide which is directly adjacent to the separation layer. The compounds/substances applied are migrated through the concentration zone with solvent front in the direction of chromatography which is focused in a narrow substance bands when these enter into separation layer.

Cyano Phase

It is a stationary phase consisting of cyanopropylsilyl group on the silica gel support. The layer is fully wettable by water.

Densitometer

This is commonly known as scanner. It is used for evaluation by recording adsorption or fluorescence in reflectance/transmittance mode. It is based on the principle that concentration of a substance is a function of the difference in amount of light reflected/emitted from the substance on the plate and the light reflected/emitted from the plate. A plot of the resulting signal against the migration distance gives an analog curve of the chromatogram which is integrated to report results as peak area or height.

Derivatization

It is a chemical reaction applied to visualize the separated components or to improve separation or selective detection. The derivatization reagent for a specific chemical reaction is applied to the plate either by spraying or by immersion of plate into reagent. Derivatization through a gas phase can also be done. Derivatization can be done before or after chromatography.

Descending Elution (Descending Development)

A mode of operation, wherein the mobile phase is supplied from the upper edge of the plate to downward movement and is governed mainly by gravity action.

Detection

It is a measurement or visualization of separated substance.

Developing Distance

It is the distance between sample application position and the final mobile phase front position.

Developing Solvent

It is a synonym to the mobile phase. But developing solvent is in fact a liquid. They may be mixture of liquids which is put into the chamber. Since the composition of the developing solvent changes during chromatography (due to the amount of interaction in the developing chamber), the composition of mobile phase moving through the stationary phase can be different from the composition of the mobile phase.

Development

It is the process of separation of the sample components which is applied to the layer by the migration of mobile phase to the end of migration path (front). The detection of migration of mobile phase can be determined by a suitable development technique, i.e. linear ascending, linear horizontal, and circular and so on. Vertical development is the most widely used method. The plate is put with the lower edge in a tank charged with developing solvent. The capillary force of the stationary phase is the main cause which allows mobile phase to move up to a desired developing distance. In case of horizontal development, it is done in a special developing chamber from one side or from both sides of the plates.

Diffusion Coefficient in the Mobile Phase(D_M or D_G)

The diffusion coefficient characterizing the diffusion in the mobile phase.

Diffusion Coefficient in the Stationary Phase (D_S)

The diffusion coefficient characterizing the diffusion in the stationary phase.

Diffusion Velocity (u_D

The term is normally used in liquid chromatography to express the reduced mobile–phase velocity. The diffusion velocity expresses the speed of diffusion into the pores of the particles:

$$U_D = D_M/d_p$$

Diol Phase

This is a modified silica gel stationary phase used in both normal and reversed phase chromatography. It consists of a diol structure (two–OH groups on adjacent carbon atoms in an aliphatic chain) and is less polar than silica gel when used in normal phase chromatography. The layer is fully wettable with water.

Dyeing Reagent

These are those reagents which can detect colourless compounds which have no UV absorbance and have no intrinsic fluorescence property. The reagents are spread on the plate through an air sprayer or the plate is dipped into the reagent. The reagent reacts in situ.

Edge Effect

These occur when a chamber is not fully or uniformly equilibrated. Evaporation from edge of the plate can cause R_f value of the compounds spotted on outer sides of the plate higher than in the middle. This is known as edge effect. This may be prevented by saturating the chamber with thick filter paper for 20 min.

Eluent

This is known as mobile phase.

Elute (Verb)

To chromatograph by elution chromatography. The elution may be stopped when all the sample components have separated on the chromatographic bed.

Equilibration

The expression indicates the level of saturation of the chromatographic bed by the mobile phase vapor before chromatography.

Fingerprint

This is a unique sequence of photodocumented zones (R_f, intensity, colour) and corresponding peaks of a chromatogram specific to the sample. Each and every fraction detected is taken into account. No fraction(s) are quantified.

Fluorescence Indicator

These are inert inorganic substances added in the layer to make colorless substance visible by fluorescence quenching at UV 254 nm. The most common fluorescence indicators in TLC are:

1. F_{254}: It is a manganese activated zinc silicate which gives green fluorescence when stimulated by short wavelength (254 nm).
2. F_{254s}: It is an acid stable alkaline with tungstate indicator gives pale–blue fluorescence emission by UV 254 nm. Suitable for acidic spray reagents.
3. F_{366}: It is known as optical brightener which, when stimulated by long wavelength, i.e. 366 nm, gives intense blue fluorescence. Rarely used.

Fluorescence Quenching

When fluorescence indicators in layer are stimulated by UV radiation (254 nm), substances absorbing UV appear as dark chromatographic zones on a pale blue or green fluorescence background.

Gas Chromatography (GC)

Gas Phase

It is the vapor phase in a developing chamber which strongly affects the chromatographic results. Particularly in TLC/HPTLC, this phase is very important but neglected.

Gradient Elution

The procedure, in which composition of the mobile phase is changed continuously or stepwise during the chromatography.

Gradient Layer

The chromatographic bed in which a gradual transition in some property occurs, i.e. composition of layer.

High-performance Thin-layer Chromatography (HPTLC)

In practice, defined as instrumentally performed thin-layer chromatography, using precoated layers with the aim of quantification.

Horizontal Elution (Horizontal Development)

A way of operation in which the plate is in a horizontal state but the movement of mobile–phase along the plate depends on capillary action only.

Immobilized Phase

A stationary phase immobilized on the support particles as in, e.g. gas chromatography.

Impregnation

The modification of the separation properties of the plate by appropriate means.

Isocratic Analysis

The composition of the mobile phase remains constant during chromatography.

Kieselguhr

This is a naturally occurring amorphous silicic acid of fossil origin which is also known as diatomaceous earth (diatoms = silicaceous algae). This is used in partition chromatography due to its advantage to a small surface area of low activity.

Liquid Chromatography (LC)

A separation technique in which the mobile phase is a liquid.

Mobile Phase

This is a mixture of solvents or pure solvent moving during development of the plate through porous TLC layer as a result of capillary forces. This phase tries to carry the sample components with it.

Mobile Phase Front

This is a term expressing the leading edge of the mobile phase as it traverses across the planar media. In all forms of development (except radial), the mobile phase front is basically a straight line parallel to the mobile phase surface. It is also known as the solvent front.

Modified Active Solid

An active solid in which the sorbtive properties changes by some treatment.

Modifier

This is an additive which changes the character of the mobile phase (e.g. hexane is a weak solvent and methanol a strong solvent can be a modifier even in very small quantities). Other modifiers can be acids or bases.

Normal Phase Chromatography

An elution procedure wherein the stationary phase is more polar than the mobile phase, e.g. silica gel.

Particle Diameter (d_p)

It is the average diameter of the solid adsorbent particles.

Particle Size

It is the average particle size in the layer. The particle size of HPTLC silica gel is 5 μm with an average distribution of materials in the range of 4 to 8 μm.

Partition Chromatography

Separation is mainly based on differences between the solubility of the components in the mobile and stationary phases (liquid chromatography) which are immiscible.

Peak

It is a part of measured analog curve corresponding to the fraction on the plate. The peak detection level is normally defined as a significant variation of the signal from the baseline (usually three times of the noise level).

Planar Chromatography

A separation technique in which the stationary phase is present as plane. The plane can be a layer of solid particles spread on a support, such as a glass plate (Thin-layer chromatography, TLC). Planar chromatography is also termed as open–bed chromatography.

Polar Bonded Phases

These are modified silica gel phases with, e.g. amino, cyano and diol functional groups, in a bound form and not coated form.

Porosity

Sorbents such as silica gel and aluminium oxide consist of solid porous particles which possess open channels with a mean opening size (pore size). Internal surface constitute some of the areas of the pores. Smaller the pore size, greater the specific area and thus stronger the adsorption effect. The internal volume of the porous sorbent particles is totally filled by the mobile phase.

Pore Radius (r_p)

It is the average radius of the pores within the solid particles.

Preconditioning

It is defined as adjusting chromatographic properties of the stationary phase by exposure to a defined gas phase (relative humidity, pH, solvent composition, etc.) before development.

Radial Elution (Radial Development) or Circular Elution (Circular Development)

In this operation, the samples are spotted near the centre of the plate and are carried outward in a circle by the mobile phase.

Relative Humidity (RH)

It is the degree of saturation of the atmosphere at a particular temperature. If sorbents are opened to an atmosphere which contains water, they adsorb certain water on their surface depending on the relative humidity. The plate can be preconditioned in a suitable vessel of defined relative humidity.

Resolution (R_s)

This is known as separation efficiency or degree of separation of two chromatographic peaks. The parameters can be measured from the chromatograph by ratio of distance between the zone and mean width. The parameters depend on interaction between the separation efficiency and the selectivity of the layer, i.e. between the band broadening (zone diffusion) and the difference between R_f value of the two substances). R_s as is accepted to be the minimum for a measurable separation to occur and for good quantitation. Value above 1.6 is desirable for rugged method.

Retention Factor (R_F)

It is a parameter used for characterizing the location of a sample compound in a chromatogram. It is a relative distance travelled by a substance as compared to distance travelled by solvent.

R_f = (peak position–application position)/solvent front position–application position). R_F values lie between 0 and 1 and best between 0.1–0.8 which depends upon the chromatographic condition, components present in a sample, etc. also known as retardation factor or reference to front.

Reversed Phase Chromatography

A procedure applied in liquid chromatography in which the mobile phase is significantly more polar than the stationary phase, e.g. silica gel RP 18 stationary phase.

Sample

The mixture consisting of a number of components for separation on the chromatographic bed, as they are carried or eluted by the mobile phase.

Sample Components

These are chemically pure constituents of the sample. They may be separated on the stationary phase by the mobile phase.

Sandwich Chamber

A chromatographic procedure, in which the layer is sandwiched between the support and a cover (sandwich) plate. This prevents the vapour from reaching the layer.

Selectivity(∞)

This is a thermodynamic factor which is a measure of relative retention of two components using a fixed stationary phase and mobile phase composition.

Separation Temperature (T_c)

The temperature of the chromatographic bed under isothermal operation.

Silanol

This is Si–OH group on the surface of the silica gel which is of different types of silanols depending upon the location and relationship with others. The strongest silanols are acidic and thus result to undesirable interactions with basic compounds during chromatography.

Silica Gel

It is the most widely used sorbent having an amorphous structure which is porous and comprises of siloxane and silanol groups.

Siloxane

It is Si–O–Si bond found in silica gel.

Solute

Sample in a solution form.

Solvent

A liquid used to dissolve solids. A term sometimes referring to the liquid phase.

Solvent Strength

This is the ability of a solvent to elute a particular solute.

Starting Point or Line

The point/line on a chromatographic layer, where the substance to be chromatographed is applied. Usually 8 mm from lower edge in band application.

Stationary Phase

One of the three phases of a chromatographic system. It can be a solid, a gel or a liquid. In case of a liquid, it may be distributed on a solid. Solid may or may not contribute to the separation process. The liquid may also be chemically bonded to the solid, e.g. bonded phase or immobilized on to it known as *immobilized* phase.

The word *chromatographic bed* or *sorbent* may be used as a general term to denote any of the different forms in which the stationary phase is used.

Stepwise Elution

The procedure in which composition of the mobile phase is changed continuously or stepwise during the chromatography.

Spot/Band

Spots in TLC refer to application of samples/standards, circular in shape. Usually 2 to 10 mm diameter. Bands are usually 6 to 8 mm long and 1 mm widened is preferred.

Spot Diameter

The width of the sample component circular spot.

Solid Support

A solid which holds the stationary phase but does not contribute to the separation process.

Two-dimensional Chromatography

A procedure in which all the separated sample components or parts there of is subjected to additional separation step.

In planar chromatography, two-dimensional chromatography states to the chromatographic process in which the components are caused to migrate in one direction first and then subsequently in a direction at right angles to the first one. The two elutions are carried out with same mobile phase to check chromatographic stability of sample or with different mobile phases to check a method or for further resolution.

Unsaturated Elution (Unsaturated Development)

The expression in chromatography which refers to an elution chamber without attaining chamber saturation.

Visualization Chamber

A device in which the planar media may be viewed under controlled-wavelength light, if necessary after spraying with chemical reagents to render the separated components as visible spots under specified conditions.

Zone

A location in the chromatographic bed where one or more components of the sample are located. The band is also used for zone.

Appendix B

LIST OF ABBREVIATIONS

A	Anisaldehyde reagent
ACN	Acetonitrile
ADC	Automatic developing chamber
AMD	Automated multiple development
AOAC	Association of analytical communities (previously Association of Official Agricultural Chemists, then Association of Official Analytical Chemists)
ASE	Accelerated solvent extraction
AVA	Acetoxyvalerenic acid
BAH	German association of pharmaceuticals manufacturers
BPC	Bonded phase chromatography
CAC	Codex Alimentarius Commission
CCD	Charged coupled device
CCMAS	Codex Committee on Methods of Analysis and Sampling
CCPR	Codex Committee on Pesticide Residue
CGMP	Current good manufacturing practices
D	Derivatization
DABITC	Dimethylaminoazobenzene isothiocyanate
DAD	Diode-array detector
DCE	1, 2-dichloroethane
DNP	Dinitrophenyl
DNS	Dinitrosulfonyl
ELSD	Evaporative light scattering detector
FBC	Flat-bottom chamber
FDA	U S Food and Drug Administration
FFDC	Forced flow development chamber
FI	Fluorescence indicator
FID	Flame ionization detector
FLI	Fluorescence intensifier
FLD	Fluorescence detection

FTIR	Fourier-transform infrared
GAP	Good agricultural practice
GC	Gas chromatography
GF	Graduated flask
GLC	Gas-liquid chromatography
GLP	Good laboratory practice
GMP	Good manufacturing practice
GSC	Gas-solid chromatography
H	Hour (s)
HC	Horizontal chamber
HETP, H	Height equivalent to a theoretical plate
HIC	Hydrophobic interaction chromatography
HPLC	High pressure liquid chromatography
HPPLC	High pressure planar liquid chromatography (circular TLC under high pressure)
HPTLC	High performance thin-layer chromatography
hR$_f$ value	Retardation factor R$_f$ multiplied by 100
IC	Ion chromatography
ICH	International Conference on Harmonization
IPC	Ion-pair chromatography
IQ	Installation qualification
IR	Infrared
I$_R$	Intensity of reflected light
ISO	International Organization for Standardization
IRS	Infrared spectrometry
I$_T$	Intensity of transmitted light
K	Flow constant or velocity coefficient
K$_A$	Coefficient of absorption per unit thickness
LC	Liquid chromatography
LIMS	Laboratory information management system
LLC	Liquid-liquid chromatography
LOD	Limit of detection
LOQ	Limit of quantitation
LSC	Liquid solid chromatography
MB	Megabytes
MD	Migration distance
MEK	Methyl ethyl ketone
Min	Minute (s)
MP	Mobile phase
MS	Mass spectrum /mass spectrometry
MWLS	Multiple wavelength scan
N	Plate number
N-Chamber	Normal developing chamber for vertical development
NF	National Formulary
NMR	Nuclear magnetic resonance
NP	Normal phase
OECD	Organization for economic Cooperation and Development

OPLC	Overpressured layer chromatography (TLC with forced flow)
OQ	Operation qualification (for analytical equipment)
PAH	Polycyclic aromatic hydrocarbon
PC	Personal computer
PC	Paper chromatography
PBP	Polar bonded phase
PEG	Polyethylene glycol
PLC	Preparative layer chromatography
PM	Photomultiplier
PQ	Performance qualification
QA	Quality assurance
R	Residue
Ref.Sol.	Reference solution
Ref.Sub.	Reference substance
rel.c.	Related compound
$\mathbf{R_F}$	Retention factor
RH	Relative humidity
RI	Refractive index
rpm	Revolutions per minute
RP	Reversed phase
$\mathbf{R_{rel}}$	Relative retardation factor
$\mathbf{R_s}$	Resolution
RSD	Relative standard deviation
RT	Room temperature
S,Sec.	Second (s)
S-chamber	Sandwich chamber
SEC	Size-exclusion chromatography
SERS	Surface-enhanced Raman spectroscopy
SFE	Supercritical-fluid extraction
SFC	Supercritical fluid chromatography
Si	Silica gel
SOP	Standard operating procedure
SPE	Solid-phase extraction
SPME	Solid-phase micro-extraction
$\mathbf{S_{rel}}$	Variation coefficient (relative standard deviation)
SRS	Secondary reference substance
SS	Solvent system
TBME	Tert-butyl methyl ether
THF	Tetrahydrofuran
TLC	Thin-layer chromatography
TLC-FID/	Combination of TLC with flame ionization detector or flame
fTID	Thermionic ionization detector
TLC scanner	Measuring equipment for the direct optical evaluation of thin-layer chromatograms
$\mathbf{\mu g}$	Microgram
$\mathbf{\mu l}$	Microliter

X_e	Proton acceptance
X_d	Proton donor
254	Short-wave UV light, reflection
366	Long-wave UV light, reflection
2D-TLC	Two-dimensional TLC
3D	Three-dimensional

Index